MATHEMATICAL THEORIES OF NONLINEAR SYSTEMS

MATHEMATICAL THEORIES OF NONLINEAR SYSTEMS

STEPHEN PAUL BANKS
Department of Control Engineering
University of Sheffield

Prentice Hall
New York London Toronto Sydney Tokyo

First published 1988 by
Prentice Hall International (UK) Ltd
66 Wood Lane End, Hemel Hempstead,
Hertfordshire, HP2 4RG
A division of
Simon & Schuster International Group

Printed and bound in Great Britain by
A. Wheaton & Co. Ltd., Exeter

Library of Congress Cataloging-in-Publication Data

Banks, Stephen P.
Mathematical theories of nonlinear systems/Stephen Paul Banks.
p. cm. (Prentice-Hall international series in systems and control engineering)
Bibliography: p.
Includes index.
ISBN 0-13-562182-8
1. Nonlinear theories. 2. System analysis. I. Title.
II. Series.
QA427.B36 1988 87-19999
003 dc19 CIP

British Library Cataloguing in Publication Data

Banks, Stephen P.
Mathematical theories of nonlinear systems. (Prentice Hall International series in systems and control engineering).
1. Nonlinear theories
I. Title
003 Q295

ISBN 0-13-562182-8

1 2 3 4 5 92 91 90 89 88

ISBN 0-13-562182-8

To my parents

CONTENTS

PREFACE

The theory of linear systems has long been regarded as a well-defined discipline in mathematics, consisting of the study of linear differential equations, with inputs, which are defined on a vector space. On the contrary, nonlinear systems have been studied from various points of view, each one depending on the type of nonlinearity involved and a coherent 'theory' of nonlinear systems did not seem possible for such diverse types of systems. However, in recent years, a large class of differentiable systems has been studied in very much the same spirit as linear systems and we can now say that a 'theory of nonlinear (differentiable) systems' exists which subsumes the linear theory. The correct objects of study are now vector (or fibre) bundles over differentiable manifolds and it is the intention of this book to present (a large part of) this theory as it now stands.

The main difficulty with this theory is that the mathematics required to understand nonlinear systems is no longer elementary vector space theory, but the theory of differential geometry. We present the basic ideas of differential geometry and functional analysis (mostly without proof) in Chapter 1. Readers with no knowledge of either discipline will almost certainly need to consult the cited textbooks on the appropriate mathematics before proceeding with the theory described in the remaining four chapters.

In Chapter 2 we discuss the basic theory of nonlinear systems on manifolds. It will be seen that many of the properties of linear systems generalize to nonlinear systems. Thus, we study reachability, controllability, observability and realization theory for such systems and obtain canonical structures in certain cases. Invertibility and decoupling are also important concepts in systems theory and these are also presented in detail. Finally, Hamiltonian systems are shown to provide a useful and important example for the general theory.

Nonlinear systems have been tackled in the past by using local linearization methods. In Chapter 3 we examine *global* linear representations and show that a large class of nonlinear systems may be represented by

M.T.N.S.—A*

(infinite-dimensional) bilinear systems. Hence we may obtain bilinear approximations to many types of nonlinear systems.

In Chapter 4 we explore the relation between bilinear representations of nonlinear systems and their Volterra series expansions. Moreover, it is shown that bilinear systems have many of the properties of linear systems (or these properties can be generalized to bilinear systems) and so it will be seen that such systems form a very important extension to the class of linear systems. We also present an approach to the frequency-domain theory of nonlinear systems which gives rise to the notion of poles and zeros. A different theory of nonlinear system zeros can be found in Nijmeijer and Schumacher, 1985.

Finally, in Chapter 5, we present an introduction to nonlinear distributed parameter systems. Because of the importance of bilinear systems in finite-dimensional space, we restrict attention mainly to this kind of system. This means that we can work on a (flat) Hilbert or Banach space. One of the main outstanding problems in nonlinear systems theory is to derive results similar to those in Chapters 2–4 for infinite-dimensional systems defined on Banach manifolds.

Since we work throughout in a differentiable category, the methods developed here do not apply to engineering systems containing hysteresis, dead zones, discontinuities, etc. Moreover we do not discuss discrete systems, although much of what we present here holds, when suitably modified, for discrete systems. The third major area of systems theory which is not covered here is that of stochastic systems. Differential-geometric methods used here for the study of deterministic systems can also be applied to stochastic systems; we refer the reader to Marcus, 1984 or Collingwood, 1985.

S.P.B.

LIST OF SYMBOLS

$\mathscr{O}$	set of open sets in a topological space
$\subseteq$	'subset of'
$\in$	'belongs to'
$\ni$	'contains the element'
$B_\varepsilon(x)$	ball of radius ε, centre x
2^X	sets of subsets of X
$X/\sim$	quotient set
$\mathbb{R}^n$	n-dimensional Euclidean space
$\lVert\cdot\rVert$	norm
T^n	n-dimensional torus
$f \circ g$	composition of functions f, g
$\triangleq$	'equal by definition'
$C^\infty(M)$ or $\mathscr{F}(M)$	set of real-valued functions on M
$f\mid U$	restriction of f to U
$\overline{\mathbb{R}^1}$	$\mathbb{R}^1 \cup \{-\infty, \infty\}$
$\square$	end of theorem, etc.
$GL(n)$	general linear group of degree n
$SL(n)$	special linear group of degree n
$O(n)$	orthogonal group of degree n
$Symm(n)$	set of n-dimensional symmetric matrices
$SO(n)$	special orthogonal group of degree n
$SO(p, q)$	special orthogonal group of type (p, q)
$Sp(n)$	symplectic group of degree n
$T_p(M)$	tangent space to M at P
$(\mathrm{d}f)_p$	differential of f at p
$(\mathrm{T}f)_p$ or f_{*p}	differential of $f: M \to N$ at p
$\otimes$	tensor product
T_p^*	cotangent space at p
f_p^*	dual map of f

$\cong$	'isomorphic to'
$\oplus$	direct sum
T_s^r	tensor space of type (r, s)
$L_r(V; F)$	set of r-linear maps of $V \times \cdots \times V$ into F
T^*	set of contravariant tensors
T_*	set of covariant tensors
$[X, Y]$	Lie bracket of X, Y
$\mathscr{T}^*_{*,p}$	tensor space at p
Ω_*	set of exterior differential forms
A_s	alternation map
$\wedge$	exterior multiplication
(E, p, B)	general bundle
$T(M)$	tangent bundle
$G_k(\mathbb{R}^n)$	Grassmann manifold
$V_k(\mathbb{R}^n)$	frame bundle
$E(\gamma_k^n)$	universal bundle
f^*E	pull-back bundle
$M_{k\times n}^k$	set of $k \times n$ matrices of rank k
$\mathbb{P}_{n-1}(\mathbb{R})$	real projective space
$\mathbb{P}_{n-1}(\mathbb{C})$	complex projective space
$\mathrm{Hom}(E, F)$	homomorphism bundle
Φ	flow of a vector field
L_X	Lie derivative with respect to x
$\lrcorner$	interior multiplication
$\mathscr{L}$, $\mathfrak{M}$, $\mathfrak{N}$	distributions
$\mathscr{T}_x^k$	subset of T_x
L_a, R_a	left and right translations
$\mathscr{I}_a$	inner automorphism of G
$\mathfrak{g}, \mathfrak{h}, \mathfrak{a}, \mathfrak{b}$	Lie algebras
ad, Ad	adjoint maps
$\mathfrak{D}\mathfrak{g}$	derived algebra of $\mathfrak{g}$
$\mathscr{C}^k\mathfrak{g}$	central series of $\mathfrak{g}$
$\mathfrak{g}^\alpha$	root space of $\mathfrak{g}$
$\triangle$	nonzero roots
$\mathbb{C}^n$	n-dimensional complex space
$\mathscr{D}_Y(p)$	discriminant variety of p with respect to Y
$\sqrt{I}$	radical of I
(p)	principal ideal
$\bar{I}$	closure of I
$\cup, \cap, \bigcup, \bigcap$	union, intersection
$\langle . , . \rangle$	inner product
X^*	dual space of X
$\langle .,. \rangle_{H^*,H}$	duality in H^*, H

$C^k(\bar{\Omega})$	functions k times differentiable on $\bar{\Omega}$
$\lVert . \rVert_k$	norm on previous space
$L^p(\Omega)$	functions with integrable p^{th} power
$\lVert . \rVert_{L^p(\Omega)}$	norm on previous space
$\lVert . \rVert_{L_\infty(\Omega)}$	norm on functions essentially bounded on Ω
$\langle . , . \rangle_{L^2(\Omega)}$	inner product on $L^2(\Omega)$
ℓ^p	pth power summable sequences
$\mathscr{D}'(\Omega)$	space of distributions on Ω
$H^{p,m}(\Omega)$	Sobolev space
$\lVert . \rVert_{p,m}$	norm on previous space
$H_0^{p,m}(\mathbb{R}^n)$	Sobolev space
$\mathscr{S}'(\mathbb{R}^n)$	space of tempered distributions
$M^\perp$	orthogonal complement of m
$\mathscr{B}(X, Y)$	bounded operators from X to Y
$\rho(A)$	resolvent set of A
$\sigma_P(A)$	point spectrum of A
$\sigma_C(A)$	continuous spectrum of A
$\sigma_R(A)$	residual spectrum of A
$\mathscr{U}$	input space
$\mathscr{Y}$	output space
$\mathfrak{M}$	space of solution trajectories
$\mathscr{L}(y)$	Laplace transform of y
$\mathscr{L}(V)$	Lie algebra generated by V
$\mathscr{L}_0(V)$	subspace of $\mathscr{L}(V)$
$\mathscr{L}'(V)$	subspace of $\mathscr{L}(V)$
$\mathscr{L}(V)$	subspace of $\mathscr{L}(V)$
$\mathbf{i} = (i_1, \ldots, i_m)$	m-tuple of integers
$\mathbf{t} = (t_1, \ldots, t_m)$	m-tuple of reals
$I_0(V, x)$	integral manifold of $\mathscr{L}_0(V)$
$I_0^t(V, x)$	integral manifold of $\mathscr{L}_0(V)$
$\bigotimes_k$	k^{th} order tensor product
$\mathscr{P}, \mathscr{I}(\mathscr{P})$	subalgebras of $\mathscr{L}(V)$
$\mathscr{R}(X; \mathscr{B}_0)$	X-radical of $\mathscr{B}_0$
$\triangle^\perp$	codistribution
$\mathscr{X}_{X_1}, \mathscr{Y}_{X_1}, \mathscr{T}_{X_1}, \mathscr{G}_{X_1}$	function spaces
$\ominus_V$	cascade operator
Φ^ξ, ξ_M^Φ	induced actions on M
Γ_ω	vector bundle mappings
$\#, \flat$	1-form to vector field translations
$\mathbb{N}^n$	n-tuples of natural numbers
ℓ_e^1	extended ℓ^1 space
$\mathscr{L}_n$	space of rank-n tensors
$\mathscr{L}_n^T$	space of simple tensors

$\mathscr{L}_{n,p}^{T}$	fibre of tensor bundle
$\mathscr{L}_{n}^{2,T}$	tensors over ℓ^2
$\mathscr{P}_k[F,G]$, $\mathscr{Q}_k[F,H]$	Grammian matrices
$\mathscr{R}[.]$	range of [.]
$\mathscr{A}, \mathscr{B}, \mathscr{C}, \mathscr{D}, \mathscr{X}, \mathscr{Y}$	subspaces of $\mathbb{R}^n$
θ	orthogonal difference
$\mathscr{H}$	subspace of $\mathscr{F}(M)$
$\mathscr{S}(x)$	ideal in $\mathscr{L}$
$\mathscr{A}_i$	tensor operator
$\rightharpoonup$	weak convergence
$\mathscr{F}$	Fréchet derivative
$\tilde{\mathscr{H}}$	abstract group of a Lie group
$\hat{i}$ or $\hat{\imath}$	means the component with value i is omitted
$\tilde{\mathbb{R}}$	$\mathbb{R}$ with a special structure
$f_{\sim}$	induced map of f under relation $\sim$
$\vee$	disjoint union
$\mathfrak{g}_{\Delta}$	direct sum of root spaces

1 MATHEMATICAL APPARATUS

1.1 DIFFERENTIABLE MANIFOLDS

1.1.1 Differentiable manifolds

We shall first recall some elementary notions from topology. A *topological space* $(X, \mathcal{O})$ is a set X, together with a set $\mathcal{O}$ of subsets of X which satisfies the properties

(a) $\emptyset, X \in \mathcal{O}$
(b) if $X_1, X_2 \in \mathcal{O}$ then $X_1 \cap X_2 \in \mathcal{O}$
(c) if $\{X_\alpha\}_{\alpha \in A} \subseteq \mathcal{O}$ then $\bigcup_{\alpha \in A} X_\alpha \in \mathcal{O}$.

The elements of $\mathcal{O}$ are called *open subsets* of X. A *neighbourhood* of a point $x \in X$ is a set N such that $x \in Y \subseteq N$ for some $Y \in \mathcal{O}$. The space X is called a *Hausdorff space* if $x_1, x_2 \in X$, $x_1 \neq x_2$ imply that there exist $X_1, X_2 \in \mathcal{O}$ with $x_1 \in X_1$, $x_2 \in X_2$ and $X_1 \cap X_2 = \emptyset$. An *open cover* of a subset $Y \subseteq X$ is a set $\mathcal{O}_1 \subseteq \mathcal{O}$ such that $Y \subseteq \bigcup \mathcal{O}_1$. The subset Y is *compact* if every open cover has a finite subcover.

A function $f\colon X \to Y$ between topological spaces X and Y is *continuous* (at $x \in X$) if $f^{-1}(W)$ is open for each open set W in Y containing $f(x)$. The function f is a *homeomorphism* if it is bijective and f and f^{-1} are continuous.

A *metric space* (X, d) is a set X together with a *distance function* $d\colon X \times X \to \mathbb{R}^+$ such that

(a) $d(x, y) = 0$ if and only if $x = y$
(b) $d(x, y) = d(y, x)$ for all $x, y \in X$
(c) $d(x, y) \leqslant d(x, z) + d(z, y)$ for all $x, y, z \in X$.

The *open ball* in X with centre x and radius ε is the set

$$B_\varepsilon(x) = \{y \in X\colon d(x, y) < \varepsilon\}.$$

Open sets in a metric space are unions of open balls and thus a metric space is a topological space.

The *distance* between a set Y and a point x in a metric space is defined as

$$d(x, Y) = \inf_{y \in Y} \{ d(x, y) \}.$$

The *distance* between subsets Y, Z is defined by

$$d(Y, Z) = \max\{ \sup_{y \in Y} d(y, Z), \quad \sup_{z \in Z} d(Y, z) \}.$$

Note that the set $(2^X, d)$ of all subsets of X with this distance function is not a metric space since

$$d(B_\varepsilon(x), \overline{B_\varepsilon(x)}) = 0$$

for any ball $B_\varepsilon(x) \triangleq \{ y \in X: d(x, y) < \varepsilon \}$. However, if we define the equivalence relation $\sim$ on 2^X by

$$Y \sim Z \quad \text{if and only if } d(Y, Z) = 0,$$

then $(2^X/\sim, d)$ is a metric space, and d is then called the *Hausdorff metric*.

Let M be a topological space. A *chart* on M is a pair (U, ϕ) where $U \subseteq M$ is open and ϕ is a homeomorphism of U onto an open subset of $\mathbb{R}^n$, for some n. If $x^i : \mathbb{R}^n \to \mathbb{R}$ denotes the standard projection on the ith component then, for any chart (U, ϕ), $x^i \circ \phi : U \to \mathbb{R}$ is called the *coordinate function*. We shall often write $x^i \circ \phi$ simply as x^i; then we are effectively identifying U with an open set in $\mathbb{R}^n$ via the homeomorphism ϕ. Two charts $\phi_1 : U_1 \to \mathbb{R}^n$, $\phi_2 : U_2 \to R^n$ are *compatible* if the mapping

$$\phi_2 \circ \phi_1^{-1} : \phi_1(U_1 \cap U_2) \to \phi_2(U_1 \cap U_2)$$

is of class C^∞ (i.e. all partial derivatives exist and are continuous). Of course, if $U_1 \cap U_2 = \emptyset$, then the charts are trivially compatible.

An *atlas* for the topological space M is a collection of compatible charts (U_i, ϕ_i) such that $\bigcup_i U_i = M$. An atlas is *complete* if it cannot be included in a larger atlas.

A *differentiable manifold* (of class C^∞) is a Hausdorff space with a complete atlas of charts. It is possible to specify a differentiable structure on M by defining any C^∞ atlas on M compatible with the given complete atlas, since the completions of both atlases are the same. Any C^∞ atlas on M will therefore define a *differentiable structure*.

It is easy to see that the number n must be constant on each connected component of M. If n is constant on the whole of M, then n is called the *dimension* of the differentiable manifold.

REMARK If we consider C^r mappings throughout the above discussion, we obtain C^r manifolds rather than C^∞ ones. Similarly, if $\mathbb{R}^n$ is replaced by $\mathbb{C}^n$

and we consider mappings which are holomorphic then we obtain *analytic manifolds* (of class C^∞). An atlas then defines the *analytic structure.*

Examples

1. $\mathbb{R}^n$ with the identity chart $id: \mathbb{R}^n \to \mathbb{R}^n$ is a differentiable manifold of dimension n.
2. The $(n-1)$-sphere $S^{n-1} \subseteq \mathbb{R}^n$ defined by $S^{n-1} = \{x \in \mathbb{R}^n : \|x\| = 1\}$ is a differentiable manifold of dimensions $n-1$, whose differentiable structure can be defined by two charts. (Note that no compact manifold can be specified by a single chart.)
3. Given two manifolds M, N with differentiable structures $\{(U_i, \phi_i)\}$, $\{(V_j, \psi_j)\}$ we can define a differentiable structure on the topological Cartesian product $M \times N$ as the collection of product charts $\{(U_i \times V_j, \phi_i \times \psi_j)\}$, these clearly being compatible. For example, the product of two circles $T^2 = S^1 \times S^1$ defines a torus; more generally, $T^n = S^1 \times \cdots \times S^1$ (n times) is an n-dimensional torus.
4. The set of $n \times m$ matrices $M_{n\times m}$ is a differentiable manifold with a single chart $\phi: M_{n\times m} \to \mathbb{R}^{n\times m}$ defined by $\phi(A) = (a_{ij})$, where A is the matrix (a_{ij}).

1.1.2 Differentiable functions

If M and N are differentiable manifolds of dimensions m and n, respectively, and $f: M \to N$ is a given function, then we say that f is *differentiable*‡ at $x \in X$ if the function

$$F \triangleq \psi \circ f \circ \phi^{-1} : \mathbb{R}^m \to \mathbb{R}^n$$

is differentiable for any charts (U, ϕ) of M and (V, ψ) of N where $x \in U, f(x) \in V$. The function $\psi \circ f \circ \phi^{-1}$ is defined on $\phi(f^{-1}(V) \cap U)$ (see Fig. 1.1). We say that f is *differentiable* if it is differentiable at each $x \in X$. Note that the definition of differentiability is independent of the choice of the coordinates ϕ, ψ, since if

$$\tilde{F} \triangleq \tilde{\psi} \circ f \circ \tilde{\phi}^{-1}$$

is another expression for f with respect to the coordinates ϕ, ψ, then the function

$$(\tilde{\psi} \circ \psi^{-1}) \circ F \circ (\phi \circ \tilde{\phi}^{-1})$$

is a restriction of $\tilde{F}$ to an open subset of the domain of F which is differentiable if and only if F is differentiable.

An injection $f: M \to N$ of a differentiable manifold M onto another

‡ If M, N are analytic manifolds, f is called *analytic*.

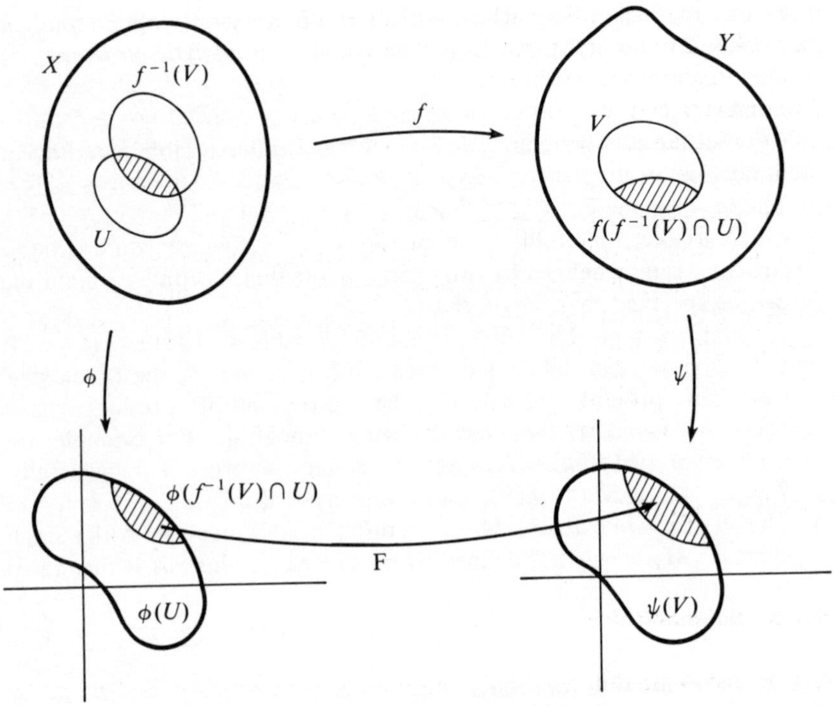

Fig. 1.1 Local representations of functions

differentiable manifold N is called a *diffeomorphism* if f and f^{-1} are differentiable. In this case M and N are said to be *diffeomorphic*.

Examples

1. If $\mathbb{R}^m$, $\mathbb{R}^n$ have their usual differentiable structures, then $f: \mathbb{R}^m \to \mathbb{R}^n$ is differentiable if it is differentiable in the usual sense.
2. Let $\mathbb{R}$ have its usual differentiable structure defined by the identity chart $\phi: \mathbb{R} \to \mathbb{R}, \phi(x) = x$, and let $\tilde{\mathbb{R}}$ be the *set* $\mathbb{R}$ with the differentiable structure defined by the single chart $\tilde{\phi}: \tilde{\mathbb{R}} \to \mathbb{R}$, $\tilde{\phi}(x) = x^3$. Then the identity map $id: \mathbb{R} \to \tilde{\mathbb{R}}$ is not a diffeomorphism since its inverse $id: \tilde{\mathbb{R}} \to \mathbb{R}$ has the coordinate expression $Id(x) \triangleq \phi \circ id \circ \tilde{\phi}^{-1}(x) = x^{1/3}$ which is not differentiable when $x = 0$. However, the map $f = \tilde{\phi}: \tilde{\mathbb{R}} \to \mathbb{R}$ is a diffeomorphism since the coordinate representations of f and its inverse $f^{-1} = (.)^{1/3}$ are both equal to the identity map $id: \mathbb{R} \to \mathbb{R}$.

REMARK Example 2 shows that a topological space can carry two differentiable structures which are not compatible, even though the resulting differentiable manifolds are diffeomorphic. The question then arises as to whether there exist truly different differentiable structures on $\mathbb{R}$. In fact, it turns out that all differentiable structures on $\mathbb{R}$ lead to diffeomorphic manifolds. Even more is true; if $n \neq 4$ then all differentiable structures on $\mathbb{R}^n$ are diffeomorphic. Milnor (1956) showed that there exists an 'exotic' differentiable structure on the seven-sphere S^7 (i.e. one which is not diffeomorphic to the standard structure). A theorem of Donaldson (1983) shows that there exists an exotic structure on $\mathbb{R}^4$; see also Freed and Uhlenbeck (1984).

It is customary to denote the set of differentiable, real-valued functions on a manifold M by $C^\infty(M)$ or $\mathscr{F}(M)$ and those real-valued functions which are defined in a neighbourhood of a point $p \in M$ and are differentiable at p by $C^\infty(p)$ or $\mathscr{F}(p)$. Note that $\mathscr{F}(M)$ (and $\mathscr{F}(p)$) are associative algebras over $\mathbb{R}$ with the operations

$$\begin{aligned} (\alpha f)(p) &= \alpha f(p), & \alpha \in \mathbb{R}, p \in M \\ (f+g)(p) &= f(p) + g(p), & p \in M \\ (fg)(p) &= f(p)g(p), & p \in M, \end{aligned}$$

for all $f, g \in \mathscr{F}(M)$.

1.1.3 Submanifolds

Let M and N be differentiable manifolds of dimensions m and n and let $f: M \to N$ be a differentiable mapping. Then we say that f is an *immersion* if for each $p \in M$ there is a neighbourhood U of p in M and a chart (V, ψ) containing $f(p)$ in N such that $\phi \triangleq \psi \circ f \mid U$ is a chart for M. The manifold M is then said to be *immersed* in N. An injective immersion is called an *embedding*. Since an immersion is clearly locally injective, it follows that an immersion is a local embedding.

A subset M of a manifold N is called a *submanifold* if the canonical injection $i: M \subseteq N$ is an embedding, for a given differentiable structure on M.

Examples

1. If M is an open subset of N, then M is a submanifold, and is called an *open submanifold*. Charts on M are just the restrictions of charts on N to M, and it is clear that the manifold dimensions of M and N are the same.
2. Let $\mathbb{R}^1$ and $\mathbb{R}^2$ have their usual differentiable structures and let $f_i: \mathbb{R}^1 \to \mathbb{R}^2$, $i = 1, 2, 3$ be the mappings shown (by their images in $\mathbb{R}^2$) in Fig. 1.2. Then f_1 is an immersion and f_2 is an embedding, so that $f_2(\mathbb{R}^1)$ (with the differentiable structure induced from $\mathbb{R}^1$) is a submanifold of $\mathbb{R}^2$. Note, however, that the topology of $f_2(\mathbb{R}^1)$ is induced from $\mathbb{R}^1$ and not that induced as a subspace of $\mathbb{R}^2$, so that any $\mathbb{R}^2$-neighbourhood of p contains points in $f_2(U)$ and $f_2(V)$, where U is

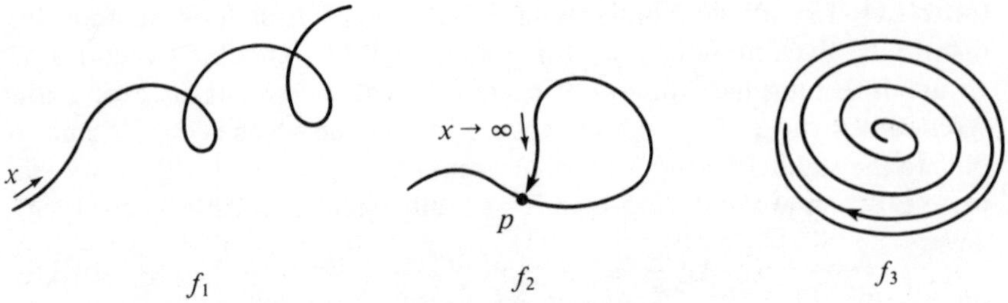

Fig. 1.2 Immersed submanifolds

a neighbourhood of $f_2^{-1}(p)$ and V is a neighbourhood of ∞ in $\bar{\mathbb{R}}^1$ such that $U \cap V = \emptyset$. $f_3 : \mathbb{R}^1 \to \mathbb{R}^2$ is topologically an unstable trajectory of a system with a limit cycle, and is an embedding.

Embedded submanifolds can be characterized as follows:

Theorem 1.1.1
A manifold M is an m-dimensional submanifold of the n-dimensional manifold N if for all $p \in X$, there exists a chart (V, ψ) of N with $p \in V$ such that

(a) $\psi(p) = 0$
(b) the set $W = \{q \in V: y^{m+1} \circ \psi(q) = \cdots = y^n \circ \psi(q) = 0\}$ and the restrictions of $y^1, \ldots, y^m$ to W form a chart of M with $p \in W$. (Here, $(y^1, \ldots, y^n)$ are the coordinate functions.)

Moreover, if $f: M \to N$ is injective and differentiable, and for every $p \in X$ there exists a chart (U, ϕ) of p in M and a chart (V, ψ) of $f(p)$ in N such that the linear map

$$D(\psi \circ f \circ \phi^{-1})(\phi(p)) : \mathbb{R}^m \to \mathbb{R}^n$$

is injective, then $f(M)$ is a submanifold of N with the differentiable structure which makes $f: M \to f(M)$ a diffeomorphism. □

Using the implicit function theorem, it can be shown that if $f: \mathbb{R}^n \to \mathbb{R}$ is a differentiable function and $M = f^{-1}(0)$, then M has a uniquely determined differentiable structure making it an $(n-1)$-dimensional submanifold of $\mathbb{R}^n$ if, for each $p \in M$, $(\text{grad } f)(p) \neq 0$. This can be generalized to functions $f: M \to N$ for manifolds M and N of dimensions m and n $(m \geqslant n)$:

Theorem 1.1.2
If $f: M \to N$ is differentiable and $q \in N$, then $f^{-1}(q)$ is a submanifold of M of dimension $(m-n)$ (or is empty) if for any $p \in f^{-1}(q)$ there are charts

(U, ϕ), (V, ψ) around p and q such h that

$$D(\psi \circ f \circ \phi^{-1})(\phi(p)) : \mathbb{R}^m \to \mathbb{R}^n$$

is surjective. □

If the condition in Theorem 1.1.2 holds for all $p \in M$, then f is called a *submersion*. Theorems 1.1.1 and 1.1.2 demonstrate that the linear map $(D(\psi \circ f \circ \phi^{-1}))(\phi(p))$ must be of full rank to determine submanifolds.

Examples

1. The *general linear group of degree n*‡ $GL(n) \subseteq M_{n\times n}$ consisting of invertible matrices of dimension n is an open submanifold of $M_{n\times n} = \mathbb{R}^{n\times n}$. Of course, if $A \in GL(n)$, then $\det A \neq 0$. The subset $SL(n)$ of $GL(n)$ of matrices of determinant $+1$ (the *special linear group of degree n*) is a subgroup of $GL(n)$ which is a submanifold since it is easy to see that the map
$$f: GL(n) \to \mathbb{R}$$
has surjective derivative $(Df)(A)$ at any $A \in GL(n)$. Hence $SL(n)$ is a manifold of dimension $n^2 - 1$.
2. The subset $O(n)$ of $GL(n)$, the *orthogonal group of degree n*, is the set of matrices A such that $AA^{\mathrm{T}} = I$. To show that $O(n)$ is a submanifold of $GL(n)$ consider the function
$$f: GL(n) \to Symm(n), \quad f(A) = AA^{\mathrm{T}} - I$$
where $Symm(n)$ is the set of symmetric matrices. Then $O(n) = f^{-1}(0)$ and since, for any $B \in M_{n\times n}$, $A \in GL(n)$,
$$\{Df(A)\}(B) = \lim_{t\to 0} \frac{1}{t}\, [f(A + tB) - f(A)] = BA^{\mathrm{T}} + (BA^{\mathrm{T}})^{\mathrm{T}}$$
the linear map $Df(A): M_{n\times n} \to Symm(n)$ is surjective. (For $C \in Symm(n)$, define $B = \frac{1}{2}C(A^{\mathrm{T}})^{-1}$.) $O(n)$ has dimension $n^2 - n(n+1)/2 = n(n-1)/2$.

   Note that $O(n)$ is not connected since $\det A = \pm 1$ if $A \in O(n)$ and the function $\det : O(n) \to \mathbb{R}\backslash\{0\}$ is continuous. The connected component
$$SO(n) = \{A \in O(n) : \det A = 1\} = O(n) \cap SL(n)$$
is a manifold of dimension $n(n-1)/2$.
3. If $0 < p < n$, $q = n - p$ and
$$B = \begin{bmatrix} -I_p & 0 \\ 0 & I_q \end{bmatrix},$$

‡ If we wish to specify the vector space $\mathbb{R}^n$ we write $GL(n; \mathbb{R})$.

where I_r is the identity matrix of dimension r, then we define

$$SO(p,q) = \{A \in GL(n) : A^{\mathrm{T}}BA = B\}.$$

Then as in (2) above it can be shown that $SO(p,q)$ is a manifold of dimension $n(n-1)/2$.

Similarly, if n is even, say $n = 2p$, and

$$C = \begin{bmatrix} 0 & \mathrm{I}_p \\ -\mathrm{I}_p & 0 \end{bmatrix},$$

then

$$Sp(n) = \{A \in GL(n) : A^{\mathrm{T}}CA = C\}$$

is called the *symplectic group of degree n* and has dimension $n(n+1)/2$.

We finish this section with a theorem of Whitney which says that any differentiable manifold may be regarded as a submanifold of some Euclidean space.

Theorem 1.1.3

Let M be an m-dimensional manifold. Then M is diffeomorphic to a submanifold of $\mathbb{R}^n$ with n at most $2m+1$. □

(See Auslander and MacKenzie, 1963.)

Note that this result is false for complex manifolds: in fact, there are no compact complex submanifolds of $\mathbb{C}^n$ of positive dimension (Wells, 1979).

1.1.4 Tangent vectors, cotangent vectors and tensors

The vector space of tangent vectors to a differentiable manifold at a point p generalizes the notion of tangent plane to an embedded submanifold of $\mathbb{R}^n$. In the latter case, if $M \subseteq \mathbb{R}^n$ is a submanifold of dimension m, then the affine plane in $\mathbb{R}^n$ tangent to M at any point has dimension m and so may be identified with $\mathbb{R}^m$. Since vectors give a sense of direction in a vector space, given any differentiable function $f: \mathbb{R}^n \to \mathbb{R}$ we can define the derivative of f in the direction of a vector $v \in \mathbb{R}^m \subseteq \mathbb{R}^n$ in the tangent space of M at p by

$$\left(\frac{\partial f}{\partial v}\right)_p \triangleq v.(\mathrm{grad}\, f)_p.$$

Now recall that $\mathscr{F}(p)$ has been defined as the set of all functions defined and differentiable in a neighbourhood of p. Then it is easy to check (using the chain rule) that $(\partial(.)/\partial v)_p$ is a linear map and that

$$(\partial(fg)/\partial v)_p = (\partial f/\partial v)_p g(p) + f(p)(\partial g/\partial v)_p.$$

Abstracting these properties of the usual directional derivative we say that, if p is any point of a differentiable manifold M of dimension n, then a linear map L from $\mathscr{F}(p)$ into $\mathbb{R}$ is called a *derivation* (on $\mathscr{F}(p)$) if

$$L(fg) = L(f)g(p) + f(p)L(g).$$

The set $T_p(M)$ (or just T_p) of all derivations on $\mathscr{F}(p)$ is called the *tangent space* to M at p.

Note that the tangent space is often defined as the set of equivalence classes of curves in M having the 'same direction' at p. The two definitions are linked by the observation that if $\alpha : (a, b) \subseteq \mathbb{R} \to M$ is a differentiable curve, then the linear map $L : \mathscr{F}(p) \to \mathbb{R}$ defined by

$$Lf = (\mathrm{d}f(\alpha(t))/\mathrm{d}t)_{t_0},$$

where $t_0 = \alpha^{-1}(p) \in (a, b)$ is a derivation, which in the case of Euclidean spaces is just $(\text{grad } f)_p.(\partial\alpha/\partial t)_{t_0}$.

$T_p(M)$ is clearly a vector space. Let (U, ϕ) be a coordinate neighbourhood of p and let x^i, $1 \leqslant i \leqslant n$, be the local coordinates in this neighbourhood. Then, expressing $f \in \mathscr{F}(p)$ in local coordinates, we have

$$\left(\frac{\mathrm{d}f(\alpha(t))}{\mathrm{d}t}\right)_t = \sum_{i=1}^{n} \left(\frac{\partial f}{\partial x^i}\right)_p \cdot \left(\frac{\mathrm{d}\alpha^i}{\mathrm{d}t}\right)_{t_0}$$

where $\alpha : (a, b) \subseteq \mathbb{R} \to M$ is a curve in M. Hence the derivations $(\partial/\partial x^i)_p$, $1 \leqslant i \leqslant n$, span T_p which therefore has dimension $\leqslant n$. However, if $\sum_{i=1}^{n} c_i(\partial/\partial x^i)_p = 0$, then the tangent vector to the curve $x^i = x^i(p) + c_i t$ is $\sum_{i=1}^{n} c_i(\partial/\partial x^i)_p$ and

$$\sum_{i=1}^{n} c_i(\partial x^j/\partial x^i)_p = 0 = c_j$$

and so the derivations $(\partial/\partial x^i)_p$ are linearly independent. Hence we have proved

Theorem 1.1.4

T_p is a vector space of dimension n with basis $(\partial/\partial x^i)_p$ in terms of the local coordinates. □

If $x = \xi(y)$ is a differentiable change of local coordinates, then we obtain the familiar transformation law for a vector, given by

$$\left(\frac{\partial}{\partial y^i}\right)_p = \sum_{j=1}^{n} \left(\frac{\partial x^j}{\partial y^i}\right)_p \left(\frac{\partial}{\partial x^j}\right)_p = \sum_{j=1}^{n} \left(\frac{\partial \xi^j}{\partial y^i}\right)_p \left(\frac{\partial}{\partial x^j}\right)_p.$$

If $f: \mathbb{R}^m \to \mathbb{R}^n$ is a differentiable map, the differential of f at $p \in \mathbb{R}^m$ is defined as the linear map $(\mathrm{d}f)_p: \mathbb{R}^m \to \mathbb{R}^n$ whose matrix representation (in the standard bases of $\mathbb{R}^m, \mathbb{R}^n$) is given by the Jacobian matrix $J_f(p) \triangleq (\partial f^i/\partial x^j)$ of f. As a linear map $\mathrm{d}f$ maps a vector v in $\mathbb{R}^m$ to a vector $w = \mathrm{d}f(v)$ in $\mathbb{R}^n$. Now any $\mathbb{R}^r$ is linearly isomorphic to its tangent space and so w defines a derivation of $\mathscr{F}(f(p))$ by

$$w\,.\,(\text{grad } g)_{f(p)}, \qquad g \in \mathscr{F}(f(p)).$$

But

$$w\,.\,(\text{grad } g) = \mathrm{d}f(v)\,.\,(\text{grad } g) = v\,.\,(\text{grad}(g \circ f)).$$

Hence we can generalize the notion of *differential* of a function $f: M \to N$ for differentiable manifolds by defining the linear map

$$(\mathrm{d}f)_p: T_p(M) \to T_{f(p)}(N)$$

by

$$\{(\mathrm{d}f)_p(L)\}(g) = L(g \circ f), \qquad L \in T_p(M), \quad g \in \mathscr{F}(f(p)).$$

It is clear that $g \circ f \in \mathscr{F}(p)$ for $g \in \mathscr{F}(f(p))$ and that $(\mathrm{d}f)_p(L)$ is in $T_{f(p)}(N)$.

The differential of f at p is also denoted by $(Tf)_p$ or f_{*p}. It is then easy to show that f_{*p} is represented locally by the Jacobian matrix of the local representation of f. Moreover we have, for any manifolds M, N, P and differentiable functions $f: M \to N$ and $g: N \to P$,

$$\begin{aligned}\{(g \circ f)_{*p}L\}(h) &= L(h \circ (g \circ f))\\ &= \{(f_{*p})L\}(h \circ g)\\ &= (g_{*f(p)}f_{*p}L)(h)\end{aligned}$$

for any $h \in \mathscr{T}((g \circ f)(p))$, and so the *chain rule*

$$(g \circ f)_{*p} = g_{*f(p)}f_{*p}$$

holds.

Note also that

$$\dim(\text{range } f_{*p}) + \dim(\ker f_{*p}) = \dim(T_pM).$$

The dimension of range f_{*p} is called the *rank* of f at p and, using local coordinates, it is seen to be just the rank of the Jacobian matrix of the local representation of f.

Examples

1. If $M_1 \times M_2$ is a product manifold and $p_i: M_1 \times M_2 \to M_i$ are the obvious projections, then the map

$$(p_{1*(a,b)}, p_{2*(a,b)}): T_{(a,b)}(M_1 \times M_2) \to T_aM_1 \oplus T_bM_2$$

is an isomorphism, for any $(a,b) \in M_1 \times M_2$. Hence identifying these spaces, if

$$f\colon M_1 \times M_2 \to L$$

is a function with values in a manifold L and $f_a\colon M_2 \to L$, $f_b\colon M_1 \to L$ are defined by $f_a(q) = f(a,q)$, $f_b(p) = (p,b)$, respectively, then

$$f_{*(a,b)}(X_1, X_2) = (f_{b*})_a(X_1) + (f_{a*})_b(X_2), \qquad X_1 \in T_aM_1, \quad X_2 \in T_bM_2.$$

This is *Leibnitz's formula*.

2. If $t\colon \mathbb{R} \to \mathbb{R}$ is the standard coordinate for $\mathbb{R}$, and $\alpha\colon (a,b) \to M$ is a (differentiable) curve in M, then the tangent vector to α at $t \in (a,b)$ is defined by

$$\dot{\alpha}(t) = (\alpha_{*t})(\mathrm{d}/\mathrm{d}t).$$

For $f \in \mathscr{F}(p)$, where $p = \alpha(t)$, we have

$$\{\dot{\alpha}(t)\}f = \alpha_{*t}(\mathrm{d}/\mathrm{d}t)f = \frac{\mathrm{d}}{\mathrm{d}t}(f \circ \alpha)$$

as seen above.

3. $f\colon M \to N$ is an immersion if and only if f_{*p} is injective for all $p \in M$, while f is a submersion if and only if f_{*p} is surjective for all $p \in M$.
4. If $f\colon M \to N$ is a submersion, then $S = f^{-1}(q)$ is a submanifold of M for any $q \in f(M)$. Then if $i\colon S \to M$ is the natural injection, the function $f \circ i$ is constant and so $f_* \circ i_* = 0$, and thus

$$i_*(T_pS) \subseteq f_{*p}^{-1}(0), \qquad p \in S$$

Since T_pS and $\ker f_{*p}$ both have the same dimension, the tangent space to S at p equals the kernel of f_{*p}.

For a manifold M and a point $p \in M$, let $T_p^*(M)$ (or just T_p^*), the *cotangent space* of M at p, denote the dual vector space of $T_p(M)$. If $f \in \mathscr{F}(p)$, then the equation

$$\langle (\mathrm{d}f)_p, X\rangle = Xf \qquad \text{for all } X \in T_p(M)$$

defines an element $(\mathrm{d}f)_p$ of $T_p^*(M)$, called the *total differential* of f at p. Given a local coordinate system $x^1, \ldots, x^n$ in M at p, it is easy to see that the cotangent vectors $(\mathrm{d}x^1)_p, \ldots, (\mathrm{d}x^n)_p$ form a basis of T_p^* dual to the basis $(\partial/\partial x^1)_p, \ldots, (\partial/\partial x^n)_p$ of T_p. We have defined above the differential $f_{*p}\colon T_p(M) \to T_{f(p)}(N)$ of $f\colon M \to N$ at p; the adjoint map $f_p^*\colon T_{f(p)}^*(N) \to T_p^*(M)$ is defined by

$$\langle f_p^*\omega, X\rangle = \langle \omega, f_{*p}X\rangle, \qquad \omega \in T_{f(p)}^*(N), \quad X \in T_p(M).$$

We shall next review the basic facts concerning tensor algebras; the results are well-known and can be found in Greub (1978) or Kobayashi and

Nomizu (1969). Let U and V be vector spaces (of finite dimension – the infinite dimensional case is similar, except for the dimension relations). The tensor product is defined (up to isomorphism) by the *universal factorization property*: it is a space $U \otimes V$, together with a bilinear map $\otimes : U \times V \to U \otimes V$, for which, given any vector space W and a bilinear map $\phi : U \times V \to W$, there exists a unique linear map $f: U \otimes V \to W$ such that the diagram

$$\begin{array}{ccc} U \times V & \xrightarrow{\ \phi\ } & W \\ {\scriptstyle\otimes}\big\downarrow & \nearrow_{f} & \\ U \otimes V & & \end{array}$$

commutes.

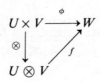

To prove the existence of $U \otimes V$ we can simply take $U \otimes V = [\mathrm{sp}(U \times V)]/N$, where $\mathrm{sp}(U \times V)$ is the vector space with basis $U \otimes V$ and N is the subspace spanned by elements of the form

$$(u_1 + u_2, v) - (u_1, v) - (u_2, v), \quad (u, v_1 + v_2) - (u, v_1) - (u, v_2)$$
$$(\alpha u, v) - \alpha(u, v), \quad (u, \alpha v) - \alpha(u, v),$$

for $u, u_1, u_2 \in U$, $v, v_1, v_2 \in V$ and scalar α. The map $\otimes : U \times V \to U \otimes V$ is defined by $\otimes(u, v) = u \otimes v$. where $u \otimes v$ is the image of (u, v) under the canonical projection $\mathrm{sp}(U \times V) \to U \otimes V$. We shall state, without proof, the following elementary facts:

(a) $U \otimes V \cong V \otimes U$, where $\cong$ means 'vector space isomorphic'.
(b) $F \otimes U \cong U \cong U \otimes F$, where F is the scalar field.
(c) $(U \otimes V) \otimes W \cong U \otimes (V \otimes W)$
(d) If $f_i : U_i \to V_i$ are linear maps, $i = 1, 2$, then there exists a unique linear map $f: U_i \otimes U_2 \to V_1 \otimes V_2$ given by $f(u_1 \otimes u_2) = f_1(u_1) \otimes f_2(u_2)$.
(e) $(U_1 \oplus \cdots \oplus U_k) \otimes V = (U_1 \otimes V) \oplus \cdots \oplus (U_k \otimes V)$.
From (e) it follows that $U \otimes V$ has a basis $u_i \otimes v_j, i = 1, \ldots, m, j = 1, \ldots, n$, where $(u_i)_{1 \leqslant i \leqslant m}$, $(v_j)_{1 \leqslant j \leqslant n}$ are bases of U and V, respectively.
(f) Let $L(U, V)$ denote the space of linear maps from U into V, and let U^* denote the dual space of U. Then $L(U^*, V) \cong U \otimes V$ where $u \otimes v$ is identified with the linear map $\langle u, . \rangle v: U^* \to V$.
(g) $U^* \otimes V^* \cong (U \otimes V)^*$ where $u^* \otimes v^* \to g \in (U \otimes V)^*$ is identified with

$$g(u \otimes v) = \langle u, u^* \rangle \langle v, v^* \rangle.$$

Now denote by $\otimes_k V$ the tensor product of k copies of V. Then the space

$$T_s^r = (\otimes_r V) \otimes (\otimes_s V^*)$$

is called the *tensor space of contravariant degree r and covariant degree s*. We often write $T^r = T^r_0$, $T_s = T^0_s$, and, by definition, $T^0_0 = F$, the scalar field. An element of T^r_s is called a *tensor of type* (r, s).

If $e_1, \ldots, e_n$ is a basis of V and $e^1, \ldots, e^n$ is the dual basis of V^*, then any $K \in T^r_s$ may be written in the form

$$K = \Sigma_{i_1, \ldots, i_r, j_1, \ldots, j_s} K^{i_1 \ldots i_r}_{j_1 \ldots j_s} e_{i_1 \ldots i_r} \otimes e^{j_1 \ldots j_s},$$

where $e_{i_1 \ldots i_r} = e_{i_1} \otimes \cdots \otimes e_{i_r}$, $e^{j_1 \ldots j_s} = e^{j_1} \otimes \cdots \otimes e^{j_s}$.

Putting $T = \oplus^{\infty}_{r,s=0} T^r_s$, we can make T into an associative algebra (the *tensor algebra over* V) by defining the product $\phi : T^r_s \times T^p_q \to T^{r+p}_{s+q}$ which is the unique bilinear map for which $\phi(K, L) = K \otimes L$ for $K \in T^r_s$, $L \in T^p_q$. In component form,

$$(K \otimes L)^{i_1 \ldots i_{r+p}}_{j_1 \ldots j_{s+q}} = K^{i_1 \ldots i_r}_{j_1 \ldots j_s} L^{i_{r+1} \ldots i_{r+p}}_{j_{s+1} \ldots j_{s+q}}$$

For $i \in \{1, \ldots, r\}$, $j \in \{1, \ldots, s\}$ the *contraction map* $C = C(i, j) = T^r_s \to T^{r-1}_{s-1}$ is the unique map for which

$$C(v_1 \otimes \cdots \otimes v_r \otimes v_1^* \otimes \cdots \otimes v_s^*) = \langle v_i, v_j^* \rangle v_1 \otimes \cdots \otimes V_{i-1} \otimes V_{i+1} \cdots \otimes V_r \otimes V_1^* \otimes \cdots \otimes V_{j-1}^* \otimes V_{j+1}^* \otimes \cdots \otimes V_s^*$$

or, in component form,

$$(CK)^{i_1 \ldots \hat{i}_k \ldots i_r}_{j_1 \ldots \hat{j}_k \ldots j_s} = \Sigma_k K^{i_1 \ldots k \ldots i_r}_{j_1 \ldots k \ldots j_s}$$

We can interpret tensors as multilinear maps in the following way:
Let $L_r(V; F)$ denote the vector space of all r-linear maps of $V \times \cdots \times V$ into F.
Then we have

(a) $T_s \cong L_s(V; F)$
(b) $T^r \cong L_r(V^*; F)$.

A *derivation* of the tensor algebra T is a linear map $D : T \to T$ such that

(a) $D(T^r_s) \subseteq T^r_s$
(b) $D(K \otimes L) = DK \otimes L + K \otimes DL$, for all $K, L \in T$
(c) $D(CK) = C(DK)$, for all $K \in T$ and any contraction C.

Finally, if $T(U)$ denotes the tensor algebra over U, then given an isomorphism of vector spaces $\phi : U \to V$, it is easy to see that ϕ extends to an isomorphism of algebras $T(\phi) : T(U) \to Y(V)$ which commutes with contractions and preserves tensor order. If $T^*(U) = \oplus^{\infty}_{r=0} T^r$, $T_*(U) = \oplus^{\infty}_{s=0} T_s$ are, respectively, the contravariant and covariant tensors over U, then any linear map $\phi : U \to V$ extends to linear maps $T^*(\phi)$, $T_*(\phi)$

such that the diagrams

$$\begin{array}{ccc} U & \xrightarrow{\phi} & V \\ \downarrow{\scriptstyle i_U} & & \downarrow{\scriptstyle i_V} \\ T^*(U) & \xrightarrow{T^*(\phi)} & T^*(V) \end{array} \qquad \begin{array}{ccc} U^* & \xleftarrow{\phi^*} & V^* \\ \downarrow{\scriptstyle i_{U^*}} & & \downarrow{\scriptstyle i_{V^*}} \\ T_*(U) & \xleftarrow{T_*(\phi)} & T_*(V) \end{array}$$

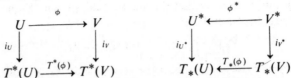

commute, where the vertical maps are the canonical injections and ϕ^* is the dual map of ϕ. We therefore see that the classical nomenclature for contravariant and covariant tensors is unfortunate, since it conflicts with the modern functorial one. In fact, the above diagrams show that T^* is a *covariant* functor, while T_* is *contravariant*.

1.1.5 Vector and tensor fields, differential forms

Let $f: X \to Y$ be a surjective function for any sets X and Y. A *section* of f is a function $s: Y \to X$ such that $f \circ s = \text{identity}$, i.e. $s(y) \in f^{-1}(y)$ for all $y \in Y$. A (differentiable) *vector field* X on a manifold M is a section of the projection $\pi: \bigcup_{p \in M} T_p(M) \to M$ such that, if we write X in local coordinates x^i, in some open set U, as

$$X = \sum_{i=1}^{n} a_i \frac{\partial}{\partial x^i}$$

then the functions a_i (defined on U) are differentiable. The set of all vector fields on M is denoted by $D(M)$. If X and Y are vector fields, then it can be checked using local coordinates that $[X, Y]$ is a vector field, where

$$[X, Y]f = X(Yf) - Y(Xf).$$

(Note that, for any vector field X, the function $Xf \in C^\infty(M)$ for each $f \in C^\infty(M)$ is defined by

$$(Xf)(p) = X_p f,$$

where $X_p \in T_p(M)$ is the tangent vector $X(p)$.)

The set of vector fields is a module over $C^\infty(M)$ if we define $(fX)_p = f(p)X_p$, for $p \in M$, $f \in C^\infty(M)$. Then we have

$$[fX, gY] = fg[X, Y] + f(Xg)Y - g(Yf)X, \qquad f, g \in C^\infty(M),$$

for, any vector fields X, Y.

Now let $f: M \to N$ be a differentiable map between the manifolds M, N and recall that f induces the map $f_{*p}: T_p(M) \to T_{f(p)}(N)$. If X is a vector field on M then we would like to use f_{*p} to define a vector field Y on N (or, at least, on $f(M)$) by the formula

$$Y_q = f_{*p} X_p, \qquad q \in f(M)$$

where $p \in f^{-1}(q)$. This is possible only if $f_{*p}X_p = f_{*p'}, X_{p'}$ whenever $f(p) = f(p')$, and so we say that the vector fields X and Y on M and N, respectively, are *f-related* if

$$Y_{f(p)} = f_{*p}X_p \qquad \text{for all } p \in M.$$

An elementary exercise shows that if the vector fields X_1, Y_1 and X_2, Y_2 are f-related, then so are $[X_1, X_2]$ and $[Y_1, Y_2]$.

If $f\colon M \to M$ is differentiable, then the vector field X is called *f-invariant* if X is f-related to itself. In this case,

$$X_{f(p)} = f_{*p}X_p \qquad \text{for all } p \in M.$$

The above ideas can be generalized to tensors in the following way. Let $T_p = T_p(M)$ be the tangent space to a manifold M at p and let

$$\mathscr{T}^*_{*,p} = \oplus\, T^r_{s,p}$$

where $T^r_{s,p}$ is the tensor space of type (r, s) over T_p. Then a *tensor field L* of type (r, s) over M is a section of the projection $\pi : \bigcup_{p \in M} T^r_{s,p} \to M$ such that, if

$$L = \Sigma L^{i_1 \ldots i_r}_{j_1 \ldots j_s} X_{i_1} \otimes \cdots \otimes X_{i_r} \otimes \omega^{j_1} \otimes \cdots \otimes \omega^{j_s}$$

is a local expression for L in a neighbourhood $U \subseteq M$, then the functions $L^{i_1 \ldots i_r}_{j_1 \ldots j_s} \colon U \to R$ are differentiable. Let $\mathscr{T}^r_s$ denote the set of tensor fields of type (r, s), and let $\mathscr{T}^r = \mathscr{T}^r_0$, $\mathscr{T}_s = \mathscr{T}^0_s$, $\mathscr{T}^* = \oplus_{r=0}^\infty \mathscr{T}^r$, $\mathscr{T}_* = \oplus_{s=0}^\infty \mathscr{T}_s$.

Using the fact that for a tensor field L of type $(0, s)$, $L_p \triangleq L(p)$ may be considered as an s-linear map of $T_p \times \cdots \times T_p \to \mathbb{R}$ it can be shown (Kobayashi and Nomizu, 1969) that L may be considered as an s-linear map of $D(M) \times \cdots \times D(M)$ into $\mathscr{F}(M)$ such that

$$L(f_1X_1, \ldots, f_sX_s) = f_1 \ldots f_s L(X_1, \ldots, X_s), \qquad f_i \in \mathscr{F}(M), \quad X_i \in D(M).$$

We simply let $L(X_1, \ldots, X_s)$ be the map which assigns the real number $L_p(X_{1p}, \ldots, X_{sp})$ to the point $p \in M$. A similar result holds for $L \in T^1_s$, but then $L : D(M) \times \cdots \times D(M) \to D(M)$.

In the final part of this section we shall define differential forms and the exterior derivative. Let Ω_s (or $\Omega_s(M)$) denote the set of $\mathscr{F}(M)$-multilinear maps of $D(M) \times \cdots \times D(M)$ into $\mathscr{F}(M)$ which are alternating; i.e. maps $\phi : D(M) \times \cdots \times D(M) \to \mathscr{F}(M)$ for which

$$\phi(X_1, \ldots, fX_i, \ldots, X_s) = f\phi(X_1, \ldots, X_s), \; f \in \mathscr{F}(M), \; X_i \in D(M)$$

and

$$\phi(X_1, \ldots, X_i, \ldots, X_j, \ldots, X_s) = 0 \qquad \text{if } X_i = X_j, \quad i \neq j.$$

If $\Omega_0 = \mathscr{F}(M)$, let $\Omega_* (= \Omega_*(M)) = \oplus_{s=0}^\infty \Omega_s$. Then Ω_* is called the set of *(exterior) differential forms* on M, and any $\phi \in \Omega_*$ is called an *s-form*. Note that Ω_s is a submodule of $\mathscr{T}_s$.

The set Ω_* can be made into an algebra in the following way. Define a linear map $A_s: \mathscr{T}_s \to \mathscr{T}_s$ by

$$A_s(d_s) = \frac{1}{s!} \sum_{\sigma \in P_s} \operatorname{sgn}(\sigma)\sigma \cdot d_s, \qquad d_s \in \Omega_s, \quad \text{if } s > 0$$

where P_s is the set of permutations of $\{1, \ldots, s\}$ and

$$\sigma \cdot d_s(X_1, \ldots, X_s) = d_s(X_{\sigma^{-1}(1)}, \ldots, X_{\sigma^{-1}(s)}), \qquad X_i \in D(M).$$

Then we can extend A_s to an $\mathscr{F}(M)$-linear map $A: \mathscr{T}_* \to \mathscr{T}_*$ by defining

$$A(d) = \bigoplus_{s=0}^{\infty} A_s(d_s), \qquad d = \bigoplus_{s=0}^{\infty} d_s, \quad d_s \in \mathscr{T}_s.$$

Clearly, if $\tau \in P_s$ and $d_s \in \mathscr{T}_s$, then $\tau . (A_s(d_s)) = \operatorname{sgn}(\tau)A_s(d_s)$, so $A_s(\mathscr{T}_s) \subseteq \Omega_s$ and $A(\mathscr{T}_*) \subseteq \Omega_*$. However, it also follows that $A_s(d_s) = d_s$ for all $d_s \in \Omega_s$ and so $A^2 = A$ and $A(\mathscr{T}_*) = \Omega_*$. Hence A is the projection of $\mathscr{T}_*$ onto Ω_*. If $N = \operatorname{Ker}(A)$, then N is a submodule of $\mathscr{T}_*$ which can be shown to be a two-sided ideal in $\mathscr{T}_*$ (see Chevalley, 1946, Helgason, 1962). Thus,

$$\Omega_* \cong \mathscr{T}_*/N$$

as associative $\mathscr{F}(M)$-algebras and we write

$$\omega_1 \wedge \omega_2 = A(\omega_1 \otimes \omega_2), \qquad \omega_1, \omega_2 \in \Omega_*.$$

Ω_* with this *exterior* multiplication is called the *Grassmann algebra* of M.

Similarly, we can define the Grassmann algebra Ω_{*p} as the $\mathbb{R}$-multilinear, alternating real-valued functions on products of the tangent space $T_p \times \cdots \times T_p$. Then

$$\omega_{1p} \wedge \omega_{2p} = (\omega_1 \wedge \omega_2)_p,$$

where $\omega_1, \omega_2 \in \Omega_*$ and ω_{1p}, ω_{2p} are well-defined, since an s-form $\omega \in \Omega_s$ can be regarded as a differentiable section of the projection $\bigcup_{p \in M} \Omega_{sp} \to M$. Note that $T_p^* \subseteq \Omega_{*p}$ and if $\omega_1, \omega_2 \in T_p^*$, then $\omega_1 \wedge \omega_2 = -\omega_2 \wedge \omega_1$ and so we have

$$\omega^1 \wedge \cdots \wedge \omega^k = \det(a_{ij})\theta^1 \wedge \cdots \wedge \theta^k,$$

where $\theta^i \in T_p^*$, $1 \leqslant i \leqslant k$ and $\omega^i = \sum_{j=1}^k a_{ij}\theta^j$, $1 \leqslant i \leqslant k$. Note that, in general,

$$\omega_1 \wedge \omega_2 = (-1)^{rs}\omega_2 \wedge \omega_1, \qquad \omega_1 \in \Omega_r, \quad \omega_2 \in \Omega_s$$

and

$$(\omega_1 \wedge \omega_2)(X_1, \ldots, X_{r+s}) = \frac{1}{(r+s)!} \sum_{\sigma \in P_{r+s}} \operatorname{sgn}(\sigma)\omega_1(X_{\sigma(1)}, \ldots, X_{\sigma(r)}) \\ \omega_2(X_{\sigma(r+1)}, \ldots, X_{\sigma(r+s)})$$

for $\omega_1 \in \Omega_r$, $\omega_2 \in \Omega_s$, $X_i \in D(M)$, $1 \leqslant i \leqslant r+s$. Note also that if $f, g \in \mathscr{F}(M)$, then $f \wedge g = fg$, $f \wedge \omega = f\omega$, $\omega \wedge f = f\omega$, $\omega \in \Omega_s$.

We next define an operator d on Ω_* which, for $f \in \mathscr{F}(M) = \Omega_{*0}$, is given by $\mathrm{d}f(X) = Xf$, for $X \in D(M)$. In general we define, for $\omega \in \Omega_p(M)$, $p \geqslant 1$, $X_i \in D(M)$,

$$\mathrm{d}\omega(X_1, \ldots, X_{p+1}) = \frac{1}{p+1} \sum_{i=1}^{p+1} (-1)^{i+1} X_i(\omega(X_1, \ldots, \hat{X}_i, \ldots, X_{p+1}))$$

$$+ \frac{1}{p+1} \sum_{i<j} (-1)^{i+j} \omega([X_i, X_j], X_1, \ldots, \hat{X}_i, \ldots, \hat{X}_j, \ldots, X_{p+1}) \quad (1.1.1)$$

In any coordinate neighbourhood U of a point $p \in M$ with coordinates $\{x^1, \ldots, x^n\}$ we then have $\mathrm{d}x^i(\partial/\partial x^j) = \delta_{ij}$ in U and so the 1-forms $\mathrm{d}x^i$ $(1 \leqslant i \leqslant n)$ provide a basis of the $\mathscr{F}(U)$ module $\mathscr{T}_1(U)$. Any element of $\mathscr{T}_*(U)$ can therefore be written

$$\Sigma T_{i_1 \ldots i_r} \mathrm{d}x^{i_1} \otimes \cdots \otimes \mathrm{d}x^{i_r}, \qquad T_{i_1 \ldots i_r} \in \mathscr{F}(U).$$

Now any form $\omega \in \Omega_*$ induces a form ω_U on U which can be written

$$\omega_U = \Sigma \alpha_{i_1 \ldots i_r} \mathrm{d}x^{i_1} \wedge \cdots \wedge \mathrm{d}x^{i_r}, \qquad \alpha_{i_1 \ldots i_r} \in \mathscr{F}(U).$$

It is easy to see that a form ω is local in the sense that if $p \in V \subseteq \bar{V} \subseteq U$ and $\bar{V}$ is compact, then

$$\mathrm{d}(\omega_V) = (\mathrm{d}\omega)_V.$$

In fact in V we have

$$(\mathrm{d}\omega)_V = \Sigma\, \mathrm{d}\alpha_{i_1 \ldots i_r} \mathrm{d}x^{i_1} \wedge \cdots \wedge dx^{i_r}$$

A simple application of the definition now shows that d is an *antiderivation*, i.e.

$$d(\omega_1 \wedge \omega_2) = \mathrm{d}\omega_1 \wedge \omega_2 + (-1)^r \omega_1 \wedge \mathrm{d}\omega_2 \qquad \text{if } \omega_1 \in \Omega_r, \quad \omega_2 \in \Omega_*.$$

Since, clearly, $\mathrm{d}(\mathrm{d}x^i) = 0$ for each i, a simple induction argument shows that $d^2 = 0$.

1.2 BUNDLES AND CONNECTIONS

1.2.1 General bundles

A *bundle* is just a topological space E and a (surjective, continuous) map $p: E \to B$ from E to a topological space B‡. We write (E, p, B) for the bundle, or just E if p and B are understood, and we call E the *total space*,

‡ A *map* $f: X \to Y$ between topological spaces X and Y is just a continuous function.

B the *base space* and p the *projection* of the bundle. If $b \in B$, $p^{-1}(b)$ is called the *fibre over b*. A topological space F is called the *fibre* of the bundle E if every fibre $p^{-1}(b)$ is homeomorphic to F.

We write $(E_1, p_1, B_1) \subseteq (E, p, B)$ and call E_1 a *subbundle* of E if $E_1 \subseteq E$, $B_1 \subseteq B$ (as topological spaces) and $p_1 = p \mid E_1$. For example, if B and F are any topological spaces, $(B \times F, p_B, B)$ (where p_B is the projection on B) is a bundle and $(B_1 \times F, p_B \mid B_1 \times F, B_1)$ is a subbundle for any subspace B_1 of B.

A *cross section* of a bundle (E, p, B) is a map $s: B \to E$ such that the diagram

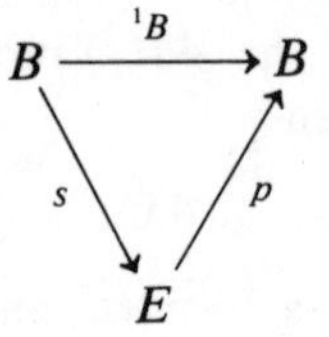

commutes, i.e. $s(b) \in p^{-1}(b)$ for each $b \in B$. Thus a section maps a point in the base space into the fibre over that point.

Examples

1. Define the *tangent bundle* $T(M)$ of a manifold M of dimension n to be the disjoint union $\bigcup_{p \in M} T_p(M)$ of tangent spaces, together with the projection map $\pi: \bigcup_{p \in M} T_p(M) \to M$ taking $X \in T_p(M)$ to p. We can topologize $T(M)$ in the following way. Let $p \in M$ and let (U, ϕ) be a coordinate system around p. Then $\phi: U \to \mathbb{R}^n$, so $\phi_{*q}(= d\phi_q)$: $T_q(M) \to T_{\phi(q)}(\mathbb{R}^n)$ for any $q \in U$. If $X \in T_q(M)$ then we can write

$$\phi_{*q}(X) = \sum_{i=1}^{n} \alpha_i(q) \left.\frac{\partial}{\partial x_i}\right|_{\phi(q)}.$$

We let

$$\psi: T(U) \to U \times \mathbb{R}^n \tag{1.2.1}$$

be defined by $\psi(X) = (q, \alpha_1(q), \ldots, \alpha_n(q)) \in U \times \mathbb{R}^n$. We then topologize $T(M)$ so that ψ is a homeomorphism for all U in an atlas of M.

2. A similar construction can be carried out for the *tensor bundle*

$$T_s^r(M) = \bigcup_{p \in m} T_{s,p}^r(M)$$

of type (r, s). We leave the details to the reader.

3. Let $G_k(\mathbb{R}^n) = \{L : L$ is a k-dimensional linear subspace of $\mathbb{R}^n\}$ and let $V_k(\mathbb{R}^n)$ denote the subspace of $S^{n-1} \times \cdots \times S^{n-1}$ (k times) consisting of k-tuples of unit $(n-1)$-vectors $(v_1, \ldots, v_k)$ such that $v_i \perp v_j$, $i \neq j$. Since S^{n-1} is compact, so is $V_k(\mathbb{R}^n)$ (as a closed subspace of a compact space). Each element of $V_k(\mathbb{R}^n)$ defines an orthonormal basis of a k-dimensional

subspace of $\mathbb{R}^n$ and so there is a projection

$$p: V_k(\mathbb{R}^n) \to G_k(\mathbb{R}^n)$$

which maps such a basis into the subspace which it spans. We give $G_k(\mathbb{R}^n)$ the quotient topology defined by p, making $G_k(\mathbb{R}^n)$ into a compact space. Let $E(G_k)$ denote the product bundle $(G_k(\mathbb{R}^n) \times \mathbb{R}^n,\ p_{G_k},\ G_k(\mathbb{R}^n))$ and let γ_k^n be the subbundle $(E(\gamma_k^n),\ p_{G_k} \mid E(\gamma_k^n),\ p_{G_k}(E(\gamma_k^n)))$ where

$$E(\gamma_k^n) = \{ (V, x) \in G_k(\mathbb{R}^n) \times \mathbb{R}^n : x \in V \},$$

γ_k^n is called the *universal bundle* on $G_k(\mathbb{R}^n)$. (Note that a similar construction holds with $\mathbb{C}$ replacing $\mathbb{R}$.)

Let (E_1, p_1, B_1) and (E_2, p_2, B_2) be two bundles. A *morphism* $(u, f): (E_1, p_1, B_1) \to (E_2, p_2, B_2)$ is a pair of maps $u: E_1 \to E_2$, $f: B_1 \to B_2$ such that the diagram

$$\begin{array}{ccc} E_1 & \xrightarrow{\ u\ } & E_2 \\ {\scriptstyle p_1}\downarrow & & \downarrow{\scriptstyle p_2} \\ B_1 & \xrightarrow{\ f\ } & B_2 \end{array}$$

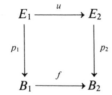

commutes. It is clear that u preserves fibres, i.e. $u(p_1^{-1}(b)) \subseteq (p_2)^{-1}(f(b))$, for all $b \in B_1$. If $B_1 = B_2 = B$ and $f = id_B$ we say that u is a *bundle morphism over B*.

Given two bundles E_1 and E_2 over B we define their *Whitney sum* or *fibre product* $(E_1 \oplus E_2, q, B)$ to be the bundle with

$$E_1 \oplus E_2 = \{ (e_1, e_2) \in E_1 \times E_2 : p_1(e_1) = p_2(e_2) \}$$

and

$$q(e_1, e_2) = p_1(e_1) (= p_2(e_2)).$$

The fibre $q^{-1}(b)$ over $b \in B$ is $p_1^{-1}(b) \times p_2^{-1}(b)$.

Two important constructions from a given bundle (E, p, B) are the *restriction* to a subset of B and the *pull-back* (or induced) bundle f^*E over B' where $f: B' \to B$ is a map. These are defined in the following way:

(a) If $A \subseteq B$, then the restriction $E \mid A$ of E to A is the bundle $(p^{-1}(A),\ p \mid p^{-1}(A),\ A)$.

(b) If $f: B' \to B$, the total space f^*E of the pull-back is the set

$$\{ (b', e) \in B' \times E : f(b') = p(e) \}.$$

The projection of the pull-back is given by $p'(b', e) = b'$. If we define

$f^*: f^*E \to E$ by $f^*(b', e) = e$, then we have a commutative diagram

$$\begin{array}{ccc} f^*E & \xrightarrow{f^*} & E \\ {\scriptstyle p'}\downarrow & & \downarrow{\scriptstyle p} \\ B' & \xrightarrow{f} & B \end{array}$$

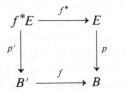

(Intuitively, for each $b \in B$, we have a copy of $p^{-1}(b)$ over each element of $f^{-1}(b)$, with all the copies for $b \in B$ being 'glued together' by the topology of $B' \times E$. See Fig. 1.3.)

The local nature of bundles is clearly very important, expecially when we require a bundle to be a manifold, and so we say that two bundles E_1, E_2 over B are *locally isomorphic* if, for each $b \in B$, there is an open neighbourhood U of b such that $E_1 \mid U$ and $E_2 \mid U$ are isomorphic over U. A bundle E over B is called *locally trivial* with fibre F if E is locally isomorphic to the product bundle $(B \times F, p, B)$.

1.2.2 Vector bundles

The most important classes of bundles for us are vector bundles and the

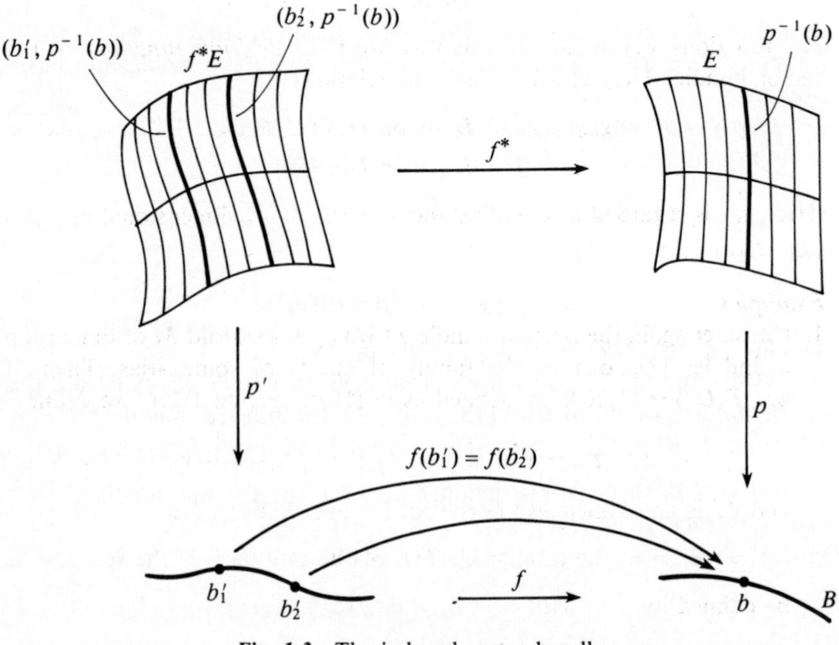

Fig. 1.3 The induced vector bundle

more general fibre bundles. We shall discuss vector bundles first. A map $p: E \to B$ between Hausdorff spaces E, B is called a *K-vector bundle of rank r* if

(a) $E_b \triangleq p^{-1}(b)$ is a vector space (over $K = \mathbb{R}$ or $\mathbb{C}$) of dimension r, for all $b \in B$

(b) For each $b \in B$, there is a neighbourhood U of b and a homeomorphism $h: p^{-1}(U) \to U \times K^r$ such that $h(E_b) \subseteq \{b\} \times K^r$, and the function $h^b: E_b \to K^r$ defined by $h^b(e) = v$, where $h(e) = (b, v)$, is a K-vector space isomorphism.

The pair (U, h) in this definition is called a *(local) trivialization*. A vector bundle is said to be *differentiable* if E and B are differentiable manifolds (of class C^∞, or more generally class C^k), p is differentiable and the local trivializations are diffeomorphisms.

For any two trivializations (U_α, h_α) and (U_β, h_β) the map

$$h_\alpha \circ h_\beta^{-1}: (U_\alpha \cap U_\beta) \times K^r \to (U_\alpha \cap U_\beta) \times K^r$$

induces a map

$$g_{\alpha\beta}: U_\alpha \cap U_\beta \to GL(r, K)$$

where

$$g_{\alpha\beta}(b) = h_\alpha^b \circ (h_\beta^b)^{-1}: K^r \to K^r.$$

The functions $g_{\alpha\beta}$ defined in this way are the *transition functions* of the vector bundle. They clearly satisfy the relations

$$\begin{aligned} g_{\alpha\beta} g_{\beta\gamma} g_{\gamma\alpha} &= I_r \qquad \text{on } U_\alpha \cap U_\beta \cap U_\gamma \\ g_{\alpha\alpha} &= I_r \qquad \text{on } U_\alpha. \end{aligned}$$

(Here $g_{\alpha\beta}$ is regarded as a matrix and I_r is the unit r-dimensional matrix.)

Examples

1. Consider again the tangent bundle $T(M)$ of a manifold M of dimension n and let $\{U_\alpha, \phi_\alpha\}$ be the family of charts of some atlas. Then, if $\psi_\alpha: T(U_\alpha) \to U_\alpha \times \mathbb{R}^n$ is defined as in (1) in Section 1.2.1, we define

$$\psi_\alpha^p: T_p(U_\alpha) \xrightarrow{\psi_\alpha} \{p\} \times \mathbb{R}^n \xrightarrow{\text{proj}} \mathbb{R}^n$$

and ψ^p is an $\mathbb{R}$-linear isomorphism for each $p \in U_\alpha$. Let

$$g_{\alpha\beta}: U_\alpha \cap U_\beta \to GL(n, \mathbb{R})$$

be defined by

$$g_{\alpha\beta}(p) = \psi_\alpha^p \circ (\psi_\beta^p)^{-1}: \mathbb{R}^n \to \mathbb{R}^n.$$

Then $g_{\alpha\beta}$ is the matrix representation of $\phi_{\alpha *} \circ \phi_{\beta *}{}^{-1}$ with respect to the basis $\{\partial/\partial x^1, \ldots, \partial/\partial x^n\}$ at $T_{\phi_\beta(p)}(\mathbb{R}^n)$ and $T_{\phi_\alpha(p)}(\mathbb{R}^n)$ and so is C^∞ on $U_\alpha \cap U_\beta$. Hence $\{(U_\alpha, \psi_\alpha)\}$ are trivializations for $T(M)$ and

$$\psi_\alpha \circ \psi_\beta^{-1}(= id \times g_{\alpha\beta}) : (U_\alpha \cap U_\beta) \times \mathbb{R}^n \to (U_\alpha \cap U_\beta) \times \mathbb{R}^n$$

is a diffeomorphism if $U_\alpha \cap U_\beta \neq \emptyset$. This makes $T(M)$ into a manifold, with the differentiable structure determined by the trivializations, such that π and ψ_α are differentiable.

2. Let $G_k(\mathbb{R}^n)$ denote the set of real k-dimensional vector subspaces of $\mathbb{R}^n$ as defined above. We can make $G_k(\mathbb{R}^n)$ into a differentiable manifold (the *Grassmann manifold*) in the following way. Let $M^k_{k\times n}$ denote the set of $k \times n$ matrices of rank k ($k \leqslant n$) and define the projection

$$\pi : M^k_{k\times n} \to G_k(\mathbb{R}^n)$$

by $\pi(A) =$ the k-dimensional subspace of $\mathbb{R}^n$ spanned by the rows of A. To define local charts on $G_k(\mathbb{R}^n)$ let $\{A_1, \ldots, A_l\}$ be the $k \times k$ minors of any given matrix $A \in M^k_{k\times n}$, and define

$$U_\alpha = \{S \in G_k(\mathbb{R}^n) : S = \pi(A) \text{ where } A_\alpha \in GL(k, \mathbb{R})\}.$$

Then the set $\{U_\alpha\}_{\alpha=1}^{l}$ covers $G_k(\mathbb{R}^n)$ and we define

$$\phi_\alpha : U_\alpha \to \mathbb{R}^{k(n-k)}$$

by

$$\phi_\alpha(\pi(A)) = A_\alpha^{-1}\bar{A}_\alpha$$

where $AP_\alpha = [A_\alpha \bar{A}_\alpha]$ for some permutation matrix P_α. Each ϕ_α is a homeomorphism and $\phi_\alpha \circ \phi_\beta^{-1}$ is a diffeomorphism for each α, β, so that $G_k(\mathbb{R}^n)$ becomes a differentiable manifold. With this differentiable structure it is easy to define trivializations for the universal bundle γ_k^n over $G_k(\mathbb{R}^n)$, in much the same way as for the tangent bundle. Details are given in Wells (1979). If $G_k(\mathbb{C}^n)$ is the complex Grassman manifold (which has holomorphic coordinate transformations), we can define the complex universal bundles in a similar way. Note that if $k = 1$ we obtain the *projective spaces* $\mathbb{P}_{n-1}(\mathbb{R}) = G_1(\mathbb{R}^n)$, $\mathbb{P}_{n-1}(\mathbb{C}) = G_1(\mathbb{C}^n)$ of one-dimensional subspaces of $\mathbb{R}^n$ or $\mathbb{C}^n$.

A *morphism* of differentiable vector bundles $f: E \to F$ is a differentiable morphism of bundles which is linear on each fibre. f is an isomorphism if f is a diffeomorphism on the total spaces and is an isomorphism on the fibres.

If $\pi_E : E \to B$ and $\pi_F : F \to B$ are vector bundles with fibres E_b, F_b respectively, then we can define a bundle

$$\pi : E \circ F \to X,$$

where $\circ$ is an operation on vector spaces and $E \circ F = \bigcup_{b \in B} E_b \circ F_b$. For example, we can define the bundles $E \oplus F$, $E \otimes F$, Hom(E, F), etc. Then $E \oplus F$ is just the Whitney sum of the vector bundles E and F.

1.2.3 Fibre bundles

Next we define the concept of a general fibre bundle, which is required in nonlinear systems theory when the fibre of a bundle is not a vector space but has some 'group-like' transformation properties. First, recall that a *topological group* G is a set G with a topological and a group structure for which the function $(g, h)\ (\in G \times G) \to gh^{-1}\ (\in G)$ is continuous. Note that all the 'classical' groups $GL(n)$, $SL(n)$, $O(n)$, $SO(n)$, $SO(p, q)$, $S_p(n)$ defined above are topological groups. A topological space X is called a *right G-space* if there is a map $X \times G \to X$ which maps (x, g) to xg, for which $x(gh) = (xg)h$ and $x1_G = x$ for all $x \in X$, $g, h \in G$.

Example

$GL(n)$ operates on $\mathbb{R}^n$ on the right by $xA \triangleq A^{\mathsf{T}}x$, $A \in GL(n)$, $x \in \mathbb{R}^n$.

A map $f\colon X \to Y$ between G spaces is a *G-morphism* if $f(xg) = f(x)g$ for all $x \in X$, $g \in G$. Given a G space X, we can define an equivalence relation on X as follows. We write $x_1 \sim x_2$ if there exists $g \in G$ such that $x_1 g = x_2$. This is clearly an equivalence relation and we let $xG = \{xg : g \in G\}$ and $X/G = X/\sim$ with the quotient topology. It follows that if $\pi : X \to X/G$ is the canonical projection, then $(X, \pi, X/G)$ is a bundle. Since $f(xG) \subseteq f(x)G$ for any G-morphism $f\colon X \to Y$, there is an induced map $f_\sim : X/G \to Y/G$ with $f_\sim(xG) = f(x)G$. Then $(f, f_\sim)$ is a bundle morphism.

A bundle (X, p, B) is called a *G-bundle* if $(1, f) : (X, p, B) \to (X, \pi, X/G)$ is a bundle isomorphism, for some G-structure on X, where $f\colon B \to X/G$ is a homeomorphism. A G-space X is *effective* (and we say that G acts *freely* on X) if $xg = x$ implies $g = 1_G$.

Suppose now that the group G also carries a differentiable structure (i.e. G is a Lie group – see below). Then a *principle fibre bundle* over a manifold M with group G consists of a manifold P and an action of G on P such that

(a) G acts freely on P. We shall write $(p, g) \in P \times G \to pg = R_g p \in P$.
(b) $M = P/G$ and $\pi : P \to M$ is differentiable.
(c) P is locally trivial; i.e. if $x \in M$ then there is a neighbourhood U of x in M such that $(\pi, \phi) : \pi^{-1}(U) \to U \times G$ is a diffeomorphism, for some G-morphism ϕ.

If U_α, U_β are open sets in M and ϕ_α, ϕ_β are associated with them as in (3) then if $u \in \pi^{-1}(U_\alpha \cap U_\beta)$ we have $\phi_\beta(ug)(\phi_\alpha(ug))^{-1} = \phi_\beta(u)(\phi_\alpha(u))^{-1}$.

Hence we can define a mapping

$$g_{\alpha\beta}: U_\alpha \cap U_\beta \to G$$

by $g_{\alpha\beta}(\pi(u)) = \phi_\beta(u)(\phi_\alpha(u))^{-1}$. These are the *transition functions* and generalize the corresponding functions for a vector bundle.

Now let P be a principal fibre bundle and F a manifold on which G acts on the left, $(g, \zeta) \in G \times F \to g\zeta \in F$. Then let G act on the right on $P \times F$ by $((u, \zeta), g) \in (P \times F) \times G \to (ug, g^{-1}\zeta) \in P \times F$ and define

$$E = P \times F/G$$

as a set. Define the projection $\pi_E: E \to M$ by $\pi_E((u, \zeta)G) = \pi(u)$ (this is clearly well-defined) and call $\pi_E^{-1}(x)$ the *fibre* over $x \in M$. Since each point $x \in M$ has a neighbourhood U such that $\pi^{-1}(U) \cong U \times G$, it follows that the action of G on the right of $\pi^{-1}(U) \times F \cong U \times G \times F$ is given by $(x, g, \xi)h = (x, gh, h^{-1}\xi)$, $x \in U$, $g, h \in G, \xi \in F$ and so $\pi_E^{-1}(U) \cong U \times F$. We can therefore make E into a differentiable manifold in which $\pi_E^{-1}(U)$ is an open submanifold of E diffeomorphic with $U \times F$ under the isomorphism above. (E, π_E, M) is called the *fibre bundle* associated with the principal fibre bundle P.

Example

The tangent bundle $T(M)$ over M is a fibre bundle with fibre $\mathbb{R}^n$. It is associated with the principal fibre bundle of *frames*, i.e. the bundle of ordered bases of $T_x(M)$ for $x \in M$ with the group action defined by $GL(n; \mathbb{R})$.

In a fibre bundle it is useful to separate the tangent space at a point into the subspace of vectors tangent to the fibre and its complement. Thus, if $\pi: E \to M$ is a fibre bundle let V_e be the subspace of $T_e(E)$ consisting of vectors tangent to the fibre through e. Then we define a *connection* H on E to be a smooth assignment of complementary subspaces of $T_e(E)$ to V_e; i.e. $T_e(E) = V_e \oplus H_e$ for each $e \in E$. Elements of V_e are called *vertical* vectors at e and elements of H_e are called *horizontal* vectors at e. Note that a vector $v \in T_e(E)$ is vertical if and only if $\pi_*(v) = 0$, since π acting on a fibre is constant, so its derivative is zero. Also, if X is a vector field on M, then there exists a unique horizontal vector field such that $\pi_* Y = X$, called the *horizontal lift* of X.

1.3 DIFFERENTIAL EQUATIONS ON A MANIFOLD

1.3.1 First-order differential equations

Recall that a *curve* in a manifold M is just a differentiable function

$\alpha: I \to M$ where $I \subseteq \mathbb{R}$ is an open interval. If t is the identity chart on R and $\partial/\partial t$ is the corresponding vector field, then

$$\dot{\alpha} \triangleq \alpha_*\left(\frac{\partial}{\partial t}\right) : \mathbb{R} \to TM, \qquad \dot{\alpha}(a) = \alpha_{*a} \left.\frac{\partial}{\partial t}\right|_a, \quad a \in \mathbb{R}.$$

is a curve in TM and $\dot{\alpha}$ satisfies

$$\pi \circ \dot{\alpha} = \alpha,$$

since $\dot{\alpha}(a) \in TM_{\alpha(a)}$. $\dot{\alpha}$ is called the *lift* of α to TM.

If X is a vector field on M, then

$$\dot{\alpha} = X \circ \alpha$$

is a *differential equation* (of first order) on M. A *solution* or *integral curve* of this equation is a curve in M with some initial condition $\alpha(0) = p \in M$ such that $X_{\alpha(t)}$ is tangent to the curve at $\alpha(t)$.

If α is a solution of such an equation and (U, x) is a chart containing p, then let $\xi = x \circ \alpha$. If $s \in \alpha^{-1}(U)$ we have

$$\begin{aligned}
\alpha_{*s} \left.\frac{\partial}{\partial t}\right|_s &= \sum_{i=1}^{n} \left(\alpha_{*s} \left.\frac{\partial}{\partial t}\right|_s\right) x^i \left.\frac{\partial}{\partial x^i}\right|_{\alpha(s)} \\
&= \sum_{i=1}^{n} \left.\frac{\mathrm{d}}{\mathrm{d}t}(x^i \circ \alpha)\right|_s \left.\frac{\partial}{\partial x^i}\right|_{\alpha(s)} \\
&= \sum_{i=1}^{n} \left.\frac{\mathrm{d}\xi^i}{\mathrm{d}t}\right|_s \left.\frac{\partial}{\partial x^i}\right|_{\alpha(s)}.
\end{aligned}$$

Hence, if $X = \Sigma f^i(x^1, \ldots, x^n)\, \partial/\partial x^i$ in U, we have

$$\frac{\mathrm{d}\xi^i}{\mathrm{d}t} = f^i(\xi^1, \ldots, \xi^n), \qquad 1 \leqslant i \leqslant n,$$

and so, locally, ξ satisfies an ordinary differential equation.

Example

On $\mathbb{R}^n$ (defined as a manifold with the identity chart) let

$$X = (Ax)\frac{\partial}{\partial x} = \sum_{i=1} \sum_{j=1} a_{ij} x^j \frac{\partial}{\partial x^i}$$

for some $n \times n$ matrix A. Then a curve $\alpha: \mathbb{R} \to \mathbb{R}^n$ with $\alpha = \xi = (\xi^1, \ldots, \xi^n)$ is an integral curve of X if

$$\frac{\mathrm{d}\xi}{\mathrm{d}t} = A\xi$$

which has solution $\xi = \mathrm{e}^{At}\xi_0$.

Note that if $X_p = 0$ then the integral curve starting from p is constant, and p is called a *critical point* (or *equilibrium point*) of the vector field X.

Clearly an integral curve can be extended to a maximal interval of definition which is open in $\mathbb{R}$. If all integral curves of X are defined for all $t \in \mathbb{R}$, then we say that the vector field X is *complete*. The existence and uniqueness of integral curves of vector fields follows from the elementary theory of differential equations – note, however, that global trajectories may not be unique in non-Hausdorff manifolds. Also it is easy to see that every vector field defined on a compact Hausdorff manifold is complete.

Now let X be a vector field on a (Hausdorff) manifold M and let $p \in M$. If $\alpha_p(t)$ is the maximal integral curve of X with $\alpha_p(0) = p$ defined on the interval I_p, then we define the function

$$\Phi : \mathbb{R} \times M \to M$$

by

$$\Phi(s, p) = \alpha_p(s)$$

for $s \in I_p$. The function Φ is called the *flow* of X and it is clear that

$$\Phi(s+t, p) = \Phi(s, \Phi(t, p)).$$

We usually write $\phi_t p = \Phi(t, p)$. It can be shown that ϕ_t is locally a diffeomorphism (on a neighbourhood V of some point p) for $t \in (-\varepsilon, \varepsilon)$ say. We then call ϕ_t a *local group of diffeomorphisms*. Note that if $q \in V$, then X_q is the tangent vector to the curve $t \to \phi_t(q)$ at $t = 0$. However, this means, as we have seen above in the general definition of a tangent vector, that

$$X_q f = \lim_{h \to 0} \frac{f(\phi_h(q)) - f(q)}{h}.$$

The next result shows that any vector field has a simple local expression:

Theorem 1.3.1
Let X be a vector field on M with $X_p \neq 0$. Then there is a coordinate system (x, U) around p such that

$$X = \frac{\partial}{\partial x^1} \quad \text{on } U.$$

Proof
Since this result is local we can assume that $M = \mathbb{R}^n$ with the usual coordinate system $t^1, \ldots, t^n$ and $p = 0$. We can assume also that $X_0 = \partial/\partial t^1 |_0$. Let ϕ_t be the flow of X and define the function ψ in a neighbourhood of 0 in $\mathbb{R}^n$ by

$$\psi(\alpha^1, \ldots, \alpha^n) = \phi_{\alpha^1}(0, \alpha^2, \ldots, \alpha^n).$$

If $\alpha = (\alpha^1, \ldots, \alpha^n)$, then

$$\psi_* \left(\frac{\partial}{\partial t^i} \bigg|_\alpha \right)(f) = (Xf)(\psi(\alpha)),$$

and

$$\psi_*\left(\frac{\partial}{\partial t^i}\bigg|_0\right)(f) = \frac{\partial f}{\partial t^i}\bigg|_0 \qquad \text{if } i > 1.$$

Since $X_0 = \partial/\partial t^1|_0$, $\psi_{*0} = I$ and so ψ^{-1} exists. Then $x = \psi^{-1}$ is the desired coordinate system. □

We write

$$L_X f = Xf$$

where

$$(Xf)(p) = \lim_{h \to 0} \frac{1}{h} [f(\phi_h(p)) - f(p)].$$

and call $L_X f$ the *Lie derivative* of f with respect to X. Moreover, if ω is a 1-form we define the *Lie derivative* of ω with respect to X by

$$(L_X\omega)(p) = \lim_{h \to 0} \frac{1}{h} [(\phi_h^*\omega)(p) - \omega(p)],$$

where we recall that ϕ_h^* is the dual operator of ϕ_{h*}, i.e.

$$\langle \phi_h^*\omega(p), X_p\rangle = \langle \omega(\phi_h(p)), \phi_{h*}X_p\rangle.$$

If Y is another vector field, the *Lie derivative* of Y with respect to X is

$$(L_X Y)(p) = \lim_{h \to 0} \frac{1}{h} [Y_p - (\phi_{h*}Y)_p].$$

Lemma 1.3.2

(a) $L_X[Y_1 + Y_2] = L_X Y_1 + L_X Y_2$
(b) $L_X[\omega_1 + \omega_2] = L_X\omega_1 + L_X\omega_2$
(c) $L_X fY = Xf.\, Y + f.\, L_X Y$
(d) $L_X f\omega = Xf.\, \omega + f.\, L_X\omega$
(e) $L_X\langle \omega, Y\rangle = \langle L_X\omega, Y\rangle + \langle \omega, L_X Y\rangle.$

Proof
These relations are obtained by elementary computation. □

If $f: M \to M$ is a diffeomorphism and the vector field X generates the one-parameter local group ϕ_t, then it is easy to see that f_*X generates the local group $f \circ \phi_t \circ f^{-1}$, where $(f_*X)_{f(p)} = f_{*p}X_p$. Hence X is f-invariant ($f_*X = X$, see above) if and only if f commutes with ϕ_t.

Lemma 1.3.3
Let X and Y be vector fields on M. If X generates ϕ_t, then

$$[X, Y] = L_X Y.$$

Proof
See Kobayashi and Nomizu (1969). □

Similarly, it can be shown that

$$(\phi_t)_*[X, Y] = \lim_{h \to 0} \frac{1}{h} [(\phi_t)_* Y - (\phi_{t+h})_* Y] .$$

for any t for which ϕ_t exists.
Writing the latter equation in the form

$$\frac{d(\phi_t)_* Y}{dt} = -(\phi_t)_*[X, Y],$$

we see that if X and Y generates the local groups ϕ_t and ψ_t, respectively, then $\phi_t \circ \psi_s = \psi_s \circ \phi_t$ for all s, t if and only if $[X, Y] = 0$. If X and Y do not commute, it follows that $[X, Y]$ is a measure of the 'mismatch' between the start and end points of the path obtained by following the flows of X, Y, X^{-1} and Y^{-1}, i.e. the path $\psi_{-t} \circ \phi_{-t} \circ \psi_t \circ \phi_t(p)$ for some $p \in M$ and $t \neq 0$ (Fig. 1.4).

We can generalize the notion of Lie derivative to general tensors in the following way. Let $\mathscr{T} = \bigoplus_{r,s=0}^{\infty} \mathscr{T}^r_s(M)$ be the set of all tensor fields on M. Then $\mathscr{T}$ is an algebra over $\mathbb{R}$ where we define the product by

$$(K \otimes L)_p = K_p \otimes L_p \qquad \text{for } p \in M, \quad K, L \in \mathscr{T}.$$

Let $f: M \to M$ be a diffeomorphism and let

$$f_*: T_p(M) \to T_{f(p)}(M)$$

be the induced isomorphism. Using the results of Section 1.1.4 it is easy to see that f_* can be extended to an isomorphism $\tilde{f}_*$ of $\mathscr{T}_p (= \bigoplus_{r,s=0}^{\infty} \mathscr{T}^r_{s,p}(M))$

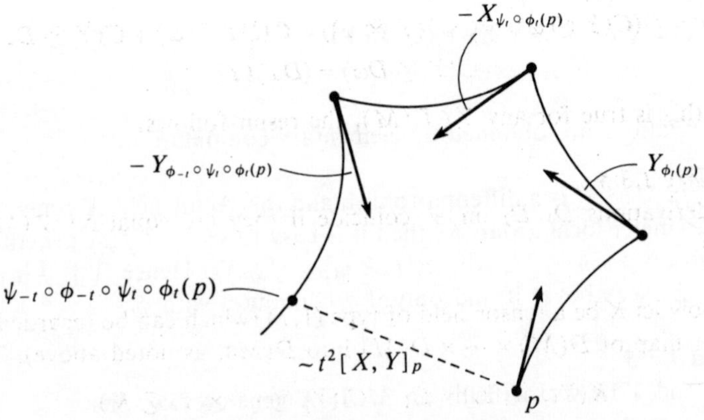

Fig. 1.4 Geometric interpretation of the Lie bracket

onto $\mathscr{T}_{f(p)}$. If $K \in \mathscr{T}$ then define

$$(\tilde{f}_* K)_{f(p)} = \tilde{f}_*(K_p), \qquad p \in M.$$

Then $\tilde{f}_*$ is an automorphism of $\mathscr{T}$ which preserves type and commutes with contractions.

Now, if X is a vector field on M which generates the local group ϕ_t, then, for simplicity, we assume that X is complete, so that ϕ_t is a diffeomorphism for each t. Then we define the *Lie derivative* of $K \in \mathscr{T}$ with respect to X by

$$(L_X K)_p = \lim_{t \to 0} \frac{1}{t} [K_p - (\tilde{\phi}_{t*} K)_p].$$

The following properties of Lie differentiation are then elementary, and generalize the above results when $K \in \mathscr{F}(M)$ or $D(M)$:

(a) L_X is a *derivation* of $\mathscr{T}$, i.e. it is linear, preserves type, commutes with contractions, and

$$L_X(K_1 \otimes K_2) = (L_X K_1) \otimes K_2 + K_1 \otimes (L_X K_2), \qquad K_1, K_2 \in \mathscr{T}.$$

(b) $L_X f = Xf$, for $f \in \mathscr{F}(M)$.

(c) $L_X Y = [X, Y]$ for $Y \in D(M)$.

Lemma 1.3.4

If D is a derivation of $\mathscr{T}$ (in the above sense) which vanishes on $\mathscr{F}(M)$ and $D(M)$, then D vanishes on $\mathscr{T}$.

Proof

By localization, it suffices to prove that $D\omega = 0$ for any one-form on M. Let $Y \in D(M)$, and let C be the contraction defined by $C(Y \otimes \omega) = \omega(Y)$. Then, since $\omega(Y) \in \mathscr{F}(M)$, we have

$$\begin{aligned} 0 = D(C(Y \otimes \omega)) &= C(D(Y \otimes \omega)) = C(DY \otimes \omega) + C(Y \otimes D\omega) \\ &= C(Y \otimes D\omega) = (D\omega)(Y). \end{aligned}$$

Since this is true for any $Y \in D(M)$, the result follows. □

Corollary 1.3.5

Two derivations D_1, D_2 on $\mathscr{T}$ coincide if they are equal on $\mathscr{F}(M)$ and $D(M)$. □

Now let K be a tensor field of type $(1, r)$ (which can be regarded as an r-linear map of $D(M) \times \cdots \times D(M)$ into $D(M)$, as noted above). Then

$$K(Y_1, \ldots, Y_r) = C_1 \ldots C_r(Y_1 \otimes \cdots \otimes Y_r \otimes K)$$

for $Y_i \in D(M)$, where the C_j are obvious contractions. Thus, if D is any derivation of $\mathscr{T}$, we have

$$D(K(Y_1, \ldots, Y_r)) = C_1 \ldots C_r D(Y_1 \otimes \cdots \otimes Y_r \otimes K)$$

$$= C_1 \ldots C_r[Y_1 \otimes \cdots \otimes Y_r \otimes DK + \sum_{i=1}^{r} Y_1 \otimes \cdots \otimes DY_i \otimes \cdots \otimes Y_r \otimes K]$$

$$= (DK)(Y_1, \ldots, Y_r) + \sum_{i=1}^{r} K(Y_1, \ldots, DY_i, \ldots, Y_r)$$

Hence if, in particular, $D = L_X$ for some $X \in D(M)$, we have

$$(L_X K)(Y_1, \ldots, Y_r) = L_X K(Y_1, \ldots, Y_r) - \sum_{i=1}^{r} K(Y_1, \ldots, L_X Y_i, \ldots, Y_r). \tag{1.3.1}$$

For any vector field X, we define the inner product $X \lrcorner \omega$ of X and the r-form ω to be the $r-1$ form defined by

$$(X \lrcorner \omega)(Y_1, \ldots, Y_{r-1}) = r \, . \, \omega(X, Y_1, \ldots, Y_{r-1}) \text{ for } Y_i \in D(M).$$

Then it can be shown that

(a) $L_X = d \circ (X\lrcorner) + (X\lrcorner) \circ d$
(b) $[L_X, Y\lrcorner] = [X, Y]\lrcorner$

where d is exterior derivation; details can be found in Kobayashi and Nomizu (1969).

In the final part of this section we shall discuss a generalization of the idea of integral curves of a differential equation. This is the theory of integral manifolds and we first need the concept of distribution. An *m-dimensional distribution* in a manifold M of dimension n is an assignment $\mathscr{L} : p \to \mathscr{L}_p (p \in M)$ where $\mathscr{L}_p$ is a linear subspace of T_pM of dimension m such that each point $p \in M$ has a neighbourhood V on which there exist vector fields $X_1, \ldots, X_m$ such that $X_{1q}, \ldots, X_{mq}$ span $\mathscr{L}_q$ for each $q \in V$. We say that a vector field X *belongs to* $\mathscr{L}$ (on an open set U) if $X_p \in \mathscr{L}_p$ for each $p \in U$. The distribution $\mathscr{L}$ is said to be *involutive* if, for any open subset U of M, given two vector fields X, Y which belong to $\mathscr{L}$ on U, then $[X, Y]$ also belongs to $\mathscr{L}$ on U.

A connected submanifold S of M is called an *integral manifold* of $\mathscr{L}$ if $T_p(S) = \mathscr{L}_p$ for all $p \in S$. The distribution $\mathscr{L}$ is said to be *integrable* if each point of M lies in some integral manifold of $\mathscr{L}$. Note that any integrable distribution $\mathscr{L}$ is involutive. For, let $p \in M$ and let X, Y be vector fields which belong to $\mathscr{L}$ on some open neighbourhood of p. Then we must show that $[X, Y]_p \in \mathscr{L}_p$. Let S be a sufficiently small integral manifold of $\mathscr{L}$ through p so that S has the induced topology from M. Then $X|_s, Y|_s$ are vector fields on S and if $i: S \to M$ is the identity map, then $X|_s, X$ and

$Y|_s$, Y are i-related pairs. Hence $[X|_s, Y|_s]$ and $[X, Y]$ are i-related, so that $[X, Y]_p \in \mathscr{L}_p$.

Now let M and N be manifolds with distributions $\mathfrak{M}$, $\mathfrak{N}$ respectively. Then we say that $\mathfrak{M}$ and $\mathfrak{N}$ are *isomorphic* if there is a diffeomorphism $f: M \to N$ such that $f_{*p}(\mathfrak{M}_p) = \mathfrak{N}_{f(p)}$ for all $p \in M$. Note that any distribution $\mathscr{L}$ on M induces a distribution $\mathscr{L} \mid U$ on an open subset U by restriction. Let $I_a^m \subseteq \mathbb{R}^m$ be the open cube defined by

$$I_a^m = \{x \in \mathbb{R}^m : |x_i| < a, i = 1, \ldots, m\}, \qquad a > 0.$$

If $t_1, \ldots, t_m$ are the standard coordinates in $\mathbb{R}^m$, then let $\mathfrak{Z}_x^k$ be the linear span of $(\partial/\partial t_1)_x, \ldots, (\partial/\partial t_k)_x$ for $x \in I_a^m$. Then $\mathfrak{Z}^k : x \to \mathfrak{Z}_x^k$ is an involutive distribution on I_a^m and the set $\{(t_1, \ldots, t_m) : t_{k+1} = a_{k+1}, \ldots, t_m = a_m\}$ is an integral manifold of $\mathfrak{Z}_x^k$ for any numbers $a_j \in (-a, a)$, $j = k+1, \ldots, m$.

Lemma 1.3.6

Let M be a manifold and let $X_1, \ldots, X_k$ be vector fields on an open set U containing $p \in M$ such that $X_{1q}, \ldots, X_{kq}$ are linearly independent for each $q \in U$ and $[X_i, X_j] = 0$ for $1 \leqslant i, j \leqslant k$. Then there exist coordinates $x_1, \ldots, x_n$ near p such that, in some neighbourhood of p,

$$X_j = \frac{\partial}{\partial x_j} + \sum_{1 \leqslant i < j} a_{ji} \frac{\partial}{\partial x_i} \qquad (1 \leqslant j \leqslant k),$$

where the a_{ji} are differentiable.

Proof

By induction on k. For $k = 1$ we have seen earlier that any vector field can be written locally in this form. Let $1 < k \leqslant n$ and assume the result is true for the vector fields $X_1, \ldots, X_{k-1}$. Then there is a connected open set V with $p \in V \cap U$ and coordinates $x_1, \ldots, x_n$ on V such that

$$X_j = \frac{\partial}{\partial x_j} + \sum_{1 \leqslant i < j} a_{ji} \frac{\partial}{\partial x_i} \qquad (1 \leqslant j \leqslant k-1), \tag{1.3.2}$$

Let $X_k = \Sigma_{1 \leqslant i \leqslant n} \xi_i \partial/\partial x_i$ and define $X_k' = \Sigma_{k \leqslant i \leqslant n} \xi_i \partial/\partial x_i$. Since $X_1, \ldots, X_k$ are linearly independent at each point of U, it follows that $X_{kq}' \neq 0$ for all $q \in V$. Now from the conditions $[X_k, X_j] = 0$ we obtain

$$\sum_{1 \leqslant i \leqslant n} \xi_i \left[\frac{\partial}{\partial x_i}, X_j\right] - \sum_{1 \leqslant i \leqslant n} (X_j \xi_i) \frac{\partial}{\partial x_i} = 0, \qquad 1 \leqslant j \leqslant k-1$$

on V. Hence $X_j \xi_i = 0$ on V for $1 \leqslant j \leqslant k-1$ and $k \leqslant i \leqslant n$, since $[\partial/\partial x_i, X_j]$ depends only on $\partial/\partial x_l$ for $1 \leqslant l \leqslant k-1$ (by (1.3.2)). By induction, and (1.3.2), $\partial/\partial x_j(\xi_i) = 0$ on V for $1 \leqslant j \leqslant k-1$, $k \leqslant i \leqslant n$, and since V is connected, ξ_i is therefore a function of $x_k, \ldots, x_n$ only. Thus we can change the coordinates $x_k, \ldots, x_n$ so that $X_k' = \partial/\partial x_k$ near p. □

Theorem 1.3.7 (Frobenius' theorem)
Let $\mathscr{L}: p \to \mathscr{L}_p$ be an involutive m-dimensional distribution on a manifold M. Then $\mathscr{L}$ is integrable.

Proof
We show that if $p \in M$ there exists a neighbourhood U of p such that $\mathscr{L} \mid U$ is isomorphic to the canonical distribution $\mathscr{Z}_x^m$ introduced above. Let $x_1, \ldots, x_n$ be coordinates near p and let $Y_1, \ldots, Y_m$ be vector fields on U such that $Y_{1q}, \ldots, Y_{mq}$ span $\mathscr{L}_q$ for each $q \in U$. Then

$$Y_j = \Sigma_{1 \leqslant i \leqslant n} a_{ji} \partial / \partial x_i$$

for some C^∞ functions a_{ji}. We may assume that, for example, $(a_{ji})_{1 \leqslant j, i \leqslant m}$ is invertible in U (by shrinking U, if necessary). Let

$$X_j = \Sigma_{1 \leqslant i \leqslant m} a'_{ji} Y_i$$

where $(a'_{ji}) = (a_{ji})^{-1}$. Then we clearly have that $X_{1q}, \ldots, X_{mq}$, span $\mathscr{L}_q$ for all $q \in U$ and

$$X_j = \partial / \partial x_j + \Sigma_{m+1 \leqslant i \leqslant n} b_{ji} \partial / \partial x_i$$

for some C^∞ functions b_{ji}. A simple argument shows that $[X_j, Y_k] = 0$ for $1 \leqslant j, k \leqslant m$. Lemma 1.3.6 now shows that $(\partial / \partial x_1)_q, \ldots, (\partial / \partial x_m)_q$ span $\mathscr{L}_q$ for all q in some neighbourhood of p. □

There is a global version of this theorem; in fact, it can be shown that any point $p \in M$ belongs to a unique maximal integral manifold through p. (See Varadarajan (1974), for example.)

1.4 LIE GROUPS AND LIE ALGEBRAS

1.4.1 Lie groups

A *Lie group* is a topological group G on which there is an analytic structure compatible with the topology of G, such that the map

$$(x, y) \to xy^{-1}, \qquad x, y \in G$$

is analytic.

The classical groups $GL(n)$, $SL(n)$, $O(n)$, $SO(n)$, $\mathrm{SO}(p, q)$ and $S_p(n)$ are Lie groups. Moreover, if $G_1, \ldots, G_n$ are Lie groups then so is $G_1 \times \cdots \times G_n$ with the direct product group structure, product topology and the product analytic structure.

Define the *left* and *right translation* on G for any $a \in G$ by

$$L_a g = ag, \quad R_a g = ga \qquad (g \in G).$$

Then L_a and R_a are analytic diffeomorphisms of G. Moreover, the map $\mathscr{I}_a : g \to aga^{-1}$ is an *inner analytic automorphism* of G.

Note that a Lie group is often defined as having only a C^∞-structure. However, it can be shown (see Montgomery and Zippin, 1955) that, even in the case of a C^0-structure, there exists a compatible analytic structure and so restricting attention to analytic manifolds is no loss of generality.

If G is a Lie group and H is a subgroup of G then H is called a *Lie subgroup* of G if it is a Lie group it is also an analytic submanifold of G. If M and N are C^k-manifolds ($k = 0, 1, \ldots, \omega$) then an embedding $f: M \to N$ is called *regular* if the topology of M is the same as that it inherits as a subset of N. The embedding f is called *quasi-regular* if, for any C^k-manifold L and any map $h : L \to M$, then h is C^k if and only if $f \circ h$ is C^k. Then we have

Theorem 1.4.1

Let G be a Lie group and H a subgroup such that there is a quasi-regular embedding $f: H \to G$. Then H, with the analytic structure defined by its embedding in G is a Lie subgroup of G which is closed in G if f is regular.

Proof

Since the map $(x, y) \to xy^{-1}$ is analytic on G it is analytic as a map from $H \times H$ to G by restriction. However, its image on $H \times H$ is H and since f is quasi-regular, this map is clearly analytic as a map from $H \times H$ into H. Hence H is a Lie subgroup of G. If f is regular, then H is locally closed in G, so that H is open in its closure $\bar{H}$ in G. But $\bar{H}$ is a subgroup of G and H is an open subgroup of $\bar{H}$, so that $H = \bar{H}$. □

Lemma 1.4.2

The component G_e of a Lie group G containing the identity element e is a connected Lie subgroup of G.

Proof

This follows from Theorem 1.4.1 since G_e is a subgroup which is a connected open submanifold of G. □

More generally, it is not difficult to prove that each component of a Lie group G is a connected open submanifold of G with a countable basis for its topology. All the components of G are analytically diffeomorphic.

1.4.2 The Lie algebra of a Lie group

Let $k = \mathbb{R}$ or $\mathbb{C}$. Then a *Lie algebra* over k is a vector space $\mathfrak{g}$ for which there is defined a bilinear map $(X, Y) \to [X, Y]$, from $\mathfrak{g} \times \mathfrak{g}$ to $\mathfrak{g}$ such that

(a) $[X, Y] + [Y, X] = 0, \ X, Y \in \mathfrak{g}$
(b) $[X, [Y, Z]] + [Y, [X, Y]] + [Z, [X, Y]] = 0, \ X, Y, Z \in \mathfrak{g}$.

The latter expression is known as the *Jacobi identity*.

If $\mathfrak{g}$ is finite dimensional and $\{X_1, \ldots, X_n\}$ is a basis of $\mathfrak{g}$, then we can write

$$[X_r, X_s] = \sum_{1 \leqslant i \leqslant n} c^i_{rs} X_i$$

for some constants c^i_{rs}, called the ***structure constants*** of $\mathfrak{g}$ relative to the basis $\{X_1, \ldots, X_n\}$. From (a) and (b) we have

(a)′ $c^i_{rs} + c^i_{sr} = 0 \qquad (1 \leqslant r, s, i \leqslant n)$

(b)′ $\sum_{1 \leqslant i \leqslant n} (c^i_{rs} c^k_{ij} + c^i_{sj} c^k_{ir} + c^i_{jr} c^k_{is}) = 0 \qquad (1 \leqslant r, s, k, j \leqslant n)$.

If $\mathfrak{a}$ and $\mathfrak{b}$ are linear subspaces of $\mathfrak{g}$, we denote by $[\mathfrak{a}, \mathfrak{b}]$ the linear space spanned by $[X, Y]$ for $X \in \mathfrak{a}$, $Y \in \mathfrak{b}$. If $[\mathfrak{h}, \mathfrak{h}] \subseteq \mathfrak{h}$ for a linear subspace $\mathfrak{h}$ of $\mathfrak{g}$, then $\mathfrak{h}$ is called a *subalgebra* of $\mathfrak{g}$, and $\mathfrak{h}$ is called an *ideal* if $[\mathfrak{g}, \mathfrak{h}] \subseteq \mathfrak{h}$.

Example

The set of vector fields on a manifold is a Lie algebra with bracket $[X, Y]$ defined as above.

After these preliminaries we can now define the Lie algebra of a Lie group. First we define an analytic vector field $X \in D(G)$, for a Lie group G, to be *left invariant* if

$$(L_{a*})_e X_e = X_a,$$

where L_a is the left translation map defined above. Note that we have defined the concept of f-invariance for a general differentiable function f on a manifold M in Section 1.1.5. Let us show that a left invariant vector field is actually L_a-invariant. In fact,

$$\begin{aligned} X_{L_a p} = X_{ap} &= (L_{ap*})_e X_e = ((L_a L_p)_*)_e X_e \\ &= (L_{a*})_p (L_{p*})_e X_e \\ &= (L_{a*})_p X_p. \end{aligned}$$

The set of left invariant vector fields $\mathscr{L}(G)$ over G form a Lie algebra with the usual bracket. Moreover, we have

Proposition 1.4.3

$\mathscr{L}(G)$ is isomorphic (as a vector space) to $T_e(G)$ (the tangent space of G at e).

Proof
First we show that if we define

$$\tilde{X}_p f = X(f \circ L_p)$$

for any real-valued analytic function f and any $X \in T_e(G)$, then the map $\tilde{X}: G \to T(G)$ taking p to $\tilde{X}_p$ is the unique left-invariant vector field on G with $\tilde{X}_e = X$. Note that $\tilde{X}_p f = (L_{p*}X)(f)$, so that $\tilde{X}_p \in T_p(G)$ and thus $\tilde{X}$ is a vector field, which is clearly left-invariant. However, L_a is a surjection and so any L_a invariant vector field is unique. To show that $\tilde{X}$ is analytic note that if $\alpha : I \to G$ is any analytic curve in G with $\dot{\alpha}(0) = X$ and $\alpha(0) = e$, then

$$\tilde{X}_p f = X(f \circ L_p) = \left(\frac{\mathrm{d}}{\mathrm{d}t}\right)_0 (f \circ L_p \circ \alpha) = \frac{\mathrm{d}}{\mathrm{d}t}\,[f(p\alpha(t))]_{t=0}.$$

Since all operations on the right are analytic, the result follows.

Now, L_p is an analytic diffeomorphism, so L_{p*} is injective. However, we have shown that any left-invariant vector field is of the form $\tilde{X}$ for some $X \in T_e(G)$ and so the result is proved. □

$\mathscr{L}(G)$ is called the *Lie algebra* of the Lie group G. If $f: \mathfrak{g}_1 \to \mathfrak{g}_2$ is a vector space homomorphism, then we say that $\mathfrak{g}_1$ and $\mathfrak{g}_2$ are *homomorphic* if $f[X, Y] = [fX, fY]$. If f is an isomorphism we say that $\mathfrak{g}_1$ and $\mathfrak{g}_2$ are *isomorphic* Lie algebras. For $X, Y \in T_e(G)$ we define

$$[X, Y] = [\tilde{X}, \tilde{Y}]_e$$

where X, Y are associated with X, Y as above. Then $T_e(G)$ becomes a Lie algebra isomorphic with $\mathscr{L}(G)$.

1.4.3 The exponential map

Let G be a (real) Lie group and $\mathfrak{g}$ its Lie algebra. (The construction for a complex Lie group is similar.) If $\mathbb{R}$ denotes the additive group of real numbers and t is the usual coordinate, then the Lie algebra of $\mathbb{R}$ is generated by $\mathrm{d}/\mathrm{d}t$. If $f: \mathbb{R} \to G$ is an analytic map, then

$$\dot{f}(\tau) \triangleq \left(\frac{\mathrm{d}f}{\mathrm{d}t}\right)_{t=\tau} = f_{*\tau}\left(\frac{\mathrm{d}}{\mathrm{d}t}\right)_\tau.$$

Then a consideration of differential equations (for details see Sagle and Walde (1973)) shows that there is a unique analytic homomorphism $f: \mathbb{R} \to G$ such that $\dot{f}(0) = X$ for any given $X \in T_e(G)$. If $\tilde{X} \in \mathfrak{g}$ is the corresponding left-invariant vector field, then

$$\dot{f}(t) = \tilde{X}(f(t))$$

for all $t \in \mathbb{R}$, and so f is the maximal integral curve of $\tilde{X}$ through $e \in G$. We then define the map

$$\exp : \mathfrak{g} \to G$$

by

$$\exp X = f(1),$$

where we have identified X and $\tilde{X}$. By uniqueness of solutions of differential equations, we have

$$\exp tX = f(t),$$

and

$$\exp(t+s)X = f(t+s) = f(t)f(s) = \exp tX \exp sX.$$

Using the exponential map we can obtain Taylor's formula for a function $f \in \mathscr{F}(G)$. In fact, we have, for $X \in \mathfrak{g}$,

$$\tilde{X}_p F = X(f \circ L_p) = \left\{\frac{\mathrm{d}}{\mathrm{d}t} f(p \exp tX)\right\}_{t=0},$$

since the homomorphism $\exp tX : \mathbb{R} \to G$ satisfies

$$\frac{\mathrm{d}}{\mathrm{d}t} (\exp tX)_{t=0} = X.$$

Hence

$$[\tilde{X}f]\,(p \exp uX) = \left\{\frac{\mathrm{d}}{\mathrm{d}t} f(p \exp uX \exp tX)\right\}_{t=0} = \frac{\mathrm{d}}{\mathrm{d}u} f(p \exp uX)$$

and so, by induction,

$$[\tilde{X}^n F]\,(p \exp uX) = \frac{\mathrm{d}^n}{\mathrm{d}u^n} f(p \exp uX).$$

If f is analytic at p, then there is a neighbourhood U of 0 in $\mathfrak{g}$ such that

$$f(p \exp X) = P(x_1, \ldots, x_n)$$

for all X in U where $X = x_1 X_1 + \cdots + x_n X_n$ for some basis $\{X_i\}$ of $\mathfrak{g}$ and P has a convergent power series

$$P(tx_1, \ldots, tx_n) = \sum_{i=0}^{\infty} \frac{1}{i!} a_i t^i,$$

for $0 \leqslant t \leqslant 1$ and fixed $x_1, \ldots, x_n$. Then,

$$a_i = [\tilde{X}^i f]\,(p),$$

and so we have the Taylor formula

$$f(p \exp X) = \sum_{i=0}^{\infty} \frac{1}{i!} [\tilde{X}^i f]\,(p), \qquad X \in U.$$

More generally, it now follows that

$$f(p\exp(t_1X_1+\cdots+t_sX_s))=\sum_{n_1,\ldots,n_s\geqslant 0}\frac{t_1^{n_1}\cdots t_s^{n_s}}{n_1!\cdots n_s!}(X^{(n_1,\ldots,n_s)}f)(p)$$

for any analytic function f on G and any $X_1,\ldots,X_s\in\mathfrak{g}$, where $X^{(n_1,\ldots,n_s)}$ is the coefficient of $t_1\cdots t_s$ in the formal expansion of

$$\frac{n_1!n_2!\ldots n_s!}{(n_1+\cdots+n_s)!}(t_1X_1+\cdots+t_sX_s)^{n_1+\cdots+n_s}.$$

Theorem 1.4.4
Let $s\geqslant 1$ and $X_1,\ldots,X_s\in\mathfrak{g}$. Then

$$\exp tX_1\ldots\exp tX_s=\exp\left\{t\sum_{1\leqslant i\leqslant s}X_i+\frac{t^2}{2}\sum_{1\leqslant i<j\leqslant s}[X_i,X_j]+O(t^3)\right\}$$

for small $|t|$.

Proof
If f is analytic near $e\in G$, let $F:(t_1,\ldots,t_s)\to f(\exp t_1X_1\cdots\exp t_sX_s)$. Then, for small $|t|$,

$$F(t,\ldots,t)=f(e)+t\sum_{1\leqslant i\leqslant s}\left(\frac{\partial}{\partial t_i}F\right)_0+\tfrac{1}{2}t^2\sum_{1\leqslant i,j\leqslant s}\left(\frac{\partial^2}{\partial t_i\partial t_j}F\right)_0+O(t^3).$$

But,

$$\left(\frac{\partial}{\partial t_i}F\right)_0=(X_if)(e),\ \left(\frac{\partial^2}{\partial t_i\partial t_j}F\right)_0=\left(\frac{\partial^2}{\partial t_j\partial t_i}F\right)_0=(X_iX_jf)(e)\quad(1\leqslant i<j\leqslant s).$$

Hence

$$F(t,\ldots,t)=f(e)+t\sum_{1\leqslant i\leqslant s}(X_if)(e)$$
$$+\frac{t^2}{2}\left\{\sum_{1\leqslant i\leqslant s}(X_i^2f)(e)+2\sum_{1\leqslant i<j\leqslant s}(X_iX_jf)(e)\right\}+O(t^3)$$

Now let $Y_1,\ldots,Y_n$ be a basis for $\mathfrak{g}$ and let x_i be maps such that

$$x_i(\exp(a_1Y_1+\cdots a_nY_n))=a_i,\qquad 1\leqslant i\leqslant n$$

for small $\|a\|$; i.e. x_i are *canonical coordinates of the first kind*. Then if

$$Z=c_1Y_1+\cdots+c_nY_n$$

we have $x_k(\exp tZ)=tc_k$ and so

$$Z^mx_k(e)=\frac{\mathrm{d}^m}{\mathrm{d}t^m}x_k(\exp tZ)\bigg|_{t=0}=\begin{cases}c_k & \text{if } n=1\\ 0 & \text{if } n\neq 1.\end{cases}$$

Let

$$X_i = \sum_{1 \leqslant k \leqslant n} c_{ik} Y_k, \quad [X_i, X_j] = \sum_{1 \leqslant k \leqslant n} d_{ijk} Y_k.$$

Hence, applying the above expression for $F(t, \ldots, t)$ with $f = x_k$, we have

$$\begin{aligned} x_k(\exp tX_1 \cdots \exp tX_s) &= t \sum_{1 \leqslant i \leqslant s} (X_i x_k)(e) \\ &\quad + \frac{t^2}{2} \sum_{1 \leqslant i < j \leqslant k} [X_i, X_j] x_k(e) + O(t^3) \\ &= t \sum_{1 \leqslant i \leqslant s} c_{ik} + \frac{t^2}{2} \sum_{1 \leqslant i < j \leqslant s} d_{ijk} + O(t^3), \end{aligned}$$

where we have used the identity

$$\sum_{1 \leqslant i \leqslant s} X_i^2 + 2 \sum_{1 \leqslant i < j \leqslant s} X_i X_j = (X_1 + \cdots + X_s)^2 + \sum_{1 \leqslant i < j \leqslant s} [X_i, X_j].$$

Now if we let $\Xi(t)$ be defined by

$$\exp \Xi(t) = \exp tX_1 \cdots \exp tX_s,$$

for small $|t|$, then

$$\Xi(t) = \sum_{1 \leqslant i \leqslant n} x_k(\exp tX_1 \cdots \exp tX_s) Y_k$$

and so

$$\Xi(t) = t \sum_{1 \leqslant i \leqslant s} X_i + \frac{t^2}{2} \left(\sum_{1 \leqslant i < j \leqslant s} [X_i, X_j] \right) + O(t^3). \qquad \square$$

In particular, we have

(a) $\exp tX \exp tY = \exp\left\{ t(X + Y) + \frac{t^2}{2}[X, Y] + O(t^3) \right\}$

(b) $\exp tX \exp tY \exp(-tX) \exp(-tY) = \exp\{t^2 [X, Y] + O(t^3)\}$.

Again, let G be a Lie group with Lie algebra $\mathfrak{g}$ and let ad X be the endomorphism of $\mathfrak{g}$ defined by

$$(\operatorname{ad} X)(Y) = [X, Y], \qquad Y \in \mathfrak{g}.$$

Then the map ad: $X \to \operatorname{ad} X$, $X \in \mathfrak{g}$ satisfies

$$\operatorname{ad}(\alpha_1 X_1 + \alpha_2 X_2) = \alpha_1 \operatorname{ad} X_1 + \alpha_2 \operatorname{ad} X_2, \qquad \alpha_1, \alpha_2 \in \mathbb{R} \text{ (or } \mathbb{C}), \quad X_1, X_2 \in \mathfrak{g}$$
$$\operatorname{ad}[X, Y] = [\operatorname{ad} X, \operatorname{ad} Y] \triangleq \operatorname{ad} X \operatorname{ad} Y - \operatorname{ad} Y \operatorname{ad} X.$$

Hence ad is a *representation* of $\mathfrak{g}$ on the space of endomorphisms of $\mathfrak{g}$, called the *adjoint representation*. For each $g \in G$, let $i_g: h \to ghg^{-1}$ be an inner automorphism of G and let $X \in \mathfrak{g}$ be regarded as an invariant vector field. Then we define

$$X^g = (i_{g*})_e(X_e),$$

and the map $\mathrm{Ad}(g): \mathfrak{g} \to \mathfrak{g}$ by

$$\mathrm{Ad}(g)X = X^g.$$

Then $\mathrm{Ad}(g)$ is an automorphism of $\mathfrak{g}$ for each $g \in G$ and the map $g \to \mathrm{Ad}(g)$ is a homomorphism of G into $GL(\mathfrak{g})$. It can be shown (Varadarajan, 1974) that

$$\mathrm{Ad}(\exp X) = e^{\mathrm{ad}\, X}, \qquad X \in \mathfrak{g}.$$

Next note that since the exponential map $\exp: \mathfrak{g} \to G$ maps $\mathfrak{g}$ into a neighbourhood of G, its derivative at $X \in \mathfrak{g}$ is a map

$$(\exp_*)_X : T_X\mathfrak{g} \to T_{\exp X}G.$$

Since $T_X\mathfrak{g} \cong \mathfrak{g}$ and $T_{\exp X}G \cong T_eG = \mathfrak{g}$ the following formula,

$$(\exp_*)_X = \sum_{n=0}^{\infty} \frac{(-1)^n}{(n+1)!}(\mathrm{ad}\ X)^n$$

makes sense and can be proved by Taylor series methods (Varadarajan, 1974). We shall not prove this result in detail here, but we shall illustrate it for the case $G = GL(n)$, with the Lie algebra $gl(n)$ = set of $n \times n$ matrices with $[X, Y] \triangleq XY - YX$. Thus, if $A, B \in gl(n)$ we have

$$(\exp_*)_A(B) = \left(\frac{\mathrm{d}}{\mathrm{d}t}\exp(-A)\exp(A+tB)\right)_{t=0}.$$

This follows from the facts that (a) the map $f: t \to \exp(-A)\exp(A+tB)$ is analytic from $\mathbb{R}$ to G with $f(0) = e$, (b) $L_{\exp A}f(t) = \exp(A+tB)$, and (c) $(L_{\exp A})_*: T_eG \cong T_{\exp A}G$. Now,

$$\begin{aligned}\exp(-A)\exp(A+tB) &= \sum_{i=0}^{\infty}\frac{(-A)^i}{i!}\sum_{j=0}^{\infty}\frac{(A+tB)^j}{j!}\\ &= \sum_{k=0}^{\infty}\sum_{i=0}^{k}\frac{(-A)^i(A+tB)^{k-i}}{i!(k-i)!}\\ &= \sum_{k=0}^{\infty}\frac{(-1)^k}{(k+1)!}t[A,[A,\ldots,[A,B]]\ldots] + o(t^2)\end{aligned}$$

as can easily be seen by induction.

Since the map exp is an analytic diffeomorphism of a neighbourhood U of $0 \in \mathfrak{g}$, there exists an analytic map $F: (X, Y) \to F(X, Y)$ of $U \times U$ into $\mathfrak{g}$ such that

$$\exp X \exp Y = \exp F(X, Y). \qquad \text{(Campbell–Hausdorff formula)}$$

Theorem 1.4.5

With the above notation,

$$F(X, Y) = \sum_{i=1}^{\infty} F_i(X, Y),$$

where $F_i(X, Y): \mathfrak{g} \times \mathfrak{g} \to \mathfrak{g}$ is a polynomial map of degree i and the series converges absolutely for $X, Y \in U$. Moreover, the polynomials F_i are given recursively by

$$F_1(X, Y) = X + Y$$

$$(i+1)F_{i+1}(X, Y) = \tfrac{1}{2}[X - Y, F_i(X, Y)] + \sum_{p \geqslant 1, 2p \leqslant n} R_{2p} \sum_{\substack{k_1, \ldots, k_{2p} > 0 \\ k_1 + \cdots + k_{2p} = n}} [F_{k_1}(X, Y), [\ldots [F_{k_{2p}}(X, Y), X + Y] \ldots],$$

where R_{2p} is the rational number given by the expansion

$$\frac{z}{1 - e^{-z}} - \tfrac{1}{2} z = 1 + \sum_{k=1}^{\infty} R_{2k} z^{2k}.$$

(For a proof, see Varadarajan (1974.)) □

Note finally that it can be shown by using the exponential map (Helgason, 1962) that if G is a Lie group and H is an abstract subgroup of G which is closed in G, then there is a unique analytic structure on H such that H is a Lie subgroup of G.

1.4.4 Lie algebras of the classical groups

We have mentioned above that the Lie algebra of $GL(n)$ is the set of matrices $gl(n)$, with the obvious bracket. Consider next $SL(n)$. Recall that $A \in SL(n)$ if $A \in GL(n)$ and $\det A = 1$. Let

$$sl(n) = \{A \in gl(n) : \operatorname{tr} A = 0\}.$$

Since $\operatorname{tr}[A, B] = 0$ if $A, B \in sl(n)$, it follows that $sl(n)$ is a (Lie) subalgebra of $gl(n)$. Moreover, if $A \in gl(n)$, we can write

$$A = \frac{1}{n} (\operatorname{tr} A) I + B,$$

where $\operatorname{tr} B = 0$. Hence $\dim\{sl(n)\} = n^2 - 1$. Finally, since for any matrix A,

$$\det(\exp A) = e^{\operatorname{tr} A}$$

it follows that the exponential map exp: $gl(n) \to GL(n)$, restricted to $sl(n)$, i.e.

$$\exp|_{sl(n)}: sl(n) \to GL(n)$$

has range $SL(n)$. Hence $sl(n)$ is the Lie algebra of $SL(n)$.

The connected component of $O(n)$ containing I is $SO(n)$. Let

$$\mathfrak{g} = \{A \in gl(n): A^{\mathsf{T}} = -A\}.$$

Then $\mathfrak{g}$ is a Lie subalgebra of $gl(n)$ and if $A \in \mathfrak{g}$, then

$$(\exp A)(\exp A)^{\mathsf{T}} = \exp A \exp A^{\mathsf{T}} = \exp(A - A) = I$$

so that $\mathfrak{g}$ is the Lie algebra of $SO(n)$ (since $\dim \mathfrak{g} = n(n-1)/2 = \dim SO(n)$).

For the group $SO(p,q)$ we have $A \in SO(p,q)$ if

$$A^{\mathsf{T}}BA = B$$

where

$$B = \begin{bmatrix} -I_p & 0 \\ 0 & I_q \end{bmatrix}.$$

Then it is easy to see that

$$A = \begin{bmatrix} A_1 & A_2 \\ A_3 & A_4 \end{bmatrix}$$

(partitioned in the obvious way) is in the Lie algebra $so(p,q)$ of $SO(p,q)$ if $A_1^{\mathsf{T}} = -A_1$, $A_4^{\mathsf{T}} = -A_4$, $A_2 = A_3^{\mathsf{T}}$, and A_2 arbitrary. Checking the dimension of $so(p,q)$ we have

$$\dim so(p,q) = p(p-1)/2 + q(q-1)/2 + pq = n(n-1)/2$$

since $p + q = n$.

Similarly, if $sp(n)$ denotes the Lie algebra of the symplectic group $Sp(n)$, then we have

$$sp(n) = \left\{A = \begin{bmatrix} A_1 & A_2 \\ A_3 & A_4 \end{bmatrix} : A_1^{\mathsf{T}} = -A_4,\ A_2^{\mathsf{T}} = A_2,\ A_3^{\mathsf{T}} = A_3\right\},$$

where each submatrix A_i is of order $n/2 \times n/2$.

1.4.5. Lie transformation groups and homogeneous spaces

If M is a Hausdorff space and G is a topological group, then we say that G is a (*topological*) *transformation group* of M if there is a homeomorphism $p \to gp$ of $M \to M$ such that

(a) $(g_1 g_2)p = g_1(g_2 p)$ for $p \in M$, $g_1, g_2 \in G$.
(b) $(g, p) \to gp$ is continuous from $G \times M$ into M

(cf. Section 1.2.3). From (a), $ep = p$ for all $p \in M$ (where e is the identity of G) and if $ap = p$ for all $p \in M$ implies $a = e$, then G is said to be *effective* on M. The group G acts *transitively* if for all $p, q \in M$, there exists $g \in G$ such that $gp = q$. If $p \in M$, then the group $G(p) = \{g \in G : gp = p\}$ is the *isotropy* subgroup of G and the set $Gp = \{gp : g \in G\}$ is called the *orbit* of $p \in M$ under G.

A *homogeneous space* M is a Hausdorff space which has a transitive group of homeomorphisms G (i.e. G acts transitively as a topological transformation group on M). The following results on transformation groups are proved in Helgason (1962).

Theorem 1.4.6
Let G be a transitive topological transformation group of a locally compact Hausdorff space M such that G is locally compact and has a countable base. Let $p \in M$ and let $G(p)$ be the isotropy group of G. Then $G(p)$ is closed and the map

$$gG(p) \to gp$$

is a homeomorphism of $G/G(p)$ onto M. □

Theorem 1.4.7
Let G be a Lie group and H be a closed subgroup of G. Then G/H (with the usual topology) has a unique analytic structure such that G is a Lie transformation group of G/H. □

Moreover, if G is a transitive Lie transformation group of a C^∞ manifold M and $p \in M$, then the map $gG(p) \to gp$ from $G/G(p)$ onto M is a diffeomorphism if it is a homeomorphism, where $G/G(p)$ is given the analytic structure in Theorem 1.4.7. In addition, if M is connected, then the identity component G_e of G is transitive on M.

1.4.6 Compact Lie algebras

Let $\mathfrak{g}$ be a real Lie algebra and let $GL(\mathfrak{g})$ denote the group of nonsingular endomorphisms of $\mathfrak{g}$. Then the Lie algebra $gl(\mathfrak{g})$ of $GL(\mathfrak{g})$ is the vector space of endomorphisms of $\mathfrak{g}$ with the usual commutator bracket. Clearly, $\mathrm{ad}(\mathfrak{g})$ is a subalgebra of $gl(\mathfrak{g})$. Let $\mathrm{Int}(\mathfrak{g})$ denote the analytic subgroup of $GL(\mathfrak{g})$ with the Lie algebra $\mathrm{ad}(\mathfrak{g})$. It can then be shown (Helgason (1962)) that

$$\mathrm{Int}(\mathfrak{g}) = \mathrm{Ad}(G).$$

Moreover, if $\mathfrak{g}$ is a real Lie algebra with centre $\{0\}$ then the centre of $\mathrm{Int}(\mathfrak{g})$ is just the identity element. The *centre* of a Lie algebra is, of course, just the kernel of the adjoint map and we say that $\mathfrak{g}$ is abelian if $\mathfrak{g} = \text{centre}(\mathfrak{g})$ i.e. $[X, Y] = 0$ for all $X, Y \in \mathfrak{g}$.

Let $\mathfrak{g}$ be a real Lie algebra and let $\mathfrak{h}$ be a subalgebra of $\mathfrak{g}$. Then $\text{ad}(\mathfrak{h})$ is a subalgebra of $\text{ad}(\mathfrak{g})$ and so there is an analytic subgroup $\mathscr{H}$ of $\text{Int}(\mathfrak{g})$ with Lie algebra $\text{ad}(\mathfrak{h})$. We say that $\mathfrak{g}$ is *compact* if $\text{Int}(\mathfrak{g})$ is compact and that $\mathfrak{h} \subseteq \mathfrak{g}$ is *compactly embedded* in $\mathfrak{g}$ if $\mathscr{H}$ is compact. Then we have (Helgason 1962)

Lemma 1.4.8
If $\tilde{\mathscr{H}}$ is the abstract group of $\mathscr{H}$ with the relative topology as a subset of $GL(\mathfrak{g})$, then $\mathscr{H}$ is compact if and only if $\tilde{\mathscr{H}}$ is compact. □

1.4.7 Structure theory of Lie algebras

The decomposition of a vector space in terms of invariant subspaces of a linear map is well known in linear algebra and gives rise to the Jordan normal form of a matrix. In fact if V is a vector space and $A \in \text{Hom}(V, V)$, then we define

$$V_i = \{v \in V : (A - \lambda_i I)^k v = 0 \text{ for sufficiently large } k\}, \qquad 1 \leqslant i \leqslant r$$

where the λ_i are the distinct eigenvalues of A. Then

$$V = \bigoplus_{i=1}^{r} V_i$$

and $AV_i \subseteq V_i$ for each i. If $d_i = \dim V_i$, then the characteristic polynomial of A is $\pi_{i=1}^{r} (\lambda - \lambda_i)^{d_i}$. Moreover, A has the decomposition

$$A = S + N$$

where S is semisimple (i.e. any invariant subspace of V for S has a complementary invariant subspace) and N is nilpotent ($N^k = 0$ for some $k > 0$). Also, $SN = NS$.

We can obtain a similar structure theory for Lie algebras by considering the linear map ad H for a particular H. There are many approaches to this theory; see, for example, Sagle and Walde (1973), Varadarajan (1974), Wan (1975). We shall follow Helgason (1962), where the proofs of the following results can be found.

Let $\mathfrak{g}$ be a Lie algebra (over $\mathbb{R}$ or $\mathbb{C}$) and let Tr denote the trace of an endomorphism of a vector space defined by

$$\text{Tr } A = \sum_{i=1}^{n} a_{ii}$$

where n is the dimension of the vector space and (a_{ij}) is a matrix representation of A with respect to any basis. This definition is, of course, basis independent.) Then we define the *Killing form* of $\mathfrak{g}$ to be the bilinear form

$$K(X, Y) = \text{Tr}(\text{ad } X \text{ ad } Y), \qquad X, Y \in \mathfrak{g}.$$

A Lie algebra $\mathfrak{g}$ is *semisimple* if its Killing form is nondegenerate, and *simple* if $\mathfrak{g} \neq 0$, it is semisimple and has no ideals other than $\{0\}$ and $\mathfrak{g}$. (An *ideal* of a Lie algebra $\mathfrak{g}$ is a vector subspace $\mathfrak{a}$ such that $[\mathfrak{a}, \mathfrak{g}] \subseteq \mathfrak{a}$.) A Lie algebra is called *reductive* if we can write $\mathfrak{g} = \mathfrak{g}_1 \oplus \cdots \oplus \mathfrak{g}_r \oplus c$ where each $\mathfrak{g}_i$ is simple and c is the centre of $\mathfrak{g}$.

Theorem 1.4.9 (Cartan)
A Lie algebra is semisimple if and only if it contains no abelian ideal other than $\{0\}$. □

Thus a semisimple Lie algebra has centre $(= \{X \in \mathfrak{g} : [X, Y] = 0 \text{ for all } Y \in \mathfrak{g}\})$ equal to $\{0\}$.

Lemma 1.4.10
A semisimple Lie algebra $\mathfrak{g}$ has a direct sum representation

$$\mathfrak{g} = \mathfrak{g}_1 \oplus \cdots \oplus \mathfrak{g}_r$$

in terms of simple ideals $\mathfrak{g}_i$. □

Next define

$$\mathfrak{D}\mathfrak{g} = \text{sp}\{[X, Y] : X, Y \in \mathfrak{g}\}$$

where sp denotes the linear span of a set. The $\mathfrak{D}\mathfrak{g}$ is called the *derived algebra* of $\mathfrak{g}$ and is clearly an ideal of $\mathfrak{g}$. Inductively, we can write

$$\mathfrak{D}^n\mathfrak{g} = \mathfrak{D}(\mathfrak{D}^{n-1}\mathfrak{g}).$$

A Lie algebra $\mathfrak{g}$ is *solvable* if $\mathfrak{D}^n\mathfrak{g} = \{0\}$ for some $n \geqslant 0$. The Lie algebra $\mathfrak{g}$ is called *nilpotent* if ad X is a nilpotent endomorphism of $\mathfrak{g}$ for each $X \in \mathfrak{g}$.

Theorem 1.4.11 (Engel)
If $V \neq \{0\}$ is a finite-dimensional vector space and $\mathfrak{g}$ is a subalgebra of $\text{gl}(V)$ containing only nilpotent elements, then

(a) $\mathfrak{g}$ is nilpotent
(b) there exists $v \in V$ such that $v \neq 0$, $Xv = 0$ for all $X \in \mathfrak{g}$.
(c) there exists a basis $e_1, \ldots, e_n$ of V in terms of which all the endomorphisms $X \in \mathfrak{g}$ have matrix representations which have zeros on or below the main diagonal. □

Corollary 1.4.12
A Lie algebra $\mathfrak{g}$ is nilpotent if and only if $\mathscr{C}^m\mathfrak{g} = \{0\}$ for $m \geqslant \dim \mathfrak{g}$, where

the *central series*

$$\mathscr{C}^0\mathfrak{g} \supseteq \mathscr{C}^1\mathfrak{g} \supseteq \mathscr{C}^2\mathfrak{g} \supseteq \cdots$$

is defined inductively by

$$\mathscr{C}^0\mathfrak{g} = \mathfrak{g}, \quad \mathscr{C}^{k+1}\mathfrak{g} = [\mathfrak{g}, \mathscr{C}^k\mathfrak{g}], \qquad k = 0, 1, 2, \ldots. \qquad \square$$

Proposition 1.4.13
A nilpotent Lie algebra is solvable □

Now let $\mathfrak{g}$ be a semisimple Lie algebra over $\mathbb{C}$. Since, by Theorem 1.4.9, a Lie algebra is semisimple if and only if it has no abelian ideals $\neq \{0\}$, it is reasonable to assume that to obtain a decomposition for $\mathfrak{g}$ in terms of invariant subspaces of ad H (for some $H \in \mathfrak{g}$) we must first remove the maximal non-semisimple subalgebra of $\mathfrak{g}$. For this reason we define a *Cartan subalgebra* of $\mathfrak{g}$ to be a subalgebra $\mathfrak{h}$ satisfying

(a) $\mathfrak{h}$ is a maximal abelian subalgebra of $\mathfrak{g}$.
(b) ad H is a semisimple endomorphism of $\mathfrak{g}$ for each $H \in \mathfrak{h}$.

Theorem 1.4.14
Every semisimple Lie algebra $\mathfrak{g}$ over $\mathbb{C}$ has a Cartan subalgebra. In fact, if we define, for any $H \in \mathfrak{g}$,

$$\mathfrak{g}(H, \lambda) = \{X \in \mathfrak{g} : (\text{ad } H - \lambda I)^k X = 0, \text{ some } k\},$$

then $\mathfrak{g}(\bar{H}, 0)$ is a Cartan subalgebra of $\mathfrak{g}$ where $\bar{H}$ is chosen so that

$$\dim \mathfrak{g}(\bar{H}, 0) = \min_{X \in \mathfrak{g}} (\dim \mathfrak{g}(X, 0)). \qquad \square$$

(Note that

$$\mathfrak{g} = \bigoplus_{i=0}^{r} \mathfrak{g}(H, \lambda_i)$$

where $0 = \lambda_0, \lambda_1, \ldots, \lambda_r$ are the distinct eigenvalues of ad H.)

Let $\mathfrak{h} \subseteq \mathfrak{g}$ be an arbitrary Cartan subalgebra. If $\alpha \in \mathfrak{h}^*$, i.e. α is a linear map from $\mathfrak{h}$ to $\mathbb{C}$, and the set

$$\mathfrak{g}^\alpha = \{X \in \mathfrak{g} : (\text{ad } H - \alpha(H))X = 0 \text{ for all } H \in \mathfrak{h}\}$$

is nonempty, then α is called a *root* of $\mathfrak{g}$ (with respect to $\mathfrak{h}$), and $\mathfrak{g}^\alpha$ is the corresponding *root space*. Since $\mathfrak{h}$ is a maximal abelian subalgebra of $\mathfrak{g}$ we have $\mathfrak{g}^0 = \mathfrak{h}$. Moreover, by the Jacobi identity,

$$[\mathfrak{g}^\alpha, \mathfrak{g}^\beta] \subseteq \mathfrak{g}^{\alpha+\beta}, \qquad \text{for all } \alpha, \beta \in \mathfrak{h}^*.$$

Let

$$\Delta = \{\alpha \in \mathfrak{h}^* : \alpha \neq 0, \quad \mathfrak{g}^\alpha \neq 0\}.$$

We can now state the main decomposition theorem for semisimple Lie algebras:

Theorem 1.4.15
Let $\mathfrak{g}$ be a semisimple Lie algebra with Killing form K and let $\mathfrak{h}$ be a Cartan subalgebra. Then

(a) $\mathfrak{g} = \mathfrak{h} \oplus \bigoplus_{\alpha \in \Delta} \mathfrak{g}^\alpha$

(b) $\dim \mathfrak{g}^\alpha = 1$ for each $\alpha \in \Delta$.

(c) If α, β are roots and $\alpha + \beta \neq 0$, then $K(X_\alpha, X_\beta) = 0$ for any $X_\alpha \in \mathfrak{g}^\alpha$, $X_\beta \in \mathfrak{g}^\beta$.

(d) $K|_{\mathfrak{h} \times \mathfrak{h}}$ is nondegenerate and for each root α there is a unique $H_\alpha \in \mathfrak{h}$ such that
$$K(H, H_\alpha) = \alpha(H) \qquad \text{for all } H \in \mathfrak{h}$$
and $\alpha(H_\alpha) \neq 0$.

(e) If $\alpha \in \Delta$ then $-\alpha \in \Delta$ and
$$[\mathfrak{g}^\alpha, \mathfrak{g}^{-\alpha}] = \mathbb{C} H_\alpha. \qquad \square$$

Let $\alpha \in \Delta$ and β be any root. Then the set of roots of the form $\beta + n\alpha$ for some integer n is called the *α-series* containing β. It can be shown that this series contains all elements of the form $\beta + n\alpha$ for $p \leqslant n \leqslant q$ for some p and q with
$$p + q = -2 \frac{\beta(H_\alpha)}{\alpha(H_\alpha)}.$$
Also, the only roots proportional to α are $-\alpha$, 0 and α and if $\alpha + \beta \neq 0$, then
$$[\mathfrak{g}^\alpha, \mathfrak{g}^\beta] = \mathfrak{g}^{\alpha+\beta}.$$
Moreover, if $X_\alpha \in \mathfrak{g}^\alpha$, $X_{-\alpha} \in \mathfrak{g}^{-\alpha}$, $X_\beta \in \mathfrak{g}^\beta$ for $\alpha, \beta \in \Delta$, then
$$[X_{-\alpha}, [X_\alpha, X_\beta]] = \frac{q(1-p)}{2} \alpha(H_\alpha) K(X_\alpha, X_{-\alpha}) X_\beta.$$

Theorem 1.4.16
If $\bar{\mathfrak{h}} = \bigoplus_{\alpha \in \Delta} \mathbb{R} H_\alpha$, then K is real and strictly positive definite on $\bar{\mathfrak{h}} \times \bar{\mathfrak{h}}$ and $\mathfrak{h} = \bar{\mathfrak{h}} + i\bar{\mathfrak{h}}$. $\square$

The vectors H_α defined in Theorem 1.4.15 determine $\mathfrak{g}$ up to isomorphism and they also have the important properties given in the next result.

Theorem 1.4.17
For each $\alpha \in \Delta$ there exists a vector $X_\alpha \in \mathfrak{g}^\alpha$ such that for any $\alpha, \beta \in \Delta$

$$[X_\alpha, X_{-\alpha}] = H_\alpha, \qquad [H, X_\alpha] = \alpha(H)X_\alpha, \quad \text{for } H \in \mathfrak{h},$$
$$[X_\alpha, X_\beta] = 0 \qquad \text{if } \alpha+\beta \neq 0 \quad \text{and} \quad \alpha+\beta \notin \Delta,$$
$$[X_\alpha, X_\beta] = N_{\alpha,\beta}X_{\alpha+\beta} \qquad \text{if } \alpha+\beta \in \Delta$$

where $N_{\alpha,\beta}$ is a constant and we have

$$N_{\alpha,\beta} = -N_{-\alpha,-\beta}.$$ □

Since the structure theory of a complex Lie algebra is given in terms of the spectral properties of ad H, for an element $H \in \mathfrak{g}$, in order to determine a similar theory for real Lie algebras, it is necessary to embed such a Lie algebra in a complex one. To do this we first consider a general real vector space V and define a *complex structure* on V to be an $\mathbb{R}$-linear endomorphism J of V such that $J^2 = -I$ (I = the identity map on V). We can then make V into a vector space $V_\mathbb{C}$ over $\mathbb{C}$ by defining

$$(a+\mathrm{i}b)X = aX + bJX \qquad \text{for } X \in V, \quad a, b \in \mathbb{R}.$$

Of course, V and $V_\mathbb{C}$ are the same when regarded as sets, but as vector spaces we have

$$\dim_\mathbb{C} V_\mathbb{C} = \tfrac{1}{2}\dim_\mathbb{R} V$$

and so $\dim_\mathbb{R} V$ must be even to support a complex structure.

Conversely, if V is a vector space over $\mathbb{C}$, then by restricting scalars to $\mathbb{R}$, we can regard V as a vector space $V_\mathbb{R}$ over $\mathbb{R}$ with dimension twice that of V (over $\mathbb{C}$). If J is defined as scalar multiplication by i on $V_\mathbb{R}$, then J is a complex structure on $V_\mathbb{R}$ and we have

$$V = (V_\mathbb{R})_\mathbb{C}.$$

A *complex structure* J in a Lie algebra $\mathfrak{g}$ over $\mathbb{R}$ is a complex structure on the underlying vector space of $\mathfrak{g}$ such that

$$[X, JY] = J[X, Y] \qquad \text{for all } X, Y \in \mathfrak{g}.$$

Then $[JX, JY] = -[X, Y]$ and $\mathfrak{g}_\mathbb{C}$ is a Lie algebra over $\mathbb{C}$, since it is easy to see that

$$[(a+\mathrm{i}b)X, (c+\mathrm{i}d)Y] = (a+\mathrm{i}b)(c+\mathrm{i}d)[X,Y].$$

Conversely, if $\mathfrak{g}$ is a Lie algebra over $\mathbb{C}$, then $\mathfrak{g}_\mathbb{R}$ has a complex structure defined by multiplication by i and $\mathfrak{g}_\mathbb{R}$ is a Lie algebra over $\mathbb{R}$.

Now let $\mathfrak{g}$ be a Lie algebra over $\mathbb{R}$. We can define a special complex structure J on the vector space $\mathfrak{g} \times \mathfrak{g}$ by defining

$$J((X, Y)) = (-Y, X).$$

Then $(\mathfrak{g} \times \mathfrak{g})_\mathbb{C}$ is called the *complexification* of $\mathfrak{g}$. (This is often defined, equivalently, by $\mathfrak{g} \otimes \mathbb{C}$.) Note that $(X, Y) = (X, 0) + \mathrm{i}(Y, 0)$ and so we can write $(X, Y) = X + iY$ and we define the bracket

$$[X_1 + iY_1, X_2 + iY_2] = [X_1, X_2] - [Y_1, Y_2] + i([Y_1, Y_2] + [X_1, Y_2]).$$

We shall write $\mathfrak{g}^{\mathbb{C}}$ for the complexification of $\mathfrak{g}$.

If $\mathfrak{g}$ is a Lie algebra over $\mathbb{C}$, then a *real form* of $\mathfrak{g}$ is a subalgebra $\mathfrak{g}_r$ of $\mathfrak{g}_{\mathbb{R}}$ such that

$$\mathfrak{g}_{\mathbb{R}} = \mathfrak{g}_r \oplus J\mathfrak{g}_r.$$

Then any element of $\mathfrak{g}$ can be written uniquely in the form $X + iY$ with $X, Y \in \mathfrak{g}_r$, and so $\mathfrak{g} \cong (\mathfrak{g}_r)^{\mathbb{C}}$.

Now let $\mathfrak{g}$ be a complex semisimple Lie algebra and let Δ be the set of nonzero roots of $\mathfrak{g}$. Then, if $\alpha \in \Delta$ let $X_\alpha \in \mathfrak{g}^\alpha$ satisfy the properties of Theorem 1.4.17, and consider the real subspace

$$\mathfrak{g}_r = \bigoplus_{\alpha \in \Delta} (iH_\alpha)\mathbb{R} \oplus \bigoplus_{\alpha \in \Delta} (X_\alpha - X_{-\alpha})\mathbb{R} \oplus \bigoplus_{\alpha \in \Delta} i(X_\alpha + X_{-\alpha})\mathbb{R}.$$

Then it can be shown that $\mathfrak{g}_r$ is a compact real form of $\mathfrak{g}$. Hence any complex semisimple Lie algebra has a compact real form and, moreover, for any two such real forms $\mathfrak{g}_1$ and $\mathfrak{g}_2$ of $\mathfrak{g}$ there is an automorphism ϕ of $\mathfrak{g}$ such that $\phi\mathfrak{g}_1 = \mathfrak{g}_2$.

If $\mathfrak{A}$ is a compact, semisimple Lie algebra, $\mathfrak{t}_0$ a maximal abelian subalgebra of $\mathfrak{A}$ and $\mathfrak{g} = \mathfrak{A}^{\mathbb{C}}$, then let $\mathfrak{h}$ be the subalgebra of $\mathfrak{g}$ generated by $\mathfrak{t}_0$. A *Weyl basis* of $\mathfrak{g}/\mathfrak{h}$ (with respect to $\mathfrak{A}$) is a basis $\{X_\alpha : \alpha \in \Delta\}$ of $\mathfrak{g}/\mathfrak{h}$ such that

(a) $X_\alpha \in \mathfrak{g}^\alpha$ and $[X_\alpha, X_{-\alpha}] = H_\alpha$ for each $\alpha \in \Delta$.
(b) $[X_\alpha, X_\beta] = N_{\alpha,\beta} X_{\alpha+\beta}$ if $\alpha, \beta, \alpha + \beta \in \Delta$ and $N_{\alpha,\beta} = -N_{-\alpha,-\beta}$.
(c) $X_\alpha - X_{-\alpha} \in \mathfrak{A}$, $i(X_\alpha + X_{-\alpha}) \in \mathfrak{A}$ for each $\alpha \in \Delta$.

It is clear from the above results that a Weyl basis always exists, and is determined up to an automorphism of $\mathfrak{g}$.

The above results enable us to prove a result of Kuranishi (1951), which is useful in the study of the controllability of bilinear systems (see also Boothby (1975) and Chapter 4).

Theorem 1.4.18
Let $\mathfrak{g}$ be a real semisimple Lie algebra of dimension n. Then there exist two elements of $\mathfrak{g}$ which generate it as an algebra.

Proof
Let $\mathfrak{g}^{\mathbb{C}}$ denote the complexification of $\mathfrak{g}$, and let $\mathfrak{h}$ be a Cartan subalgebra of $\mathfrak{g}^{\mathbb{C}}$ with nonzero root set $\Delta = \{\alpha\}$ and corresponding root spaces $\mathfrak{g}^\alpha$. Let $\{X_\alpha\}$ be a Weyl basis of $\mathfrak{g}^{\mathbb{C}}$, so that

$$\mathfrak{g}^{\mathbb{C}} = \mathfrak{h} \oplus \bigoplus_{\alpha \in \Delta} \mathfrak{g}^\alpha = \mathfrak{h} \oplus \mathfrak{g}_\Delta$$

where

$$\mathfrak{g}_{\Delta} = \bigoplus_{\alpha \in \Delta} \mathfrak{g}^{\alpha}.$$

Choose $H_0 \in \mathfrak{h}$ so that the minimal polynomial of the operator $\mathrm{ad}(H_0)$ and its characteristic polynomial are equal. This can be done by choosing H_0 so that

$$\alpha_1(H_0) \neq \alpha_2(H_0) \qquad \text{for all } \alpha_1 \neq \alpha_2 \text{ in } \Delta.$$

Then $\mathrm{ad}(H_0)$ is cyclic on $\mathfrak{g}_{\Delta}$ and so

$$X_0,\ \mathrm{ad}(H_0)X_0,\ \mathrm{ad}(H_0)^2 X_0, \ldots$$

span $\mathfrak{g}_{\Delta}$ for some $X_0 \in \mathfrak{g}_{\Delta}$. Hence H_0 and X_0 span $\mathfrak{g}^{\mathbb{C}}$; in particular, there is a set of n monomial brackets P_i in H_0 and X_0 which form a basis of $\mathfrak{g}^{\mathbb{C}}$, i.e.

$$P_i(H_0, X_0) = [\ldots[H_0, X_0]\ldots], \qquad i = 1, \ldots, n.$$

This proves the result for the complexification $\mathfrak{g}^{\mathbb{C}}$ of $\mathfrak{g}$,

To prove the result for the real algebra $\mathfrak{g}$ let $Y_1, \ldots, Y_n$ be a basis of $\mathfrak{g}$. Then $Y_1, \ldots, Y_n$ is also a basis for $\mathfrak{g}^{\mathbb{C}}$ (over $\mathbb{C}$). Let

$$P_i\left(\sum_{k=1}^{n} z^k Y_k, \sum_{l=1}^{n} w^l Y_l\right) = \left[\ldots\left[\sum_{k=1}^{n} z^k Y_k, \sum_{l=1}^{n} w^l Y_l\right], \ldots\right]$$

$$= \sum_{j=1}^{n} \lambda_i^j(z, w) Y_j,$$

where λ_i^j $(i \leqslant i, j \leqslant n)$ is a polynomial in the $2n$ complex variables $(z, w) = (z^1, \ldots, z^n, w^1, \ldots, w^n)$. The function

$$f\colon (z, w) \to \det \lambda_i^j(z, w)$$

is a polynomial in $2n$ variables with real coefficients and does not vanish at the value (z_0, w_0) for which

$$H_0 = \Sigma z_0^k Y, \qquad X_0 = \Sigma w_0^l F_l.$$

Hence f is not identically zero and so there must exist $2n$ real numbers $(a, b) = (a^1, \ldots, a^n, b^1, \ldots, b^n)$ such that $f(a, b) \neq 0$. Hence we can write $Y_1, \ldots, Y_n$ in terms of the values $P_i(A, B)$ where

$$A = \Sigma a^k Y_k, \qquad B = \Sigma b^l Y_l. \qquad \square$$

1.5 ALGEBRAIC GEOMETRY OF CURVES

1.5.1 Projective space

In this section we shall discuss the topological structure of projective varieties in $\mathbb{P}^2(\mathbb{C})$ which we shall use later to develop a spectral theory for nonlinear systems. A complete introduction to this subject is given by Kendig (1977); we shall merely outline the main results.

The topological space $\mathbb{C}^n = \{(z_1, ..., z_n) : z_i \in \mathbb{C}\}$ is called *affine n-space* over $\mathbb{C}$. Note that, in general algebraic geometry, we can replace $\mathbb{C}$ by any field, but $\mathbb{C}$ will suffice here because of our interest in spectral theory. The set $\mathbb{P}^n(\mathbb{C})$ of one-dimensional (complex) subspaces of $\mathbb{C}^{n+1}$ is the n-dimensional (complex) projective space. Note that $\mathbb{P}^n(\mathbb{C})$ is just $G_1(\mathbb{C}^n)$, the Grassmann manifold introduced above.

Two nonzero $(n+1)$-tuples $(x_1, ..., x_{n+1})$, $(y_1, ..., y_{n+1}) \in \mathbb{C}^{n+1}$ define the same element of $\mathbb{P}^n(\mathbb{C})$ if there exists $\alpha \in \mathbb{C}\backslash\{0\}$ such that $x_i = \alpha y_i$, $1 \leqslant i \leqslant n+1$. It follows that an alternative description of $\mathbb{P}^n(\mathbb{C})$ is as the 'completion' of $\mathbb{C}^n$ by the addition of a point for each 1-subspace of $\mathbb{C}^n$. In fact, if H is any hyperplane in $\mathbb{C}^{n+1}$ and H_a is any affine space parallel to H, then all one-subspaces not lying entirely in H intersect H_a in a single point. This gives rise to a one-to-one correspondence between the points of $H = \mathbb{C}^n$ and the one-subspaces of $\mathbb{P}^n(\mathbb{C})$ not parallel to H. The remaining one-subspaces lying lying in H account for the rest of $\mathbb{P}^n(\mathbb{C})$. Such a hyperplane H in $\mathbb{C}^{n+1}$ defines a projective subspace of $\mathbb{P}^n(\mathbb{C})$ denoted by $\mathbb{P}^{n-1}(\mathbb{C})$ which is called the *hyperplane at infinity relative to H*. $\mathbb{P}^n(\mathbb{C})\backslash\mathbb{P}^{n-1}(\mathbb{C})$ is called the corresponding *affine part* of $\mathbb{P}^n(\mathbb{C})$. Choosing different hyperplanes H allows us to visualize the 'hidden' part of $\mathbb{P}^n(\mathbb{C})$ 'at infinity'.

1.5.2 Projective varieties

If $\{p_\alpha\}$ is a set of polynomials in $\mathbb{C}[X_1, ..., X_n]$, then the set

$$V(\{p_\alpha\}) = \{x = (x_1, ..., x_n) \in \mathbb{C}^n : p_\alpha(x) = 0 \text{ for each } \alpha\}$$

is the *affine variety* (or just the *variety*) defined by the set $\{p_\alpha\}$. If V is a variety in $\mathbb{C}^n$, then the closure of V in $\mathbb{P}^n(\mathbb{C})$ is called the *projective completion* of V and is denoted by $\bar{V}$. Since $\mathbb{P}^n(\mathbb{C})$ is the set of complex lines in $\mathbb{C}^{n+1}$, $\bar{V}$ is a subset of $\mathbb{C}^{n+1}$ such that

$$x \in \bar{V} \Rightarrow cx \in \bar{V} \qquad \text{for all } c \in \mathbb{C}.$$

Any subset of $\mathbb{C}^{n+1}$ having this property is called a *homogeneous subset* of $\mathbb{C}^{n+1}$. A *homogeneous variety* in $\mathbb{C}^{n+1}$ is an affine variety which is a homogeneous set. Hence a projective variety in $\mathbb{P}^n(\mathbb{C})$ can be regarded as a homogeneous variety in $\mathbb{C}^{n+1}$. Homogeneous varieties (in $\mathbb{C}^{n+1}$) are defined by homogeneous polynomials, i.e. polynomials in which all terms have the same total order. Any polynomial

$$p = p(X_1, ..., X_{n+1}) \in \mathbb{C}[X_1, ..., X_{n+1}]$$

is of the form

$$p = p_0 + \cdots + p_s$$

where p_i is homogeneous of degree i.

Now recall that any affine hyperplane H in $\mathbb{C}^{n+1}$ can be regarded as the hyperplane at infinity in $\mathbb{P}^n(\mathbb{C})$ and we may observe the affine parts of any projective variety by studying the intersection of the homogeneous set in $\mathbb{C}^{n+1}$ defined by the variety and those affine hyperplanes H which do not pass through the origin. The most obvious affine hyperplanes H are those parallel to the axes:

$$H_i : X_i = 1, \qquad 1 \leqslant i \leqslant n+1,$$

although we may choose any other one. Hence, if $\{p_\alpha\}$ is a set of homogeneous polynomials defining a projective variety in $\mathbb{P}^n(\mathbb{C})$, then we can obtain the corresponding 'canonical' affine parts by substituting $X_1 = 1, X_2 = 1, \ldots, X_{n+1} = 1$ successively in each p_α. If

$$p = p_0 + p_1 + \cdots + p_d \in \mathbb{C}[X_1, \ldots, \hat{X}_k, \ldots, X_{n+1}]$$

is a polynomial in which p_i is of degree i, then the *homogenization of p at X_k* is defined as the polynomial

$$H_k(p) = p_0 X_k^d + p_1 X_k^{d-1} + \cdots + p_d.$$

Similarly, if $p = p(X_1, \ldots, X_{n+1})$ is a homogeneous polynomial, then the *dehomogenization* of p at X_k is defined by

$$D_k(p) = p(X_1, \ldots, X_{n+1}) \mid_{X_k = 1}.$$

Theorem 1.5.1

Let $V = V(p_1, \ldots, p_r) \subseteq \mathbb{C}^n$ be an affine variety. Then $\bar{V} \subseteq \mathbb{P}^n(\mathbb{C})$ is represented by the homogeneous variety

$$V(H_{n+1}(p_1), \ldots, H_{n+1}(p_r)) \subseteq \mathbb{C}^{n+1}. \qquad \square$$

Note that Theorem 1.5.1 is not true if we replace $\mathbb{C}$ by $\mathbb{R}$. Also, for any polynomial $p \in \mathbb{C}[X_1, \ldots, X_n]$, we have

$$D_{n+1}(H_{n+1}(p)) = p$$

but it may happen that

$$H_{n+1}(D_{n+1}(p)) \neq p.$$

(Consider the example $p(X_1, X_2) = X_1 X_2$.)

As a simple example of the process of projective completion and dehomogenization, consider the circle $V(X^2 + Y^2 - 1) \subseteq \mathbb{R}^2$. By homogenization this curve must be a part of the homogeneous variety

$V(X^2+Y^2-Z^2)\subseteq \mathbb{R}^3$. The latter variety is just a cone and by dehomogenization with respect to various affine hyperplanes we obtain the standard conic sections.

We shall now proceed to describe the topological structure of a complex algebraic curve; i.e. the topological structure of $C=V(p(X,Y))\subseteq \mathbb{P}^2(\mathbb{C})$, the projective variety defined by a single polynomial p. Firstly note that by the implicit function theorem, it is clear that C is locally the graph of an analytic function (near (x_0, y_0)) if

$$\frac{\partial p}{\partial X}(x_0, y_0)\neq 0 \quad \text{or} \quad \frac{\partial p}{\partial Y}(x_0, y_0)\neq 0.$$

Hence, at such a point, C looks locally like a disc. At the points where $\partial p/\partial X=\partial p/\partial Y=0$, C may not be locally a topological disc. To describe the behaviour of C at such points let $D_1, \ldots, D_n$ be topological discs with base points $p_1, \ldots, p_n$ and let $D=\vee D_i/\sim$ where $\vee$ denotes the disjoint union and $\sim$ is the relation obtained by identifying base points. Then D is called a *one-point union of discs.*

Lemma 1.5.2

Let C be a complex algebraic curve. Then C is locally a disc at all but finitely many points where C is locally a one-point union of finitely many discs.

Proof

To show that C is locally a disc at all but finitely many points it is sufficient to show that there are only finitely many points $(x_0, y_0)\in\mathbb{C}^2$ such that

$$p(x_0, y_0)=\frac{\partial p}{\partial Y}(x_0, y_0)=0.$$

First note that if $n=\deg p$, then by a change of coordinates (if necessary) we can write

$$p(X,Y)=Y^n+a_1(X)Y^{n-1}+\cdots+a_n(X),$$

where $\deg a_i(X)\leqslant i$ or $a_i(X)=0$. Moreover, we can assume that p has no repeated factors, for if $p_1^{n_1}\cdots p_r^{n_r}=p$, then

$$V(p)=V(p_1^{n_1})\cdots V(p_r^{n_r})=V(p_1\cdots p_r).$$

Now let $\mathscr{D}_Y(p)$ be the discriminant of p with respect to Y. This is a polynomial in X which is zero at x_0 if p and $\partial p/\partial Y$ have a common zero at x_0. $\mathscr{D}_Y(p)\not\equiv 0$ since p has no repeated factors and so the polynomial $\mathscr{D}_Y(p)\in\mathbb{C}[X]$ has only a finite number of zeros. Hence p and $\partial p/\partial Y$ have a common zero at a finite number of points x_i, $1\leqslant i\leqslant k$ say. At each such point the equations

$$p(x_i, Y) = \left(\frac{\partial p}{\partial Y}\right)(x_i, Y) = 0$$

can only have a finite number of solutions, since $p(x_i, Y)$ is a nonzero polynomial of degree n.

To prove the second part, on the topological structure of C in a neighbourhood of the points where C is not a topological disc we merely apply the classical Puisseaux series argument. In fact, if

$$p(X, Y) = Y^n + a_1(X)Y^{n-1} + \cdots + a_n(X),$$

then in a neighbourhood of $(x_0, y_0) \in C$ we can write, for some N,

$$p(X, Y) = \prod_{j+1}^{N} (Y - Y_j),$$
$$Y_j = y_0 + a_{j1}(X - x_0)^{1/m_j} + a_{j2}(X - x_0)^{2/m_j} + \cdots,$$

where $m_1 + \cdots + m_N = r$ and r is the multiplicity of the zero y_0 in $p(x_0, Y)$. Then C is the disjoint union of N discs in a neighbourhood of (x_0, y_0). □

REMARK We recall that if p and q are polynomials given by

$$p(X) = a_0X^m + a_1X^{m-1} + \cdots + a_m$$
$$q(X) = b_0X^n + b_1X^{n-1} + \cdots + b_n,$$

then the *resultant* of p and q is the determinant

$$\left|\begin{array}{cccccccc} a_0 & a_1 & a_2 & & a_m & & & \\ & a_0 & a_1 & \cdots & & a_m & & \\ & & a_0 & \cdots & & & a_m & \\ & & & \cdots & & & & \\ & & & & a_0 & \cdots & & a_m \\ b_0 & b_1 & b_2 & \cdots & & b_n & & \\ & b_0 & b_1 & \cdots & & & b_n & \\ & & \cdots & & & & & \\ & & & & b_0 & \cdots & & b_n \end{array}\right| \begin{array}{l} \left.\vphantom{\begin{array}{c} a\\a\\a\\a\\a \end{array}}\right\} n \text{ rows} \\ \left.\vphantom{\begin{array}{c} b\\b\\b\\b \end{array}}\right\} m \text{ rows} \end{array}$$

The *discriminant* $\mathcal{D}(p)$ of p is the resultant of p and $\mathrm{d}p/\mathrm{d}X$. □

Coming now to the global structure of a complex algebraic curve we must consider covering surfaces of a topological 2-manifold M. A locally compact space N is a *covering surface* of M if there is a surjective continuous map $\pi: N \rightarrow M$ such that

(a) for each $p \in M$ there is an open disc $D \subseteq M$ with $p \in M$ such that each component of $\pi^{-1}(D)$ is an open disc D_α, $\alpha \in A$, for some set A.
(b) $\pi \mid D_\alpha : D_\alpha \to D$ is a homeomorphism for each $\alpha \in A$.

If $\#(A) = n$, then (N, M, π) is called an *n-sheeted cover*. If $P = \{p_1, \ldots, p_n\}$ is a finite subset of M and $\bar{M} = M \backslash P, \bar{N} = N \backslash \pi^{-1}(P)$, $\bar{\pi} = \pi \mid N$ and if $(N, M, \bar{\pi})$ is an n-sheeted cover, then (N, M, π) is called a *near n-sheeted cover*.

Lemma 1.5.3
If $p(X, Y) = a_0(X)Y^n + a_1(X)Y^{n-1} + \cdots + a_n(X)$ where $a_i(X) \in \mathbb{C}[X]$ and $a_0 \neq 0$, and p has no repeated factors, then

$$(V(p), \mathbb{C}, \pi)$$

is a near n-sheeted cover, where $V(p)$ is regarded as an affine variety in $\mathbb{C}^2$ and $\pi : \mathbb{C}^2 \to \mathbb{C}$ is the map $\pi(X, Y) = X$. □

We can compactify the variety $V(p)$ in Lemma 1.5.3 and regard it as a covering of the sphere $\mathbb{P}^1(\mathbb{C})$ by taking its projective completion in $\mathbb{P}^2(\mathbb{C})$. Then we have that if C is a projective algebraic curve, then $(C, \mathbb{P}^1(\mathbb{C}), \pi)$ is a near n-sheeted cover of $\mathbb{P}^1(\mathbb{C})$.

The exceptional points of this cover are of two types; the points where C is locally a disc and the points where C is locally a one-point union of finitely many discs. The first type of point is called *topologically nonsingular* and the second type is called *topologically singular*. We have seen that C is locally a disc at $p \in C$ if $\partial P/\partial X \neq 0$ or $\partial P/\partial Y \neq 0$ at p, where P is a polynomial describing C in some dehomogenization. However the converse is not true. We may have $\partial P/\partial X = 0$ and $\partial P/\partial Y = 0$ at p and still C is a topological disc at p. At such a point, C is not locally the graph of a smooth function and we say that C is *singular* at a point where $\partial P/\partial X = 0$ and $\partial P/\partial Y = 0$. We say that C is a *nonsingular* curve if it is not singular anywhere. We can now state the classification theorem for algebraic curves.

Theorem 1.5.4
If $p \in \mathbb{C}[X, Y]$ is not a constant and is irreducible, then the projective variety $V(p)$ is obtained from a real two-dimensional, compact, connected, orientable manifold by identifying finitely many points to finitely many points (corresponding to the one-point union of discs). If $p \in \mathbb{C}[X, Y] \backslash \mathbb{C}$ is any polynomial, then $V(p)$ is a finite union of such varieties which touch each other at finitely many points. □

Recall that a real two-dimensional, compact, connected, orientable manifold is just a Riemann surface and, as such, is a topological sphere with g handles. If $C \subseteq \mathbb{P}^2(\mathbb{C})$ is a nonsingular projective curve defined by an irreducible polynomial $p(X, Y)$ of degree n, then it can be shown that

$$g = (n-1)(n-2)/2.$$

In generalizing some of the above ideas to higher-dimensional varieties we are led to consider ideals in the ring of polynomials $\mathbb{C}[X_1, \ldots, X_n]$. An *ideal I* in a commutative ring R is a subset which is closed under subtraction and for which

$$ar \in I \qquad \text{for all } a \in I, r \in R.$$

Now, if V is an algebraic variety in $\mathbb{C}^n$ defined by polynomials $p_1, \ldots, p_k \in \mathbb{C}[X_1, \ldots, X_n]$, i.e.

$$V = \{x \in \mathbb{C}^n : p_i(x) = 0 \text{ for all } i\}$$

then the set

$$I_V = \{p \in \mathbb{C}[X_1, \ldots, X_n] : p(x) = 0 \text{ for all } x \in V\}$$

is an ideal in $\mathbb{C}[X_1, \ldots, X_n]$. Hence, if V is a variety defined by a set of polynomials, then we associate with V the ideal consisting of all polynomials which vanish on V.

If $p \in \mathbb{C}[X_1, \ldots, X_n]$, then the set

$$(p) = \{pq : q \in \mathbb{C}[X_1, \ldots, X_n]\}$$

is an ideal in $\mathbb{C}[X_1, \ldots, X_n]$ called the *principal ideal generated by p*. Generally, if $\{p_\alpha\}$ is a set of polynomials in $\mathbb{C}[X_1, \ldots, X_n]$, then the *ideal generated by* $\{p_\alpha\}$ is the set

$$\{r_1 p_1 + \cdots + r_s p_s : r_i \in \mathbb{C}[X_1, \ldots, X_n],\ p_i \in \{p_\alpha\}\}.$$

We can now define the product $I_1 I_2$ of the ideals I_1 and I_2 as the smallest ideal containing the products pq with $p \in I_1, q \in I_2$. Then

$$(p) . (q) = (pq),$$

for principal ideals. It is clear that

$$(p) . (q) = (p) \cap (q)$$

if p and q have no common factors.

For any ideal I in a ring R we define the *radical* $\sqrt{I}$ of I to be the set

$$\sqrt{I} = \{p \in R : q^m \in I \text{ for some integer } m > 0\}.$$

We have introduced above the ideal I_V generated by a variety V. In the same way an ideal $I \subseteq \mathbb{C}[X_1, \ldots, X_n]$ generates a variety V_I given by

$$V_I = \{x \in \mathbb{C}^n : p(x) = 0 \text{ for all } p \in I\}.$$

Clearly, we have

$$I \subseteq J \Rightarrow V_J \subseteq V_I$$

and

$$V_1 \subseteq V_2 \Rightarrow I_{V_2} \subseteq I_{V_1}.$$

Moreover, it is easy to show that, for any ideals I, J,

$$V_{I \cap J} = V_{I.J} = V_I \cap V_J$$

and

$$V_{I+J} = V_I \cap V_J$$

where $I + J$ is the ideal

$$I + J = \{p + q : p \in I, q \in J\}.$$

Note however, that two different ideals may define the same variety. Nevertheless, for any ideal I,

$$I \subseteq I_{V_I}$$

and I_{V_I} is the largest defining ideal of V. If $\bar{I} = I_{V_I}$, then we clearly have

$$V_{\bar{I}_1 \cap \bar{I}_2} = V_{\bar{I}_1} \cup V_{\bar{I}_2}$$
$$V_{\overline{I_1 + I_2}} = V_{\bar{I}_1} \cap V_{\bar{I}_2}.$$

For any ideal $I \subseteq \mathbb{C}[X_1, ..., X_n]$, $\bar{I}$ is called the *closure* of I and if $I = \bar{I}$ we say that I is *closed*.

An ideal $I \subseteq R$ is *irreducible* if $I = I_1 \cap I_2$ implies that $I = I_1$ or $I = I_2$, and I is called a *prime ideal* if $r_1 . r_2 \in I$ implies that $r_1 \in I$ or $r_2 \in I$. A variety $V \subseteq \mathbb{C}^n$ (or $\mathbb{P}^2(\mathbb{C})$) is *irreducible* if $V = V_1 \cup V_2$, where V_1 and V_2 are varieties, implies $V = V_1$ or $V = V_2$. Then it is easy to prove that any variety V can be written

$$V = V_1 \cup \cdots \cup V_k$$

where the V_i are irreducible varieties, $V_i \neq V_j$, for $i \neq j$ and the decomposition is unique up to the order of the V_i. Similarly, any closed ideal $I \subseteq \mathbb{C}[X_1, ..., X_n]$ may be written as an irredundant intersection of irreducible closed ideals, uniquely up to order:

$$I = I_1 \cap \cdots \cap I_k.$$

In fact, each I_i in this expression is a prime ideal, since any prime ideal is irreducible and closed. The proof of the last statement requires Hilbert's 'Nullstellensatz' which states that every proper ideal I of $\mathbb{C}[X_1, ..., X_n]$ has a zero $x \in \mathbb{C}^n$, i.e. $p(x) = 0$ for all $p \in I$. Note also that it can be shown that the closure $\bar{I}$ of an ideal I in $\mathbb{C}[X_1, ..., X_n]$ is precisely the radical $\sqrt{I}$ of I. Moreover, we have

$$\bar{I} = \sqrt{I} = \bigcap_{P \supseteq I} P$$

where the intersection is taken over all prime ideals containing I. Hence we have seen that irreducible varieties are defined by prime ideals and vice versa. These irreducible varieties are the higher-dimensional analogues of the irreducible projective curves studied above. However, the topological classification of the higher-dimensional irreducible varieties is not a simple matter, as in the case of curves; in fact it has been proved (Markov, 1958) that it is impossible to devise an algorithm producing a complete set of topological invariants for closed, connected, compact manifolds of (real) dimension $\geqslant 4$.

The final concept from elementary algebraic geometry which we shall need is that of the coordinate ring of a variety. First note that $\mathbb{C}[X_1, \ldots, X_n]$ is the set of all polynomial functions defined on $\mathbb{C}^n$ and is generated by the polynomials $X_1, \ldots, X_n$. The only identically zero polynomial in $\mathbb{C}[X_1, \ldots, X_n]$ is the zero polynomial. Regarding $\mathbb{C}^n$ as the irreducible variety defined by the zero polynomial we call $\mathbb{C}[X_1, \ldots, X_n]$ the *affine coordinate ring* of $\mathbb{C}^n$. If I is a prime ideal in $\mathbb{C}[X_1, \ldots, X_n]$, then V_I is an irreducible variety and we can consider the ring of polynomials defined on V_I. Since two polynomials p, q in $\mathbb{C}[X_1, \ldots, X_n]$ are equal on V_I if $p - q \in I$ it is natural to define the *affine coordinate ring* of V_I as the quotient ring $\mathbb{C}[X_1, \ldots, X_n]/I$. We can write this as $\mathbb{C}[x_1, \ldots, x_n]$, where

$$x_i = X_i + I, \qquad 1 \leqslant i \leqslant n.$$

1.6 FUNCTIONAL ANALYSIS

1.6.1 Hilbert and Banach spaces

In the last section of this chapter we shall present a brief review of some elementary functional analysis which we shall need later. The proofs of the results can be found in Yosida (1974), Taylor (1958), Takesaki (1979).

Recall first that a *Banach space* is a normed linear space which is *complete*, i.e. any Cauchy sequence converges in the space. A *Hilbert space* is a Banach space whose norm is defined by an inner product. The basic structure theory of a Hilbert space H is similar to that of $\mathbb{R}^n$ in that there exists an *orthonormal basis* $\{e_i\}_{i \in I} \subseteq H$; i.e.

$$\langle e_i, e_j \rangle = \delta_{ij}, \qquad i, j \in I$$

and H is the closed linear span of $\{e_i\}$. If I is a countable index set, then H is said to be *separable* and then we have *Parseval's relation*

$$\| x \|^2 = \sum_{i=0}^{\infty} |\langle x, e_i \rangle|^2, \qquad \text{for all } x \in H,$$

where $I = \{0, 1, 2, \dots\}$.

A *linear form* x^* on a Banach space X is a continuous linear map $x^*: X \to F$, where F is the ground field of X and is equal to $\mathbb{R}$ or $\mathbb{C}$ in our case. The *dual space* X^* of X is the set of all forms on X. The space X is *reflexive* if the map $J: X \to X^{**}$ $(=(X^*)^*)$ defined by

$$\langle J(x), x^* \rangle = \langle x, x^* \rangle$$

is onto X^{**}. The dual space X^* is a Banach space under the norm

$$\| x^* \| = \sup_{\| x \| = 1} | \langle x^*, x \rangle |.$$

If H is a Hilbert space then for any $h^* \in H^*$ there is a unique $h \in H$ such that

$$\langle h^*, x \rangle_{H^*, H} = \langle x, h \rangle_H \qquad \text{for all } x \in H$$

and $\| h^* \| = \| h \|$. ($\langle .,. \rangle_{H^*, H}$ denotes the duality pairing with respect to H^*, H.) Thus the map $j_H: h \to h^*$ is an isometric isomorphism.

Examples

We shall now present some examples of Hilbert and Banach spaces which we shall need later.

1. Let $\Omega \subseteq \mathbb{R}^n$ be open and bounded and let $C^k(\bar{\Omega})$, $k \in \mathbb{N}$ be the set of all functions defined on Ω whose derivatives to order k exist and are uniformly continuous. Define a norm on $C^k(\bar{\Omega})$ by

$$\| f \|_k = \sup_{|p| \leqslant k} \left\{ \sup_{x \in \Omega} |(\partial/\partial x)^p f(x)| \right\}, \qquad f \in C^k(\bar{\Omega})$$

where $p \in \mathbb{N}^n$, $|p| = p_1 + \cdots + p_n$ and $(\partial/\partial x)^p = (\partial/\partial x_1)^{p_1} \cdots (\partial/\partial x_n)^{p_n}$. Then $C^k(\bar{\Omega})$ is a Banach space. We write

$$C(\bar{\Omega}) = C^0(\bar{\Omega}) \qquad \text{and} \qquad C^\infty(\bar{\Omega}) = \bigcap_{k=0}^{\infty} C^k(\bar{\Omega}).$$

2. Let $\Omega \subseteq \mathbb{R}^n$ and define the spaces $L^p(\Omega)$, $p \in \mathbb{R}^+$, as the sets of (Lebesgue) measurable functions on Ω such that

$$\| f \|_{L^p(\Omega)} \triangleq \left(\int | f(x) |^p dx \right)^{1/p} < \infty, \qquad p < \infty$$

$$\| f \|_{L^\infty(\Omega)} \triangleq \operatorname{ess\,sup}_{x \in \Omega} | f(x) | < \infty.$$

If $1 \leqslant p \leqslant \infty$ then $L^p(\Omega)$ is a Banach space. If $p = 2$, then $L^2(\Omega)$ is a Hilbert space with inner product

$$\langle f, g \rangle_{L^2(\Omega)} = \int_\Omega f(x) \bar{g}(x) \, dx$$

3. Let ℓ^p, $1 \leqslant p \leqslant \infty$ be the space of sequences $x = \{x_i\}_{i=1,2,\ldots}$ of elements of F ($= \mathbb{R}$ or $\mathbb{C}$) such that

$$\| x \|_p = \left(\sum_{i=1}^{\infty} | x_i |^p \right)^{1/p} < \infty, \qquad 1 \leqslant p < \infty$$

$$\| x \|_\infty = \sup_{1 \leqslant i < \infty} | x_i | < \infty, \qquad p = \infty.$$

ℓ^p is a Banach space and ℓ^2 is a Hilbert space.

4. Let $\mathscr{D}'(\Omega)$ denote the space of distributions on the open set $\Omega \subseteq \mathbb{R}^n$. Then $L^p(\Omega) \subseteq \mathscr{D}'(\Omega)$, $1 \leqslant p \leqslant \infty$ and so we can define

$$H^{p,m}(\Omega) = \{ f \in \mathscr{D}'(\Omega) : (\partial/\partial x)^\alpha f \in L^p(\Omega), \qquad |\alpha| \leqslant m \}$$

with the norm

$$\| f \|_{p,m} = \left\{ \sum_{|\alpha| \leqslant m} \int_\Omega \left| \left(\frac{\partial}{\partial x} \right)^\alpha f(x) \right|^p \mathrm{d}x \right\}^{1/p}.$$

$H^{p,m}(\Omega)$ is a Banach space and $H^{2,m}(\Omega)$ is a Hilbert space with the obvious inner product.

5. $H_0^{p,m}(\Omega)$ $(1 \leqslant p \leqslant \infty, m \geqslant 1)$ is the closure of $\mathscr{D}(\Omega)$ (the space of infinitely differentiable functions with compact supports in Ω) in $H^{p,m}(\Omega)$. Put

$$H^{q,-m}(\Omega) = (H_0^{p,m}(\Omega))^*, \qquad q = p/(p-1), \quad 1 \leqslant p < \infty.$$

If $\Omega = \mathbb{R}^n$, then

$$H_0^{p,m}(\mathbb{R}^n) = H^{p,m}(\mathbb{R}^n).$$

6. Define, for any $s \in \mathbb{R}$,

$$H^s = \{ f \in \mathscr{S}'(\mathbb{R}^n) : (1 + |\xi|^2)^{s/2} \hat{f} \in L^2(\mathbb{R}^n_\xi) \}$$

where $\mathscr{S}'(\mathbb{R}^n)$ is the space of tempered distributions and $\hat{f}$ is the Fourier transform. The spaces $H^{p,m}(\Omega)$, $H_0^{p,m}$ and H^s are called the *Sobolev spaces*.

1.6.2 Operator theory

Now let X and Y be Banach spaces and let A be a function from a linear manifold $D(A)$ of X into Y such that

$$A(\alpha x_1 + \beta x_2) = \alpha A x_1 + \beta A x_2, \qquad x_1, x_2 \in D(A), \quad \alpha, \beta \in F.$$

Then A is a *linear operator* with *domain* $D(A)$. The *kernel* of A is the linear manifold

$$\ker A = \{ x \in D(A) : Ax = 0 \},$$

and the *range* $R(A)$ of A is the range of A as a function.

If $D(A) = X$ and

$$\| A \| = \sup_{\| x \| = 1} \| Ax \| < \infty$$

then we say that A is *bounded*. The space of bounded operators on X is denoted by $\mathscr{B}(X, Y)$ or $\mathscr{B}(X)$ if $X = Y$. An operator $P \in \mathscr{B}(X)$ is a *projection* (on $R(P)$) if $P^* = P$. If H is a Hilbert space and M is a closed subspace, then

$$H = M \oplus M^{\perp}$$

where $M^{\perp} = \{h \in H: \langle h, m \rangle = 0, \text{ for all } m \in M\}$, and the map

$$P_M(h) = m, \ h = m + m^{\perp}, \qquad m \in M, \ \ m^{\perp} \in M^{\perp}$$

is a projection such that $\| P_M \| \leqslant 1$.

If $A \in \mathscr{B}(X, Y)$ then we define the *dual* operator $A' \in \mathscr{B}(Y^*, X^*)$ by

$$\langle A' y^*, x \rangle_{X^*, X} = \langle y^*, Ax \rangle_{Y^*, Y}, \qquad x \in X, \ \ y^* \in Y^*.$$

When X and Y are Hilbert spaces, we also define the *adjoint* operator A^* by

$$\langle A^* h_1, h_2 \rangle_X = \langle h_1, A h_2 \rangle_Y, \qquad h_2 \in X, \ \ h_1 \in Y.$$

Clearly, we have

$$A^* = j_X^{-1} A' j_Y.$$

We shall not preserve the distinction between A^* and A', but shall refer simply to the adjoint A^*.

If $\sup_{\| x \| = 1} \| Ax \| = \infty$, then A is said to be *unbounded*. An important class of unbounded operators is the class of densely defined closed ones. Then we can extend the notion of adjoint operator to this class and we have

$$\langle y^*, Ax \rangle = \langle A^* y^*, \ x \rangle, \qquad x \in D(A), \ \ y^* \in D(A^*),$$

and $\overline{D(A^*)} = Y^*$. An operator A is *closed* if $x_n \to x$ and $Ax_n \to y$ imply that $x \in D(A)$ and $y = Ax$.

We now consider a special kind of unbounded operator defined on a Sobolev space. This operator is given by

$$L = \sum_{|\alpha| \leqslant m} a_\alpha(x) D^\alpha = \sum_{k=0}^{m} \left\{ \sum_{\alpha_1 + \ldots + \alpha_n = k} a_{\alpha_1 \ldots \alpha_n}(x) D_1^{\alpha_1} \cdots D_n^{\alpha_n} \right\}$$

where $D_i = \partial / \partial x_i$, $\alpha \in \mathbb{N}^n$ and $a_\alpha(x)$ is defined in a bounded open set $\Omega \subseteq \mathbb{R}^n$. The operator L is, of course, a partial differential operator and is called (*uniformly*) *strongly elliptic* in Ω if m is even and

$$(-1)^{m/2} \operatorname{Re} \left\{ \sum_{|\alpha| = m} a_\alpha(x) \xi^\alpha \right\} \geqslant C |\xi|^m$$

for all $x \in \Omega$, $\xi \in \mathbb{C}^n$ and some constant C.

Consider the *Dirichlet problem* given by

$$(L+c)u = f \qquad \text{in } \Omega$$

$$\frac{\partial^j u}{\partial \nu^j} = g_j \qquad \text{on } \partial\Omega, \quad 0 \leqslant j \leqslant m-1$$

for some constant c, where ν is the (outward) normal to Ω. If $g_j \in C^{2m}(\partial\Omega)$ then it can be shown that the above problem is equivalent to the problem

$$(L+c)u = f, \qquad f \in L^2(\Omega)$$

$$\frac{\partial^j u}{\partial \nu^j} = 0.$$

A *solution* of the latter problem is an element $u \in H_0^{2,m}(\Omega)$ such that

$$B(\phi,u) = \langle \phi, f \rangle_{L^2} \qquad \text{for all } \phi \in \mathscr{D}(\Omega),$$

where B is the bilinear form given by

$$B(v,u) = \langle v, Lu \rangle_{L^2} = \sum_{0 \leqslant |\beta|,|\gamma| \leqslant q} \langle D^\beta v, a^{\beta\gamma} D^\gamma u \rangle_{L^2}$$

and the coefficients $a^{\beta\gamma}$ are defined by L in the 'divergence form':

$$Lu = \sum_{0 \leqslant |\beta|,|\gamma| \leqslant q} (-1)^{|\beta|} D^\beta (a^{\beta\gamma}(x) D^\gamma u).$$

Then a standard result in the theory of partial differential equations is the following (Friedman, 1969):

Theorem 1.6.1
Let Ω be a bounded domain in $\mathbb{R}^n$ with smooth boundary and let L be strongly elliptic in Ω, with coefficients $a_\alpha \in C^j(\bar{\Omega})$, $j = \max(0, |\alpha| - m)$. Then the operator A defined by

$$(Au)(x) = Lu(x), \qquad u \in D(A)$$

is closed in $L^2(\Omega)$ and the *resolvent* $R(\lambda; A) \triangleq (\lambda I - A)^{-1} : L^2(\Omega) \to L^2(\Omega)$ exists for all $\lambda \in \mathbb{C}$ such that λ is in the sector

$$\{\lambda : \tfrac{1}{2}\pi < \arg(\lambda + k) < \tfrac{3}{2}\pi, \qquad \text{for some } k \geqslant 0\}$$

and

$$\| (\lambda I - A)^{-1} \|_{\mathscr{B}(L^2(\Omega))} \leqslant \frac{C}{|\lambda| + 1}$$

for some $C > 0$. □

In particular, the Dirichlet problem has a unique solution for all c greater than some number.

1.6.3 Spectral theory

Next we shall mention, briefly, the spectral theory of operators. Let A be an operator with domain and range in a complex Banach space X. We partition the complex plane into four regions depending on the nature of the operator $(\lambda I - A)$, $\lambda \in C$. We define

$$\rho(A) = \{\lambda \in \mathbb{C} : (\lambda I - A)^{-1} \in \mathscr{B}(X)\}$$
$$\sigma_P(A) = \{\lambda \in \mathbb{C} \backslash \rho(A) : (\lambda I - A) \text{ is not } 1\text{–}1\}$$
$$\sigma_C(A) = \{\lambda \in \mathbb{C} \backslash \rho(A) : \overline{(\lambda I - A)^{-1}} = X\}$$
$$\sigma_R(A) = \{\lambda \in \mathbb{C} \backslash \rho(A) : \overline{(\lambda I - A)^{-1}} \neq X\}.$$

Then

$$\mathbb{C} = \rho(A) \cup \sigma_P(A) \cup \sigma_C(A) \cup \sigma_R(A)$$

and $\rho(A)$ is called the *resolvent set* of A and $\sigma(A) = \sigma_P(A) \cup \sigma_C(A) \cup \sigma_R(A)$ is called the *spectrum* of A.

The main result from spectral theory which we shall need is the spectral theorem for a self-adjoint operator (see Helmberg, 1969):

Theorem 1.6.2
Let A be a self-adjoint operator (i.e. $A = A^*$) on a Hilbert space H. Then there exists a family of projections $\{P(\lambda) : \lambda \in \mathbb{R}\}$ such that

(a) $P(\lambda) \leqslant P(\lambda')$, for $\lambda \leqslant \lambda'$
(b) $(s)\lim_{\lambda \to -\infty} P(\lambda) = 0$, $(s)\lim_{\lambda \to +\infty} P(\lambda) = I$
(c) $P(\lambda + 0) = P(\lambda)$, for all $\lambda \in \mathbb{R}$
(d) $h \in D(A)$ if and only if $\int_{-\infty}^{\infty} \lambda^2 d \| P(\lambda)h \|^2 < \infty$.

We also have

$$Ah = \int_{-\infty}^{\infty} \lambda \, dP(\lambda)h, \qquad \text{for all } h \in D(A). \qquad \square$$

In this theorem (s)lim means 'strong limit', so that $(s) \lim_{\lambda \to -\infty} P(\lambda) = 0$ is equivalent to $\lim_{\lambda \to -\infty} P(\lambda)h = 0$ for all $h \in H$.

1.6.4 Semigroup theory

In generalizing the exponential solution of a finite dimensional linear differential equation

$$\dot{x} = Ax$$

to the case where A is an unbounded operator, such as the partial differential operator above, we are faced with the difficulty that the usual series for $\exp(At)$ does not converge and so we must seek to define '$\exp(At)$' in a different way. If the equation $\dot{x} = Ax$ (defined on a Hilbert space H) has global continuous solutions which are uniquely defined for $t \in [0, \infty)$, then if $x(t; x_0)$ denotes the solution through x_0 at $t = 0$, we define the operator $T: H \to H$ by

$$T(t)x_0 = x(t; x_0).$$

Then we have

(a) $T(0) = I$ (the identity operator)
(b) $T(t_1 + t_2) = T(t_1)T(t_2),\ t_1, t_2 \geqslant 0$
(c) $\lim_{t \to 0+} T(t)x = x, \qquad$ for all $x \in H$.

An operator-valued function $T: \mathbb{R}^+ \to \mathscr{B}(H)$ satisfying axioms (a)–(c) is called a *strongly continuous (or C^0) semigroup of operators*. Given such a semigroup, let $D(A)$ be the set of $x \in H$ such that

$$\lim_{t \to 0+} \left\{ \frac{T(t)x - x}{t} \right\} = Ax$$

exists. Then the operator A is called the (*infinitesimal*) *generator* of T. The following standard results concerning C_0 semigroups can be found in Yosida (1974) or Banks (1983).

Theorem 1.6.3
If T is a C^0 semigroup then there exists $\omega_0 \in \mathbb{R}^+$ such that

$$\| T(t) \| \leqslant M e^{\omega t},$$

for all $\omega > \omega_0$ and some $M > 0$. Moreover,

(a) $T(t): D(A) \to D(A)$ for $t \geqslant 0$

(b) $\dfrac{d^n}{dt^n}(T(t)x) = A^n T(t)x = T(t)A^n x,\ x \in D(A^n),\ t > 0$

(c) A is closed and $\overline{D(A)} = X$. □

Theorem 1.6.4 (Hille–Yosida)
In order that a closed linear operator A with dense domain in H generates a strongly continuous semigroup $T(t)$, it is necessary and sufficient that there exist real numbers M, ω such that for all $\sigma > \omega$, we have $\sigma \in \rho(A)$ and

$$\| R(\sigma; A)^m \| \leqslant M(\sigma - \omega)^{-m}, \qquad m \geqslant 1.$$

Then

$$\| T(t) \| \leqslant M e^{\omega t}.$$ □

An operator A defined on $D(A) \subseteq H$ is *dissipative* if $\mathrm{Re}\langle Ax, x\rangle \leqslant 0$, for all $x \in D(A)$.

Theorem 1.6.5
If $\overline{D(A)} = H$ then A generates a contraction semigroup $T(t)$ (i.e. $\| T(t) \| \leqslant 1$ for all $t \geqslant 0$) if A and A^* are dissipative. □

Another type of operator which is important in the theory of parabolic partial differential equations is the sectorial class. If A is closed and $\overline{D(A)} = H$ then if, for some $\phi \in (0, \pi/2)$,

(a) $S_{\phi+\pi/2} \triangleq \{\lambda \in \mathbb{C} : |\arg \lambda| < \pi/2 + \phi\} \subseteq \rho(A)$
(b) $\| R(\lambda; A) \| < C/|\lambda|, \quad$ if $\lambda \in S_\phi, \quad \lambda \neq 0$

where C is a constant, we call A a *sectorial operator.*

Theorem 1.6.6
If A is sectorial, then A generates a strongly continuous semigroup $T(t)$ such that

(a) $T(t)$ can be continued analytically into the sector
$S_\phi = \{t \in \mathbb{C} : |\arg t| < \phi, t \neq 0\}$
(b) $AT(t)$ and $\mathrm{d}T(t)/\mathrm{d}t$ are bounded for each $t \in S_\phi$ and

$$\frac{\mathrm{d}T(t)}{\mathrm{d}t} x = AT(t)x, \qquad x \in H$$

(c) For any $0 < \varepsilon < \phi$, there exists C' such that

$$\| T(t) \| < C', \; \| AT(t) \| < C'/|t|, \text{ if } t \in S_{\phi-\varepsilon}.$$ □

From this result it follows that the partial differential operator $L + c$ defined above generates a C^0 semigroup. Moreover, from Theorem 1.6.5 it follows that if we consider the wave-type equation

$$\frac{\partial^2 \phi}{\partial t^2} = A\phi, \; \phi|_{\partial\Omega} = 0$$

where A is a strongly elliptic operator, then defining the operator

$$A' = \begin{bmatrix} 0 & I \\ A & 0 \end{bmatrix}$$

on the Hilbert space $H = H_0^1(\Omega) \oplus L^2(\Omega)$ with domain $(H_0^1(\Omega) \cap H^2(\Omega)) \oplus H_0^1(\Omega)$, A' generates a C^0 semigroup. (For details see Curtain and Pritchard, 1978 or Banks, 1983.)

1.6.5 Fixed point theory

In this section we shall discuss some fixed point theorems which are useful in nonlinear systems theory. First, however, we shall state a result which is important in the application of certain of the fixed point theorems to nonlinear differential equations.

Theorem 1.6.7 (Arzela–Ascoli)
Let (S, μ) be a compact metric space and $C(S)$ the Banach space of complex-valued continuous functions $x(s)$ with the sup norm

$$\| x \| = \sup_{s \in S} | x(s) |.$$

Then a sequence $\{x_n\} \subseteq C(S)$ is relatively compact in $C(S)$ if:

(a) $\{x_n\}$ is equibounded, i.e. $\sup_{n \geqslant 1} \sup_{s \in S} | x_n(s) | < \infty$.

(b) $\{x_n\}$ is equicontinuous, i.e. $\lim_{\delta \to 0} \sup_{n \geqslant 1, \mu(s, s') \leqslant \delta} | x_n(s) - x_n(s') | = 0.$ □

(For a proof see Yosida, 1974.)

The simplest fixed point theorem is the *contraction mapping theorem* (see Holtzman, 1970). To state the theorem, let (X, d) be a metric space and let $f: K \to K$, where K is closed in X, be a *contraction mapping* on K with *contraction constant* $\gamma < 1$, i.e.

$$d(f(x), f(y)) \leqslant \gamma d(x, y), \qquad \text{for all } x, y \in K.$$

Then we have

Theorem 1.6.8
If $f: K \to K$ is a contraction mapping on a closed subset K of a complete metric space, then f has a unique fixed point x^*; i.e. $f(x^*) = x^*$. □

The other fixed point theorem which we shall need is due to Schauder and is stated in terms of compact mappings. A continuous map $f: X \to Y$ where X and Y are Banach spaces is said to be *compact* if it maps bounded sets into relatively compact sets.

Theorem 1.6.9
Let K be a closed convex subset of a Banach space X. Then if $f: K \to K$ is compact, it has a fixed point. □

(For a proof, see Schwartz (1969).)

1.6.6 Tensor products of Hilbert and Banach spaces

We shall conclude this chapter by discussing briefly the tensor product of Hilbert and Banach spaces; a more detailed exposition is given by Takesaki (1979). Consider first the case of two Hilbert spaces H_1 and H_2, and let H denote the usual algebraic tensor product of H_1 and H_2. Then any $h \in H$ may be written in the form

$$h = \sum_{i=1}^{n} x_i \otimes y_i, \qquad x_i \in H_1, \quad y_i \in H_2, \quad 1 \leqslant i \leqslant n.$$

If $g = \sum_{j=1}^{m} v_j \otimes w_j$ is another element of H, then we define

$$\langle h, g \rangle = \sum_{i=1}^{n} \sum_{j=1}^{M} \langle x_i , v_j \rangle \langle y_i, w_j \rangle$$

and it is clear that $\langle .,. \rangle$ is an inner product on H. Note that for any *simple* tensor in H (i.e. one of the form $x \otimes y$, $x \in H_1, y \in H_2$) we have

$$\| x \otimes y \| = \| x \| . \| y \|.$$

Such a norm on an algebraic tensor product of spaces is called a *cross-norm*. The *tensor product* $H_1 \otimes H_2$ of H_1 and H_2 is the completion of H in the norm defined by this inner product.

If we now consider the case of Banach spaces B_1, B_2 then let B denote the algebraic tensor product of B_1 and B_2. In general, there are many cross-norms on B, two important ones being defined by

$$\| x \|_\lambda = \sup \left\{ \left| \sum_{i=1}^{n} f(x_i) g(y_i) \right| : f \in B_1^*, \ g \in B_2^*, \ \| f \|, \ \| g \| \leqslant 1 \right\}$$

for $x = \sum_{=1}^{n} x_i \otimes y_i$ and

$$\| x \|_\gamma = \inf \left\{ \sum_{i=1}^{n} \| x_i \| \ \| y_i \| : x = \sum_{i=1}^{n} x_i \otimes y_i \right\}.$$

$\| . \|_\lambda$ is called the *injective cross-norm* and $\| . \|_\gamma$ is called the *projective cross-norm*. Clearly,

$$\| x \|_\gamma \geqslant \| x \|_\lambda$$

and both are, in fact, norms. We denote by $B_1 \otimes_\lambda B_2$, $B_1 \otimes_\gamma B_2$ the completions of B under the respective cross-norms. However, when we are not concerned with any particular cross-norm, we shall simply write $B_1 \otimes B_2$.

2 NONLINEAR SYSTEMS

2.1 DEFINITIONS OF NONLINEAR SYSTEMS

2.1.1 Systems with state-independent input spaces

Any physical system is usually regarded as being specified in one of two basic ways. In the first we consider the system to be a 'black box' into which inputs may be injected and from which we may obtain outputs in the form of measurements. This is illustrated in Fig. 2.1, where S is the system and $u(t)$ and $y(t)$ are the input and output functions, respectively. The input functions are assumed to belong to some function space $\mathscr{U}$ and the output functions will then belong to some other function space $\mathscr{Y}$. We can then regard S as a function from $\mathscr{U}$ to $\mathscr{Y}$, i.e.

$$S: \mathscr{U} \to \mathscr{Y}$$

and we shall denote this system by the triple $(\mathscr{U}, \mathscr{Y}, S)$. When we specify a system in this way we say that it is in the *input–output* form.

The second method of describing a system is via its internal dynamics. By this we mean that we associate with S a dynamical system $\phi_t(p, u(\,.\,))$, defined on some differentiable manifold M, which depends on the input function $u(\,.\,)$, and a function $h: \mathscr{M} \to \mathscr{Y}$,‡ called the *output map*, where $\mathscr{M}$ is a space of functions on M containing the solution trajectories of the dynamical system. We then require that

$$h[\phi_t(p, u(\,.\,))] = S(u(\,.\,))$$

for each $u(\,.\,) \in \mathscr{U}$ and some $p \in M$.

Suppose that $\mathscr{U}$ and $\mathscr{Y}$ are spaces of functions which are defined on some time interval, J say, with values in U and Y, respectively, where U and Y are differentiable manifolds. ($\mathscr{U}$ and $\mathscr{Y}$ may have more structure, of

‡More generally, h may also depend on $\mathscr{U}$, so that $h: \mathscr{M} \times \mathscr{U} \to \mathscr{Y}$

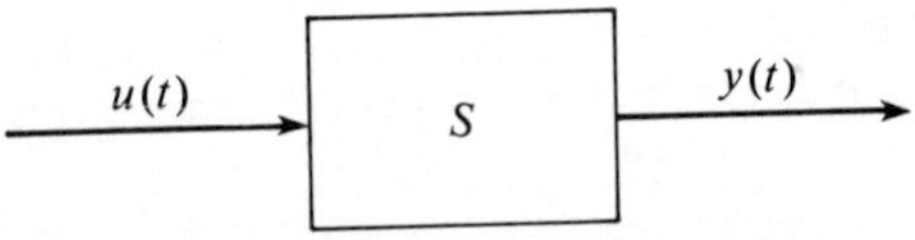

Fig. 2.1 System block diagram

course, as we shall see later.) The spaces M, U and Y may be finite- or infinite-dimensional; we shall discuss the infinite-dimensional case in Chapter 5, and so for the present we shall assume that M, U and Y are finite-dimensional. Now let

$$X: U \to D(M) \tag{2.1.1}$$

be a differentiable map from U to the space of (differentiable) vector fields on M and assume that for any given $u(.) \in \mathscr{U}$, $\phi_t(p, u(.))$ is the flow of the time-dependent vector field $X(u(.))$ for some initial value $p \in M$. Then we shall refer to X as the *state-space realization* of S with output map h. Note that h is often memoryless in the sense that there exists a function h' such that

$$\eta(t) = h(\xi(.))(t) = h'(\xi(t))$$

for any $\xi(.) \in \mathscr{M}$, $\eta(.) \in \mathscr{Y}$. We shall not preserve the distinction between h and h' in this case and denote both functions by h.

In terms of local coordinates the system $X: U \to D(M)$ may be written in the form

$$\dot{x} = f(x, u)$$

for some differentiable function f, and the output equation is

$$y = h(x), \tag{2.1.2}$$

where x and y are local coordinates for the manifolds M, Y.

Examples

1. Linear systems defined on the state space $\mathbb{R}^n$ with controls in $\mathbb{R}^m$ take the form

$$\dot{x} = Ax + Bu, \qquad A, B \in \mathbb{R}^{n \times n}$$

with the output equation

$$y = Cx, \qquad C \in \mathbb{R}^{k \times n}.$$

(In the more general situation mentioned in the above footnote, $y = Cx + Du$.) The input–output map is given by the familiar variation of constants formula

$$y(t) = Ce^{At}x_0 + \int_0^t Ce^{A(t-s)}Bu(s)\,ds.$$

Note also that in the linear case we may obtain the (complex) frequency domain input–output map by using Laplace transforms (denoted by $\mathscr{L}$). Then,

$$Y(s) = \mathscr{L}(y)(s) = C(sI - A)^{-1}x_0 + C(sI - A)^{-1}BU(s).$$

This relation is the cornerstone of classical control theory (where $x_0 = 0$, in the usual case). We shall see later how to generalize the frequency domain ideas to nonlinear systems.

2. It will become clear later that the next most important class of systems after the linear ones is that consisting of the bilinear systems. If the state space is $\mathbb{R}^n$, then a typical bilinear system takes the form

$$\dot{x} = Ax + u_1B_1x + u_2B_2x + \cdots + u_mB_mx + Bu,$$
$$y = Cx,$$

$$A, B_1, \ldots, B_m \in \mathbb{R}^{n\times n}, \quad B \in \mathbb{R}^{n\times m}$$
$$C \in \mathbb{R}^{k\times n}.$$

The input–output map can be written as a convergent power series in the functions $u(.)$ with operator-valued coefficients and will be discussed in detail subsequently.

3. As an example of a nonlinear system whose global behaviour is naturally defined on a differentiable manifold other than $\mathbb{R}^n$, consider the simple pendulum (Fig. 2.2). The equation of motion is, of course,

$$ml\ddot{\theta} = -mg\sin(\theta) - k\dot{\theta}$$

or, in phase plane coordinates,

$$\dot{\theta} = \omega$$

$$\dot{\omega} = \ddot{\theta} = -\frac{g}{l}\sin(\theta) - \frac{k}{ml}\omega$$

The phase plane trajectories are shown in Fig. 2.3 in the strip $[-\pi, \pi] \times (-\infty, \infty)$; the orbit structure is periodic in θ with period 2π.

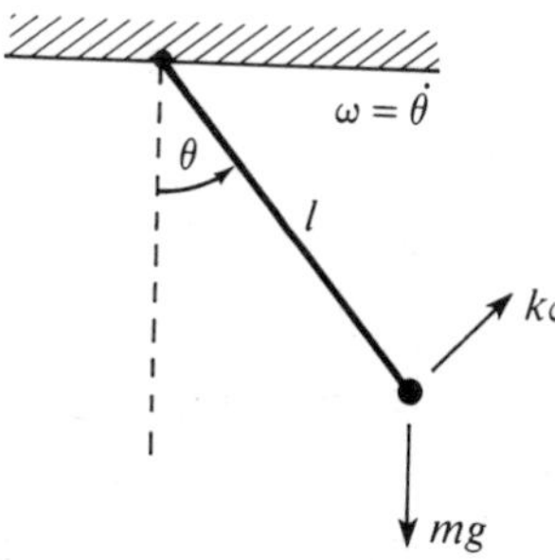

Fig. 2.2 Damped pendulum

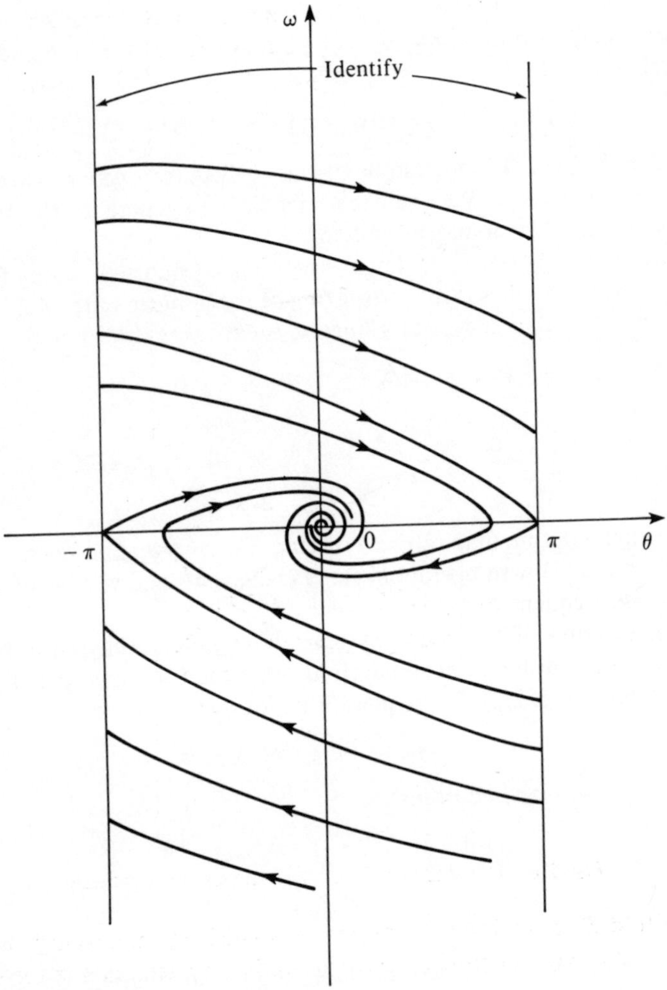

Fig. 2.3 Phase-plane trajectories of the damped oscillator

However, since $\theta = -\pi$ and $\theta = +\pi$ correspond to the same physical point it is natural to identify the lines $-\pi \times (-\infty, \infty)$ and $\pi \times (-\infty, \infty)$ giving rise to a dynamical system defined on an infinite cylinder (Fig. 2.4). We can even compactify the manifold on which this dynamical system is defined. In fact, if we identify the ω-axis $(-\infty, \infty)$ with $(-1, 1)$ in the obvious way, then we obtain a system defined on a finite open cylinder (Fig. 2.5(a)) and then identifying opposite sides of this cylinder we obtain a Klein bottle K with an unstable limit cycle on the identified edge (Fig. 2.5(b)). Hence we see that, even in this simplest of systems, the global structure is defined on a nonorientable compact

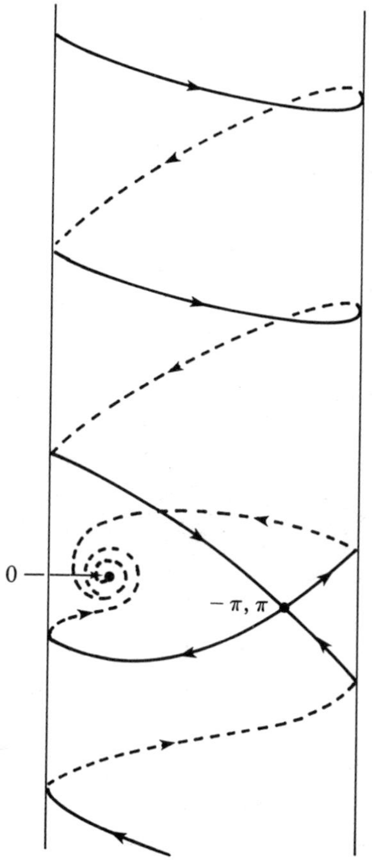

Fig. 2.4 Damped oscillator trajectories on a cylinder

manifold K. A multiple pendulum is defined on the product manifold $K \times \cdots \times K$ with a copy of the Klein bottle for each mass in the system.

We shall discuss more general Hamiltonian systems at the end of this chapter. If we define the simple pendulum on the cylinder, then the state space is a product of the circle group and the ω-axis. For a study of dynamical systems defined on more general rotation groups (for example, $SO(3)$) see Bonnard (1984) where the attitude control of a satellite is considered. In this case (as shown by Brockett, 1972) the orientation of a rigid body relative to fixed axes is described by a 3×3 orthogonal matrix A for which

$$\mathring{A} = \begin{bmatrix} 0 & \omega_3(t) & -\omega_2(t) \\ -\omega_3(t) & 0 & \omega_1(t) \\ \omega_2(t) & -\omega_1(t) & 0 \end{bmatrix} A.$$

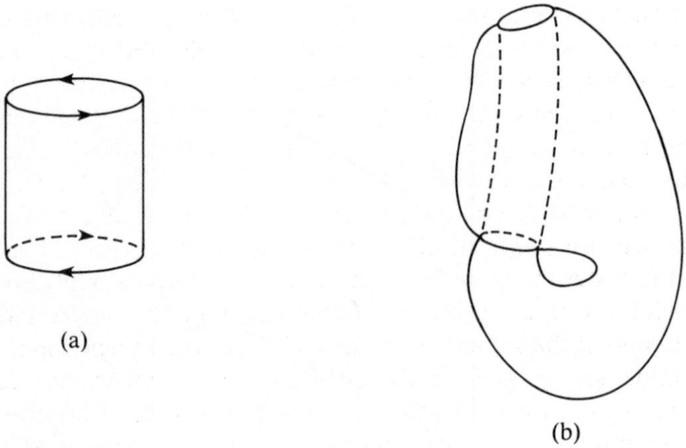

Fig. 2.5 The Klein bottle

The angular velocities ω_i are controlled by angular accelerations about each axis and so we have the equations

$$\dot{\omega}_1(t) = [(I_2 - I_3)/I_1]\,\omega_2(t)\omega_3(t) + u_1(t)/I_1$$
$$\dot{\omega}_2(t) = [(I_3 - I_1)/I_2]\,\omega_1(t)\omega_3(t) + u_2(t)/I_2$$
$$\dot{\omega}_3(t) = [(I_1 - I_2)/I_3]\,\omega_1(t)\omega_2(t) + u_3(t)/I_3.$$

(For a derivation of Euler's equations of motion, see Banks, 1986a.) (A, ω) is then defined on the state space $SO(3) \times \mathbb{R}^3$.

2.1.2 Systems with state-dependent input spaces

In the last section we introduced a class of nonlinear systems with local representatives of the form

$$\begin{aligned} \dot{x} &= f(x, u) \\ y &= h(x) \end{aligned} \tag{2.1.3}$$

in which the input space U is independent of the state x. This describes many kinds of physical systems very accurately, but there are cases when one is forced to consider state-dependent input spaces. Following Willems (1979) and Van der Schaft (1982a) we shall now discuss the extension of the nonlinear systems considered above to the situation when the input spaces may depend on the state.

For many types of physical system it is obvious what are the inputs and what are the outputs. However, as pointed out by Willems (1979), there are cases when it is not clear which variables should be regarded as inputs and which should be considered as outputs. For example, an ideal diode has a characteristic of the form shown in Fig. 2.6 (where we have ‘smoothed out’ the corner to make the characteristic differentiable). When $v < 0$ then it is natural to consider v as the input and i as the output, and when $i > 0$ then i should be regarded as the input. When $v = 0$, $i = 0$ then there is no neighbourhood around this point in which we can regard either v or i as the input. By changing coordinates, however, we can consider, for example, the input or output to be one of the variables $i - v$, $i + v$. From this example it is clear that, for some physical systems, we can only expect a local separation of the ‘external variables’ into inputs and outputs. Only when there exists a global separation of the external variables into inputs and outputs can we represent the system in the form (2.1.2) with a single input space U.

Coming now to the formal definition of a system with state-dependent input spaces note that to each point $x \in M$ (the state space) we want to assign an input space, i.e. we require a bundle of spaces over M. Thus, let B denote a fibre bundle over M with projection $\pi : B \to M$. In the case where we have a global input space U then B is the trivial bundle $\pi : M \times U \to M$ and we can reformulate the system (2.1.1) as a commutative diagram:

$$\begin{array}{ccc} M \times U & \xrightarrow{\quad F \quad} & TM \\ & {\scriptstyle \pi} \searrow \qquad \swarrow {\scriptstyle \pi_M} & \\ & M & \end{array} \qquad (2.1.4)$$

where π_M is the tangent bundle projection, and F has the local representative f. We can include the output map h given by (2.1.2) or, more generally by $y = h(x, u)$, in this diagram in the obvious way to obtain the commutative diagram

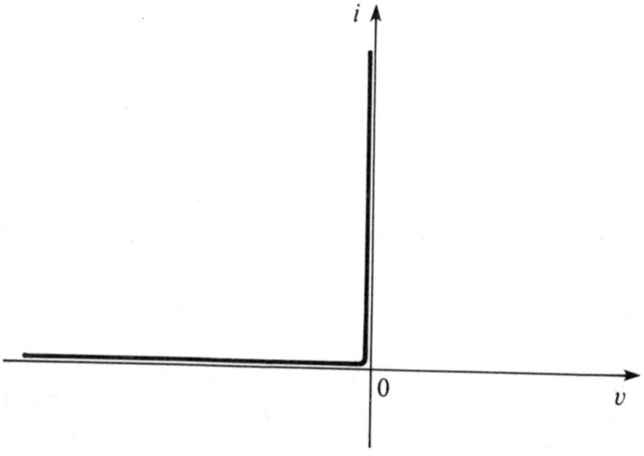

Fig. 2.6 ‘Smoothed’ diode characteristic

$$\begin{array}{ccc} M \times U & \xrightarrow{F \times H} & TM \times Y \\ & {\scriptstyle \pi} \searrow \quad \swarrow {\scriptstyle \pi_M} & \\ & M & \end{array} \tag{2.1.5}$$

where, by abuse of notation, we write $\pi_M = \pi_M \circ pr_1$, with pr_1 equal to the projection on the first component.

In the general situation with $M \times U$ replaced by B, since we do not have a global separation of external variables into inputs and outputs, we must replace the output space Y by the space W of external variables containing both inputs and outputs. (The nature of the space W will depend on the local input and output spaces, but it will certainly be assumed to be a manifold.) We can thus replace the above commutative diagram by the following one:

$$\begin{array}{ccc} B & \xrightarrow{F \times H} & TM \times W \\ & {\scriptstyle \pi} \searrow \quad \swarrow {\scriptstyle \pi_M} & \\ & M & \end{array} \tag{2.1.6}$$

We then say that a *smooth nonlinear system* is a commutative diagram of the form (2.1.6), which in local coordinates has the form

$$\dot{x} = f(x, u)$$
$$w = h(x, u).$$

Consider now the sense in which W is the set of both inputs and outputs. Suppose first that H, restricted to the fibres of B, is an immersion into W. Then the matrix $\partial h / \partial u$ is injective and so the implicit function theorem implies that, if $(x', u') \in B$, $w' \in W$ satisfy $h(x', u') = w'$, then there are neighbourhoods O_1 of (x', u') in B and O_2 of w' in W such that we can choose coordinates (x, u) and (w_1, w_2) in these respective neighbourhoods such that

$$h(x, u) = (w_1, w_2) = (h_1(x, u), u), \qquad (x, u) \in O_1, \quad (w_1, w_2) \in O_2$$

for some function h_1 defined on O_1. This implies that locally we can identify w_1 as the output with output equation

$$w_1 = h_1(x, u)$$

and w_2 $(= u)$ as the input.

In the case where a system has an output which is independent of explicit mention of u, i.e. $y = h(x)$, then we can include such systems in the above formulation by assuming that W is a fibre bundle over a globally defined output space Y. We then have

Lemma 2.1.1

If a system has a globally defined output space Y, $p: W \to Y$ is a fibre bundle over Y and $H: B \to W$ is a bundle morphism which is diffeomorphic on the fibres of B to those of W, then choose $x' \in X$ and $y' \in Y$ such that

$$h(\pi^{-1}(x')) = p^{-1}(y').$$

Let $(x', v') \in \pi^{-1}(x')$ and $(y', u') \in p^{-1}(y)$ be related by

$$H(x', v') = (y', u').$$

Then there are local coordinates v around v' in the fibres of B and u around u' in the fibres of W such that

$$h(x, v) = (y, u) = (h_1(x), v),$$

for some function h_1.

Proof

Since W is a fibre bundle we can find a trivializing chart (O, ϕ) of W in a neighbourhood of y'. If U is the fibre of W, then $\phi: p^{-1}(O) \to O \times U$. Choosing coordinates u near $u' \in U$, y near $y' \in Y$ and (x, v) near (x', v') in B we can write h in the form

$$h(x, v) = (y, u) = (h_1(x), h'(x, v))$$

since H is a bundle morphism. However, H restricted to fibres is a diffeomorphism and therefore so is any local representative h'; i.e. for each fixed x we can choose the coordinates v to satisfy

$$v = (h'_x)^{-1}(u)$$

where $h'_x(v) = h'(x, v)$. The map h then has the form

$$h(x, v) = (h_1(x), u).$$

□

If W and B are trivial bundles over Y and M, respectively, then we can write $W = Y \times U$ and $B = M \times U$ where we have identified the diffeomorphic fibres of W and B to U. We are then in the situation where the inputs and outputs are globally defined and we have the diagram

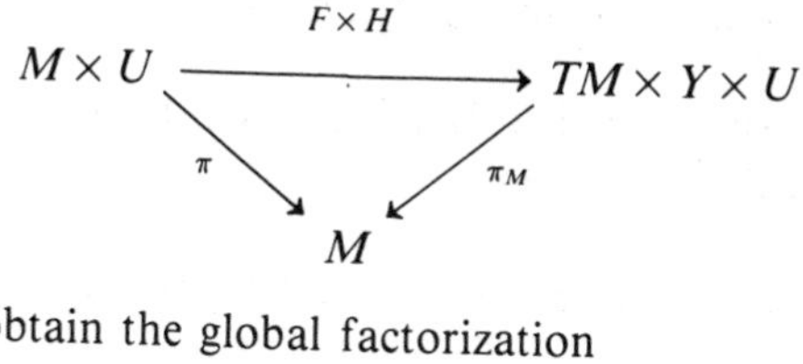

from which we obtain the global factorization

$$\begin{array}{ccc} M \times U & \xrightarrow{F} & TM \\ & \searrow^{\pi} \quad {}^{\pi_M}\swarrow & \\ & M & \end{array}$$

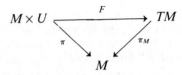

which is the same as (2.1.4).

Example (Takens, 1976)
As an example of a smooth nonlinear system of the form (2.1.6), consider the Lagrangian dynamics of a mechanical system. (An elementary introduction to Lagrangian dynamics is given in Banks (1986a).) The state space M is defined by the phase space of the system corresponding to some system of generalized coordinates $q, \dot{q}$. Let Q denote the appropriate manifold on which q is defined. Then $M = TQ$ (the tangent bundle of Q); in the example of the simple pendulum Q is the circle and TQ is a cylinder. The external forces on the system are specified by the variation of the work w in the form

$$\mathrm{d}w = F_{q_i}\, \mathrm{d}q_i.$$

Thus the forces appear as the coefficients in the local representation of the cotangent bundle T^*Q which we may take as the space W. The bundle B which we require in the general system formulation is locally a product of the state space TQ and the inputs, which are contained in the bundle T^*Q. Hence we choose B to be the pull-back bundle $\pi_Q^* T^*Q$ of T^*Q with respect to the projection map $\pi_Q : TQ \to Q$. (See Chapter 1 for the definition of a pull-back bundle.) Thus we have the diagram

$$\begin{array}{ccc} \pi_Q^* T^*Q & \longrightarrow & T^*Q \\ \Big\downarrow & & \Big\downarrow \\ TQ & \xrightarrow{\pi_Q} & Q \end{array}$$

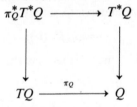

Finally let $L : TQ \to \mathbb{R}$ be the Lagrangian of the system. Then we associate a map $F: B \to TTQ$ with L whose local representation gives Lagrange's equations of motion.

2.2 CONTROLLABILITY AND OBSERVABILITY

2.2.1 Nonlinear controllability

We now come to one of the most important properties of a system, namely that of controllability. Following Sussmann and Jurdjevic (1972), let a system of the form (2.1.4) (or (2.1.1)) be given on an analytic manifold M. By abuse of notation we shall write the system in the form

$$\dot{x}(t) = F(x(t), u(t)) \tag{2.2.1}$$

where $F(., u)$ is an analytic vector field on M for each u. The control functions are assumed to belong to an 'admissible' control set $\mathscr{U}$, consisting of piecewise continuous functions with finite one-sided limits defined on $[0, \infty)$ with values in a set $U \subseteq \mathbb{R}^m$ (which is locally path connected). Note that $\mathscr{U}$ contains the piecewise constant functions. On $\mathscr{U}$ we put the topology of uniform convergence on compact intervals (in which a sequence $u_i(t) \in \mathscr{U}$ converges to $u(t) \in \mathscr{U}$ if and only if

$$\sup_{t \in K} | u_i(t) - u(t) | \to 0$$

as $i \to \infty$, for any compact set $K \subseteq \mathbb{R}$). We also assume that $F: M \times U \to TM$ is jointly continuously differentiable and that $F(., u)$ is a *complete* analytic vector field for each $u \in U$. Thus, for each $x \in M$, $u(.) \in \mathscr{U}$, the equation

$$\dot{x}(t) = F(x(t), u(t)), \; x(0) = x$$

has a unique solution on some interval $[0, \delta)$, $\delta > 0$ which we denote by $\phi_t(x, u(.))$, $t \in [0, \delta)$. We then assume that $\phi_t(x, u)$ is defined for all $t \in [0, \infty)$.

Definition
We say that $y \in M$ is *reachable* (or *attainable*) *from* $x \in M$ *at time* $t \geqslant 0$ if there is a control $u(.) \in \mathscr{U}$ such that

$$y = \phi_t(x, u(.)).$$

Define the set

$$R(x, t) = \{ y : y \text{ is reachable from } x \text{ in time } t \}$$

for each $x \in M$ and let

$$R(x, [0, t]) = \bigcup_{s \leqslant t} R(x, s), \; R(x) = \bigcup_{t \geqslant 0} R(x, [0, t]).$$

Then the system (2.2.1) is *controllable from* $x \in M$ if $R(x) = M$ and is *controllable* if it is controllable from any $x \in M$.

Another property which is very important in the discussion of controllability is that of accessibility – intuitively this means the capability of reaching a full neighbourhood of some point. Formally, we say that the system (2.2.1) has the *accessibility property from* x‡ if interior $(R(x)) \neq \emptyset$,

‡It may be that $x \notin \text{int}(R(x))$

and has the *accessibility property* if it has the accessibility property from any $x \in M$. The system is said to have the *strong accessibility property from* x if interior$(R(x, t)) \neq \emptyset$ for some $t > 0$ and the *strong accessibility property* if it has the strong accessibility property from any $x \in M$.

Now let $V \subseteq D(M)$ be a set of analytic vector fields on M and let $\mathscr{L}(V)$ denote the Lie subalgebra of $D(M)$ generated by V. Then, by Frobenius' theorem, there exists a maximal integral submanifold of $\mathscr{L}(V)$ through any point $x \in M$, which we denote by $I(V, x)$. It is clear that $\alpha(t) \in I(V, x)$ for any integral curve $\alpha: [t_0, t_1] \to M$ of V such that $\alpha(\tau) = x$ for some $\tau \in [t_0, t_1]$. The crucial result for reachability, which is a converse of the last remark, is provided by a theorem of Chow (1939) which may be stated as follows:

Theorem 2.2.1
Let V be symmetric in the sense that $-X \in V$ if $X \in V$, and let $x \in M$. Then for each $y \in I(V, x)$, there exists an integral curve $\alpha: [0, T] \to M$ of V, with $T \geqslant 0$, such that $\alpha(0) = x$ and $\alpha(T) = y$.

Proof (After Hermann, 1968)
It is sufficient to assume that $\mathscr{L}(V) = D(M)$ and $M = I(V, x)$. First let $X, Y \in D(M)$ and let ϕ_t denote the flow of X. Then, from the basic property of the Lie derivative (Chapter 1), we have

$$\frac{\mathrm{d}}{\mathrm{d}t}((\phi_t)_* Y) = -[X, (\phi_t)_* Y]$$

and since all functions are analytic we have

$$(\phi_t)_* Y = \sum_{j=0}^{\infty} \frac{t^j \mathrm{Ad}^j(X)(Y)}{j!} \tag{2.2.2}$$

(the Lie series).

Now if $[V, V] \subseteq V$, then the result follows, since

$$M_p = \mathscr{L}(V)_p = V_p (= \{X(p): X \in V\}).$$

Suppose, therefore, than $[X, Y] \notin V$ for some vector fields $X, Y \in V$. Then, by (2.2.2), we must have

$$(\phi_t)_* Y \notin V$$

for some t, say $t = \tau$. Let V_1 be the subspace of $D(M)$ spanned by V and $(\phi_\tau)_* Y$. Now, if ψ_t denotes the flow of a vector field Y, then $\phi_t \circ \psi_s$ is the flow of the vector field $\phi_{t_*}(Y)$ (for fixed t) and so we have

$$\phi_t \circ \psi_s = \xi_{t,s} \circ \phi_t$$

where $\xi_{t,s}$ is the flow of $\phi_{t_*}(Y)$, for t fixed. It follows that every point which can be reached along an integral curve (see remark below) of V_1 can also along an integral curve of V. If $[V_1, V_1] \subseteq V_1$, then the result is proved, since then $\mathscr{L}(V) = V_1$. If $[V_1, V_1] \nsubseteq V_1$, choose $X_1, Y_1 \in V_1$ such that $\phi_{1\tau_*} Y_1 \in V_1$ and then repeat the argument. ($\phi_{1\tau}$ is the flow of X_1.) The process must terminate and so the theorem is proved. □

REMARK An *integral curve* of a set of vector fields V is a curve $\alpha: [0, T] \to M$ which is an integral curve of some $X \in V$ on some subinterval of $[0, T]$.

We can extend the notion of reachability defined above to the case of an arbitrary family $V \subseteq D(M)$ of vector fields. We say that $y \in M$ is *V-reachable from x at time T* if there exists an integral curve α of V on $[0, T]$ such that $\alpha(0) = x$, $\alpha(T) = y$. The set of such points y is denoted by $R(x, t; V)$. The sets $R(x, [0, t]; V)$ and $R(x; V)$ are then defined in the obvious way.

Chow's theorem can be used to prove an accessibility result for vector fields:

Lemma 2.2.2

If $V \subseteq D(M)$ is symmetric, then $R(x; V)$ has a nonempty interior in M if and only if $\dim \mathscr{L}(V)(x) = \dim M$ $(= n)$.

Proof

If $\dim \mathscr{L}(V)(x) = n$, then $I(V, x)$ is an n-dimensional submanifold of M and so is open in M. However, by Chow's theorem, we have $R(x; V) = I(V, x)$ and so $R(x; V)$ is open in M. If, conversely, $\dim \mathscr{L}(V)(x) < n$, then $I(V, x)$ is a connected submanifold of M of dimension $< n$. It is clear that such a submanifold of M has empty interior in M and so the lemma is proved. □

Lemma 2.2.2 can be generalized to the case where V is not symmetric. Then we have

Theorem 2.2.3

Let M be an n-dimensional analytic manifold with $V \subseteq D(M)$ a given system of complete vector fields. Then $R(x; V)$ has nonempty interior in M if and only if $\dim \mathscr{L}(V)(x) = n$. In this case, interior $(R(x, [0, T]; V))$ is dense in $R(x, [0, T]; V)$ for each $T > 0$. (For a proof of this result see Lobry (1970) and Sussmann and Jurdjevic (1972).) □

In Theorem 2.2.3 we allow the time taken for reachability to be anything from 0 to T (or ∞). To derive a similar result for the set $R(x, T; V)$ we must replace $\mathscr{L}(V)$ by a similar but smaller subspace $\mathscr{L}_0(V)$ of $D(M)$ which effectively removes those points reachable in arbitrary time and leaves only those points which can be reached in some precisely specified time. If we allow reverse time in our model (merely for the purpose of describing the space $\mathscr{L}_0(V)$), then an obvious solution is to include directions in $\mathscr{L}_0(V)(x)$, for any point $x \in M$, which are limiting directions of 'reachable curves' of zero time. To make this statement more precise we consider the geometrical meaning of the Lie algebra $\mathscr{L}_0(V)$. Let $X, Y \in V$ and let ϕ_t, ψ_t be their corresponding flows. Then the reachable curve

$$t \to \phi_{-t}(\psi_{-t}(\phi_t(\psi_t(x))))$$

has time duration zero (adding the time on each flow algebraically) and as $t \to 0$ has limiting direction along $[X, Y]$, as seen from the geometric interpretation of the Lie bracket given in Fig. 1.4. Thus all brackets of all elements of V should be included in $\mathscr{L}_0(V)$, as should all higher brackets, e.g. $[X, [Y, Z]]$ etc. by the same argument. Similarly, the limiting direction of the curve

$$t \to \phi_{-t}(\psi_t(x))$$

is $Y - X$ and so all the differences of elements of V should be in $\mathscr{L}_0(V)$. Since $\mathscr{L}_0(V)$ is, in particular, a vector space, this means that $\mathscr{L}_0(V)$ should contain the vector space $\mathscr{A}(V)$ generated by all finite linear combinations $\sum_{i=1}^k \lambda_i X_i$ of all elements $X_i \in V$ for which $\sum_{i=1}^k \lambda_i = 0$. Hence we define

$$\mathscr{L}_0(V) = \mathscr{L}'(V) + \mathscr{A}(V)$$

where $\mathscr{L}'(V)$ is the Lie algebra generated by the set

$$\{X \in D(M) : X = [Y, Z] \text{ for some } Y, Z \in V\}$$

and is called the *derived algebra* of V.

Note that we do not include those elements of V in $\mathscr{L}_0(V)$ which are not already in $\mathscr{L}'(V) + \mathscr{A}(V)$, since, for any such $X \in V$, X is not the limiting direction of a reachable curve of zero time duration.

In order to generalize Theorem 2.2.3 to the set $R(x, T; V)$ we must consider the time variable in a more explicit way. In fact, we associate with V the set of vector fields V^* on the product manifold $M \times \mathbb{R}$ given by

$$V^* = \{X \oplus \frac{\partial}{\partial t} : X \in V\}.$$

Since $[X \oplus \partial/\partial t, Y \oplus \partial/\partial t] = [X, Y] \oplus 0$, it follows that

$$\mathscr{L}'(V^*) = \{X \oplus 0 : X \in \mathscr{L}'(V)\},$$

and so

$$\mathscr{L}(V^*) = \left\{ Z \in D(M \times \mathbb{R}) : Z = \sum_{i=1}^k \lambda_i \left(X_i \oplus \frac{\partial}{\partial t} \right) + Y \oplus 0,\ X_i \in V,\ Y \in \mathscr{L}'(V),\ \lambda_i\text{s scalars} \right\} \quad (2.2.3)$$

Lemma 2.2.4
$\dim[\mathscr{L}(V^*)(x,0)] = n+1$ if and only if $\dim[\mathscr{L}_0(V)(x)] = n$.

Proof
If $\dim[\mathscr{L}(V^*)(x,0)] = n+1$ and $v \in T_xM$, then $(v,0) \in \mathscr{L}(V^*)(x,0)$ and so $(v,0) = Z(x,0)$ for some vector field $Z \in \mathscr{L}(V^*)$. Hence by (2.2.3),

$$Z(x,0) = (v,0) = \left(\Sigma\lambda_i X_i(x) + Y(x), \Sigma\lambda_i \frac{\partial}{\partial t}(0)\right)$$

for some scalars λ_i. Thus, $\Sigma\lambda_i = 0$ and so the vector field $\Sigma\lambda_i X_i + Y \in \mathscr{L}_0(V)$, whence $v \in \mathscr{L}_0(V)(x)$ and $T_xM \subseteq \mathscr{L}_0(V)(x)$. Since $\dim T_xM = n$ we have $\dim[\mathscr{L}_0(V)(x)] = n$.

To prove the converse, let $v \in T_xM$ so that $v \in \mathscr{L}_0(V)(x)$. Thus $v = (\Sigma\lambda_i X_i + Y)(x)$ where $X_i \in V$, $Y \in \mathscr{L}'(V)$, $\Sigma\lambda_i = 0$. It follows that

$$(v,0) = \left(\Sigma\lambda_i\left(X_i \oplus \frac{\partial}{\partial t}\right) + Y \oplus 0\right)(x,0) \in \mathscr{L}(V^*)(x,0).$$

If $X \in V$, then $(X \oplus \partial/\partial t)(x,0) \in V^*(x,0) \subseteq \mathscr{L}(V^*)(x,0)$ and $(X \oplus 0)(x,0) \in \mathscr{L}(V^*)(x,0)$. Hence $(0, \partial/\partial t(0)) \in \mathscr{L}(V^*)(x,0)$ (which is a vector space). Since v is arbitrary in T_xM the result follows. □

We can now prove

Theorem 2.2.5
Under the assumptions of Theorem 2.2.3, $R(x,T;V)$ has nonempty interior in M if and only if $\dim[\mathscr{L}_0(V)(x)] = n$ for any $x \in M$. Then, interior $(R(x,T;V))$ is dense in $R(x,T;V)$.

Proof
First note that there is a one-to-one correspondence between the integral curves α of V such that $\alpha(0) = x$ and the integral curves β of V^* such that $\beta(0) = (x,0)$ defined by the map

$$\alpha \leftrightarrow \beta \qquad \text{with } \beta(t) = (\alpha(t), t).$$

Thus, $y \in R(x,T;V)$ if and only if $(y,T) \in R((x,0),[0,T],V^*)$. Hence the trick of adding in the vector field $\partial/\partial t$ allows us to consider the set $R((x,0),[0,T],V^*)$ rather than $R(x,T;V)$, which puts us in the situation of Theorem 2.2.3, which states that $R((x,0),[0,T];V^*)$ has nonempty interior if and only if $\dim[\mathscr{L}(V^*)(x,0)] = n+1$. However, by Lemma 2.2.4 this is the case if and only if $\dim[\mathscr{L}_0(V)(x)] = n$ and so if we can show that $R(x,T;V)$ has nonempty interior in M if and only if $R((x,0);V^*)$ has nonempty interior in $M \times \mathbb{R}$, then the first part of the theorem will follow.

Hence, suppose that $R(x,T;V)$ has nonempty interior (in M) and let $U \neq \emptyset$ be an open set in $R(x,T;V)$. If $X \in V$, define the map

$F: U \times \mathbb{R} \to M \times \mathbb{R}$ by

$$F(u, t) = (\phi_t(u), T + t)$$

where ϕ_t is the flow of X. Then $\mathrm{d}F$ has rank $n + 1$ everywhere and so maps open sets to open sets. However, $F(U \times (0, \infty)) \subseteq R((x, 0), V^*)$ and so $R((x, 0); V^*)$ has nonempty interior in $M \times \mathbb{R}$. If, conversely, $R((x, 0), V^*)$ has nonempty interior in $M \times \mathbb{R}$, then so does $R((x, 0), [0, t]; V^*)$ for each $t \in (0, T)$, by Theorem 2.2.3. Let $U \times I \neq \emptyset$ be an open set in $R((x, 0), [0, T], V^*)$ and let $s \in I$. Since $U \times \{s\} \subseteq R((x, 0), [0, T]; V^*)$ it is clear that $U \subseteq R(x, s; V)$ and so if $X \in V$ has flow ϕ_t, then the map $y \to \phi_{T-s}(y)$ defined on U is open and has image in $R(x, T; V)$ and the result follows.

To prove the second part of the theorem note that there is no loss of generality in assuming that V is a finite set and so we let $V = \{X_1, \ldots, X_l\}$, say. Let $\phi_{i,t}(.)$ be the flow of X_i and let

$$\phi_{\mathbf{i},\mathbf{t}}(x) \triangleq \phi_{i_1, t_1}(\phi_{i_2, t_2}(\ldots(\phi_{i_m, t_m}(x))\ldots))$$

where $\mathbf{i} = (i_1, \ldots, i_m)$, $1 \leqslant i_j \leqslant l, \mathbf{t} = (t_1, \ldots, t_m) \in \mathbb{R}^m$ for any m. Now, if $y \in \phi_{\mathbf{i}}(\mathbf{t}, x)$, with $t_i > 0$ for $i = 1, \ldots, m$ and $\| \mathbf{t} \| \triangleq t_1 + \cdots + t_m = T$, then let $\{s_k\} \subseteq (0, t_m)$ be a sequence with $\lim_{k \to \infty} s_k = 0$. If $R(x, T; V)$ has nonempty interior in M then so does $R(x, t; V)$ for all $t > 0$ since, by the first part, this is equivalent to $\dim[\mathscr{L}_0(V)(x)] = n$, which is independent of T. Hence, for each $k > 0$ there exists x_k in the interior of $R(x, s_k; V)$ and so if $\mathbf{t}_k = (t_1, \ldots, t_{m-1}, t_m - s_k)$ and $y_k = \phi_{\mathbf{i}, \mathbf{t}_k}(x_k)$, then y_k is in the interior of $R(x, T; V)$ since $\phi_{\mathbf{i}}$ is a diffeomorphism. Since V is finite and $s_k \to 0$, $R(x, s_k; V) \to \{x\}$ as $k \to \infty$. Moreover, ϕ_i is continuous and so $y_k \to y$ as $k \to \infty$ (and $\mathbf{t}_k \to \mathbf{t}$). □

The dimension of $\mathscr{L}_0(V)(x)$ is equal to k or $k - 1$ where $k = \dim \mathscr{L}(V)(x)$ and when $k < n$ we can obtain results similar to those above. Note, first, that it can be shown that if S is a maximal integral manifold of $\mathscr{L}(V)$, then $\dim \mathscr{L}_0(V)(x)$ is constant on S. Let $I_0(V, x)$ denote the maximal integral manifold of $\mathscr{L}_0(V)$. Then we have

Theorem 2.2.6
Let $V \subseteq D(M)$ be a set of complete vector fields, and let $x \in M$. Define

$$I_0^t(V, x) = \phi_t(I_0(V, x))$$

where ϕ_t is the flow of a vector field $X \in V$. Then $I_0^t(V, x)$ is independent of the particular vector field $X \in V$ chosen in the definition. Moreover, for each $T > 0$, $R(x, T; V) \subseteq I_0^t(V, x)$ and the interior of $R(x, T; V)$ in the relative topology of $I_0^t(V, x)$ is dense in $R(x, T; V)$.

(For a proof, see Sussmann and Jurdjevic, 1972). □

We can now return to the study of the controllability of the system

$$\dot{x}(t) = F(x(t), u(t)), \qquad x(0) = x. \tag{2.2.4}$$

where $F: M \times U \to TM$. For each $u \in U$, let $X_u = F(.\,, u)$. Then, by assumption, X_u is a complete analytic vector field on M. We can apply the above results by defining $V = \{X_u : u \in U\}$, giving the following two theorems:

Theorem 2.2.7
Let $x \in M$. Then $R(x) = R(x; V) \subseteq I(V, x)$ and for each $T > 0$ the interior of $R(x, [0, T])$ relative to $I(V, x)$ is dense in $R(x, [0, T])$. □

Theorem 2.2.8
Let $x \in M$. Then, for each $t > 0$, $R(x, t) \subseteq I_0^t(V, x)$ and the interior of $R(x, t)$ relative to $I_0^t(V, x)$ is dense in $R(x, t)$. □

Hence we obtain the following two corollaries:

Corollary 2.2.9
The system (2.2.4) has the accessibility property from x if and only if $\dim[\mathscr{L}(V)(x)] = \dim M$. In this case $R(x, [0, T])$ has nonempty interior for every $T > 0$.
□

Corollary 2.2.10
The system (2.2.4) has the strong accessibility property from x if and only if $\dim[\mathscr{L}_0(V)(x)] = \dim M$, and then $R(x, T)$ has nonempty interior for all $T > 0$.
□

Clearly the strong accessibility property implies the accessibility property but not conversely, in general. The reverse implication does hold, however, in certain cases. In fact, Sussmann and Jurdjevic (1972) show that if M has a compact universal covering space (see Spanier, 1966) then a system with the accessibility property must also have the strong accessibility property. Moreover, it can be shown that if M has a fundamental group with no elements of infinite order (e.g. if M is simply connected) then every controllable system on M has the strong accessibility property. (Of course, controllability implies accessibility, but not strong accessibility, in general.)

Examples

1. For linear systems of the form

$$\dot{x} = Ax + Bu \tag{2.2.5}$$

we have the well-known controllability criterion

$$\operatorname{rank}[B, AB, A^2B, \ldots, A^{n-1}B] = n \tag{2.2.6}$$

To see how this relates to the accessibility property above let $M = \mathbb{R}^n$, $U = \mathbb{R}^m$ and $F: M \times U \to TM$ be given by

$$F(x, u) = Ax + Bu.$$

Clearly, $V = \{A \cdot + Bu : u \in \mathbb{R}^m\}$ and it is easy to check that $\mathscr{L}(V)(x)$ contains the vectors $Ax \pm b_i, \pm Ab_i, \ldots, \pm A^{n-1}b_i$, $1 \leqslant i \leqslant m$ where b_i is the ith column of B. Moreover, these vectors generate $\mathscr{L}(V)(x)$. Hence Corollary 2.2.9 implies that the linear system (2.2.5) is accessible from the origin (i.e. $R(0, [0, t])$ has nonempty interior in $\mathbb{R}^n$) if and only if the rank condition (2.2.6) holds.

We also have $\mathscr{L}_0(V)(0) = \mathscr{L}(V)(0)$ and so accessibility, strong accessibility and controllability are equivalent in this case.

2. Consider the 'bilinear system'

$$\dot{x} = (A_0 + \sum_{i=1}^{m} A_i u_i)x \tag{2.2.7}$$

where $A_0, \ldots, A_m$ are $n \times n$ matrices. Here, $V = \{(A_0 + \sum_{i=1}^{m} u_i A_i)(.) : u \in \mathbb{R}^m\}$. Let L_0, L denote the Lie algebras spanned by $\{A_0, \ldots, A_m\}$, $\{A_1, \ldots, A_m\}$, respectively. Then it is easy to check that

$$\mathscr{L}(V)(x) = \{Px : P \in L_0\}, \; \mathscr{L}_0(V)(x) = \{Px : P \in L\}$$

and so the accessibility properties of the bilinear system (2.2.7) are determined by the Lie algebras L_0, L.

In the case where the manifold M is a Lie group G it is possible to obtain stronger controllability results. Consider the system

$$\dot{x}(t) = X_0(x(t)) + \sum_{i=1}^{m} u_i(t) X_i(x(t)), \tag{2.2.8}$$

where $x(t) \in G, t \geqslant 0$ and $X_0, \ldots, X_m$ are right-invariant vector fields on G. By right-invariance it is clear that

$$R(g) = R(e)g, \qquad g \in G$$

where e is the identity of G. Hence we can just consider the reachable set from e. The following results are proved in Jurdjevic and Sussmann (1972):

Theorem 2.2.11

A necessary condition for the system (2.2.8) to be controllable is that G is connected and that $\{X_0, \ldots, X_m\}_{\mathrm{LA}}$ equals the Lie algebra of G, where $\{X_0, \ldots, X_m\}_{\mathrm{LA}}$ is the Lie algebra generated by $X_0, \ldots, X_m$. If G is compact or $X_0 = 0$, then the condition is also sufficient. □

Theorem 2.2.12

If G is compact and connected, $\{X_0, \ldots, X_m\}_{\mathrm{LA}}$ equals the Lie algebra of G and $X_0 \in \{X_1, \ldots, X_m\}_{\mathrm{LA}}$ then (2.2.8) is controllable. □

We shall next describe some results of Hirschorn (1975) which provide explicit expressions for the reachable set of a nonlinear system

$$\dot{x}(t) = F(x(t), u(t)), \tag{2.2.9}$$

where the control functions $u(.)$ are restricted to be piecewise constant functions from $[0, \infty)$ into $\mathbb{R}^m$. The results are based on certain properties of Lie transformation groups with finite-dimensional Lie algebras and so we assume that the Lie algebra $\mathscr{L}(V)$, where V is the set of vector fields of the form $F(., u)$, is finite-dimensional. Note first that Chow's theorem already specifies the reachable set if V is symmetric, for then $R(x) = I(V, x)$.

In the case where M is a Lie group G and V is a collection of right-invariant vector fields on G, then $\mathscr{L}(V)$ is automatically finite-dimensional. To describe the reachable set in this case consider the following sequence of Lie algebras: if $\mathscr{P}$ is a Lie subalgebra of $\mathscr{L}(V)$ and $X \in V$ is a right-invariant vector field on G we have the sequence

$$\mathscr{P} \subseteq \mathscr{I}(\mathscr{P}) \subseteq \{\mathscr{P}, X\}_{\mathrm{LA}}$$

where $\mathscr{I}(\mathscr{P})$ is the ideal generated by $\mathscr{P}$ in $\{\mathscr{P}, X\}_{\mathrm{LA}}$. If $\mathscr{P}$ is an ideal in $\mathscr{I}(\mathscr{P})$, then this sequence is called an *X-chain*, and if $\mathscr{P} \subseteq \mathfrak{k} \subseteq \mathscr{L}(V)$, where $\mathfrak{k}$ is a Lie subalgebra of $\mathscr{L}(V)$, then it is called an *X-chain from* $\mathfrak{k}$. In this case the Lie algebra $\{\mathfrak{k}, \mathscr{I}(\mathscr{P})\}_{\mathrm{LA}}$ is said to be *X-generated from* $\mathfrak{k}$. Now, if $\mathscr{B}_0$ and $\mathscr{B}_n$ are Lie subalgebras of $\mathscr{L}_0(V)$ and $\mathscr{B}_0 \subseteq \mathscr{B}_n$, then a chain of Lie subalgebras

$$\mathscr{B}_0 \subseteq \mathscr{B}_1 \subseteq \cdots \subseteq \mathscr{B}_{n-1} \subseteq \mathscr{B}_n$$

is called an *X-series for* $\mathscr{B}_0$ *terminating at* $\mathscr{B}_n$ if $\mathscr{B}_{i+1}$ is X-generated from $\mathscr{B}_i$ for $i = 0, 1, \ldots, n-1$. We define the *X-radical* for $\mathscr{B}_0$, $\mathscr{R}(X; \mathscr{B}_0)$, as the largest Lie subalgebra $\mathfrak{k}$ of $\mathscr{L}_0(V)$ such that there exists an X-series for $\mathscr{B}_0$ terminating at $\mathfrak{k}$. Then we have

Theorem 2.2.13 (Hirschorn, 1973)
Consider the system

$$\dot{x} = A(x) + \sum_{i=1}^{m} u_i B_i(x), \qquad x(0) = x_0$$

defined on a Lie group G, with piecewise constant controls and $A, B, \ldots, B_m \in \mathscr{L}(G)$. Let $\mathscr{B}$ be the Lie algebra

$$\mathscr{B} = \{r_1 X_1 + \cdots + r_n X_n : r_i \in \mathbb{R}, \Sigma r_i = 0, \ X_i \in V, \ n = 1, 2, \ldots\}_{\mathrm{LA}}$$

and suppose that $\mathscr{R}(A; \mathscr{B}) = \mathscr{L}_0(V)$. Then

$$R(x_0, t) = \{\exp \mathscr{L}_0(V)\}_H \exp(tA) x_0,$$

where H is the subgroup

$$H = \{ X_{t_1}^{u_1} \circ X_{t_2}^{u_2} \circ \cdots \circ X_{t_n}^{u_n} : u_i \in \mathbb{R}^m, t_i \in \mathbb{R}, n = 1, 2, \ldots \}$$

of diff(M). (We have denoted the flow of $F(\,.\,, u)$ by X_t^u). Also, $\{S\}_H$ denotes the Lie subgroup of H generated by S. □

In the case of a system of the form (2.2.9) on a Hausdorff differentiable manifold M we can obtain a similar result by introducing a Lie transformation group G on M by defining

$$G = \{\exp t_1 L_1 \exp t_2 L_2 \cdots \exp t_n L_n : t_i \in \mathbb{R},\ L_i \in \mathscr{L}(V),\ n = 1, 2, \ldots\}$$

and similar groups G_0 and B related to $\mathscr{L}_0(V)$, $\mathscr{B}$ in the same way. Then we have

Theorem 2.2.14 (Hirschorn, 1975)
Let $A \in \mathscr{L}(V)$ denote the vector field $F(\,.\,, 0)$ and let A^0 be the isomorphic image of A in $\mathscr{L}(G)$. Suppose that the A-radical for $\mathscr{B}$ is $\mathscr{L}_0(V)$ and that for all $t \in \mathbb{R}^+$, $B \exp tA^0 \subseteq \bar{H}_t$ where

$$H_t = \{ X_{t_1}^{u_1} \circ \cdots \circ X_{t_n}^{u_n} : u_i \in \mathbb{R}^m, t_i \in \mathbb{R}^+, \Sigma t_i = t, n = 1, 2, \ldots \}.$$

Then, for all $t > 0$,

$$R(x, t) = G_0.\ \exp(tA^0)x.$$ □

Corollary 2.2.15
Consider the nonlinear system

$$\dot{x}(t) = A(x(t)) + u_1(t)B_1(x(t)) + \cdots + u_m(t)B_m(x(t)),$$

on M where $u_i(\,.\,)$ is piecewise constant, $A, B_1, \ldots, B_m$ are complete vector fields in $D(M)$ and $\{A, B_1, \ldots, B_m\}_{\text{LA}}$ is finite-dimensional. Then $\mathscr{B} = \{B_1, \ldots, B_m\}_{\text{LA}}$, $\mathscr{L}(V) = \{A, B_1, \ldots, B_m\}_{\text{LA}}$ and if $\mathscr{R}(A; \mathscr{B}) = \mathscr{L}_0(V)$, then for all $x \in M$, $t \in \mathbb{R}^+$, we have

$$R(x, t) = G_0 \exp(tA^0)x,$$

where A^0 is the isomorphic image of A in $\mathscr{L}(G)$. □

Example
Consider the system

$$\dot{x} = A(x(t)) + u_1(t)B_1(x(t)) + u_2(t)B_2(x(t))$$

in $\mathbb{R}^4$, where

$$A(x) = (x_2^2 \quad 1 \quad x_1 \quad x_2)^{\mathsf{T}}, B_1(x) = (x_2 \quad -1 \quad 1 \quad 0)^{\mathsf{T}}$$
$$B_2(x) = [4 \quad 0 \quad -2x_2 \quad 0]^{\mathsf{T}}.$$

A, B_1, B_2 are complete vector fields and we have

$$(\text{ad } A)B_1(x) = \left(x_2^2 \frac{\partial}{\partial x_1} + \frac{\partial}{\partial x_2} + x_1 \frac{\partial}{\partial x_3} + x_2 \frac{\partial}{\partial x_4}\right)\left(x_2^2 \frac{\partial}{\partial x_1} - \frac{\partial}{\partial x_2} + \frac{\partial}{\partial x_3}\right)$$
$$- \left(x_2^2 \frac{\partial}{\partial x_1} - \frac{\partial}{\partial x_2} + \frac{\partial}{\partial x_3}\right)\left(x_2^2 \frac{\partial}{\partial x_1} + \frac{\partial}{\partial x_2} + x_1 \frac{\partial}{\partial x_3} + x_2 \frac{\partial}{\partial x_4}\right)$$
$$= 4x_2 \frac{\partial}{\partial x_1} - x_2^2 \frac{\partial}{\partial x_3} + \frac{\partial}{\partial x_4}$$
$$\equiv (4x_2 \quad 0 \quad -x_2^2 \quad 1)^{\mathsf{T}}.$$

Similarly,

$$(\text{ad } A)^2 B_1(x) = (4 \quad 0 \quad -6x_2 \quad 0)^{\mathsf{T}},$$

and

$$B_3(x) \triangleq [B_1, B_2](x) = (0 \quad 0 \quad 2 \quad 0)^{\mathsf{T}}.$$

Furthermore,

$$(\text{ad } A)B_3 = [\text{ad } A \ B_1, B_3] = [\text{ad } A \ B_1, B_2] = [B_1, B_3] = [B_2, B_3] = 0$$
$$\text{ad } A \ B_2 = [(\text{ad } A)^2 B_1, B_1] = -3B_3, \ [\text{ad } A \ B_1, B_1] = B_2$$
$$(\text{ad } A)^3 B_1 = -5B_3.$$

Thus $\mathscr{L}(V)$ has a basis $\{A, B_1, B_2, B_3, \text{ad } A \ B_1, (\text{ad } A)^2 B_1\}$, $\mathscr{B}$ has a basis $\{B_1, B_2, B_3\}$ and $\mathscr{L}_0(V)$ has a basis $\{B_1, B_2, B_3, \text{ad } A \ B_1, (\text{ad } A)^2 B_1\}$. Since $\mathscr{B}$ is clearly an ideal in $\mathscr{L}_0(V)$ we have $\mathscr{R}(A; \mathscr{B}) = \mathscr{L}_0(V)$, and since, moreover, $\mathscr{L}(V)$ is finite-dimensional, Corollary 2.2.15 applies, giving

$$R(x_0, t) = G_0 \exp(tA^0)x_0.$$

Now $G_0 z_0 = \mathbb{R}^4$ for any $z_0 \in \mathbb{R}^4$, for if $y = (y_1, y_2, y_3, y_4) \in \mathbb{R}^4$ then follow the flow of $(\text{ad } A)B_1$ from z_0 for a time $y_4 - z_4$ to get to $p = (p_1, p_2, p_3, y_4)$ and then the flow of B_1 from p for a time $p_2 - y_2$ to get to $q = (q_1, y_2, q_3, y_4)$, and then the flow of B_2 from q for a time $(y_1 - q_1)/4$ to get to $r = (y_1, y_2, r_3, y_4)$ and finally the flow of B_3 from r for a time $(y_3 - r_3)/2$ to get to y. Hence the reachable set from any $x_0 \in \mathbb{R}^4$ is $\mathbb{R}^4$.

Consider finally the 'bilinear' system

$$\dot{x} = (A + u_1(t)B_1 + \cdots + u_m(t)B_m)x(t) + Ev(t) \qquad (2.2.10)$$

where $A, B_1, \ldots, B_m$ are $n \times n$ matrices, $v(t) \in \mathbb{R}^l$, $E = (e_1, \ldots, e_l)$ with $e_i \in \mathbb{R}^n$. As usual, let

$$\mathscr{L}(V) = \mathscr{L} = \{A, B_1, \ldots, B_m\}_{\text{LA}}$$

and

$$\mathscr{L}^i = \{ L_{k_1} L_{k_2} \dots L_{k_i} : L_{k_j} \in \mathscr{L}, j = 1, \dots, l \}_{LS}$$

where $\{S\}_{LS}$ is the linear space spanned by S. Then Hirschorn, 1975, proves

Theorem 2.2.16
The reachable set for the system (2.2.10) is given by

$$R(x, t) = \exp tA \{\exp \mathscr{L}_0(V)\}_H x + \{\mathscr{L}^i e_j : i = 0, 1, \dots, j = 1, \dots, l\}_{LS}$$

for all $x \in \mathbb{R}^n$, $t \in \mathbb{R}^+$, provided $\mathscr{R}(A; \mathscr{B}) = \mathscr{L}_0(V)$. □

If $x = 0$ in this result we have

$$R(0, t) = \{\mathscr{L}^i e_j : i = 0, 1, \dots, j = 1, \dots, l\}_{LS}$$

which reduces to the classical case $R(0, t) = \text{range } (E, AE, \dots, A^{n-1}E)$ for linear systems if $B_1 = \cdots = B_m = 0$. This was shown by Brockett (1973).

Further results on accessibility are given by Sussmann (1976) and Grasse (1984) where the stability of accessibility and controllability under perturbations of the vector field are studied. Rebhuhn (1977) obtains conditions on sets of vector fields for the attainable set to be contained in the closure of its interior. For the application of Lie theory methods to the controllability of delay systems, see Hermes (1979). Local controllability for systems with the control appearing linearly is discussed by Sussmann (1983).

2.2.2 Controllability by fixed point theory

Nonlinear controllability can be studied analytically as well as by the algebraic methods above. We shall illustrate this approach by following the work of Davison and Kunze (1970). Thus, consider a nonlinear system of the form

$$\dot{x}(t) = A(t, x(t))x(t) + B(t, x(t))u(t), \qquad t_0 \leqslant t < \infty \tag{2.2.11}$$

where $x(t) \in \mathbb{R}^n$, $u(t) \in \mathbb{R}^m$ for each t, A is an $n \times n$ matrix function and B is an $n \times m$ matrix function. The elements $a_{ij}(t, x)$, $b_{ik}(t, x)$ of A and B are assumed to be continuous in x (for fixed t) and piecewise continuous in t (for fixed x) and to be bounded:

$$| a_{ij}(t, x) | \leqslant M, \qquad 1 \leqslant i \leqslant n, 1 \leqslant j \leqslant n$$
$$| b_{ik}(t, x) | \leqslant N, \qquad 1 \leqslant i \leqslant n, 1 \leqslant k \leqslant m$$

for each $x \in \mathbb{R}^n$ and $t \geqslant t_0$.

We say that the system (2.2.11) is *completely state controllable* at t_0 in the controllability domain $D \subseteq \mathbb{R}^n$, if each initial state $x(t_0)$ in D can be transferred to any final state x_f in D in some time $t_f < \infty$. If $D = \mathbb{R}^n$ the controllability is said to be *global*, and *local* otherwise. If the system is completely state controllable on any interval $[t_0, t_f]$, $t_f > t_0$, it is said to be *totally state controllable on D*.

The controllability criterion which we shall prove below is stated in terms of the linear system

$$\dot{x}(t) = A(t, z(t))x(t) + B(t, z(t))u(t), \tag{2.2.12}$$

where $z(t)$ is any given function in $C^n[t_0, t_f]$ (the Banach space of continuous $\mathbb{R}^n$-valued functions on $[t_0, t_f]$). The advantage of fixing the arguments of A and B in $C^n[t_0, t_f]$ is that (2.2.12) is a linear system with solution

$$x(t) = \phi(t, t_0; z(.))x_0 + \int_{t_0}^{t} \phi(t, \tau; z(.))B(\tau, z(\tau))u(\tau)\, d\tau, \tag{2.2.13}$$

where ϕ is the state-transition matrix of the system

$$\dot{x}(t) = A(t, z(t))x(t).$$

We define

$$H(t_0, \tau; z(.)) = \phi(t_0, \tau; z(.))B(\tau, z(\tau))$$

and

$$G(t_0, t; z(.)) = \int_{t_0}^{t} H(t_0, \tau; z(.))H^{\mathrm{T}}(t_0, \tau; z(.))\, d\tau.$$

Then the following well-known result (see Kreindler and Sarachik (1964)) specifies the controllability of the system (2.2.12).

Lemma 2.2.17

The system (2.2.12) is completely state controllable at t_0 if and only if there exists a finite time $t_f > t_0$ such that the rows of the matrix $H(t_0, \tau; z(.))$ are linearly independent functions of τ on $[t_0, t_f]$. (2.2.11) is totally state controllable if and only if the rows of $H(t_0, \tau; z(.))$ are linearly independent functions of τ on $[t_0, t_f]$ for all $t_f > t_0$. □

Thus, if (2.2.12) is controllable on $[t_0, t_f]$ it follows that $G(t_0, t_f; z(.))$ is positive definite and hence invertible. We may therefore consider the control

$$u(t_0, t, t_f; z(.)) = H^{\mathrm{T}}(t_0, t; z(.))G(t_0, t_f; z(.))^{-1}[\phi(t_f, t_0; z(.))^{-1}x_f - x_0]$$

in (2.2.13). Substituting, we have

$$x(t) = \phi(t, t_0; z(.))x_0 + \int_{t_0}^{t} \phi(t, \tau; z(.))B(\tau, z(\tau))H^{\mathrm{T}}(t_0, t; z(.)) \cdot G(t_0, t_f; z(.))^{-1}[\phi(t_f, t_0; z(.))^{-1}x_f - x_0]\, d\tau.$$

However, by the transition property,

$$\phi(t, \tau; z(.)) = \phi(t, t_0; z(.))\phi(t_0, \tau; z(.))$$

and so

$$x(t) = \phi(t, t_0; z(.))\{x_0 + G(t_0, t; z(.))G(t_0, t_f; z(.))^{-1} [\phi(t_f, t_0; z(.))^{-1}x_f - x_0]\}. \quad (2.2.14)$$

Hence

$$x(t_0) = x_0, \; x(t_f) = x_f.$$

Write the right-hand side of (2.2.14) as $P(z)(t)$. Then (2.2.14) may be expressed in the form

$$x = P(z). \quad (2.2.15)$$

If we can find a solution of (2.2.15) for z in terms of x, then we will clearly have solved the controllability problem for the nonlinear system (2.2.11).

Theorem 2.2.18

Consider the system (2.2.11) for which the matrices A, B satisfy the above assumptions. Then, if there exists a constant $c > 0$ such that

$$\inf_{z \in C^n[t_0, t_f]} \det G(t_0, t_f; z(.)) \geqslant c$$

for some $t_f > t_0$ (or for all t_0 and $t_f > t_0$) then (2.2.11) is globally completely state controllable at t_0 (or totally state controllable).

Proof

As we have remarked, we must prove that (2.2.15) has a solution. We shall do this by applying Schauder's fixed point theorem (see Chapter 1). Put the norm

$$\| z \| = \max_{t_0 \leqslant t \leqslant t_f} \sum_{i=1}^{n} | z_i(t) |$$

on $C^n[t_0, t_f]$. Since each element of A is bounded by M, we have

$$| \phi(t, t_0; z(.))x | \leqslant e^{nM(t_f - t_0)} | x |, \qquad t_0 \leqslant t \leqslant t_f,$$

where $| x | = \Sigma_{i=1}^{n} | x_i |$. Similarly,

$$\| G(t_0, t; z(.)) \| \triangleq \max_j \sum_{i=1}^{n} | g_{ij}(t_0, t; z(.)) | \leqslant nmN^2(t_f - t_0)e^{2nM(t_f - t_0)},$$

where $G = (g_{ij})$. Hence, by (2.2.14), we have

$$\| x \| \leqslant \{(1 + C) | x_0 | + C | x_f | e^{nM(t_f - t_0)}\}e^{nM(t_f - t_0)} = K, \quad \text{say},$$

where

$$C = \sup_{z \in C^n[t_0, t_f]} \| G(t_0, t_f; z(\,.\,))^{-1} \| \, nmN^2(t_f - t_0)e^{2nM(t_f - t_0)}.$$

Let $S = \{z : z \in C^n[t_0, t_f],\ \|z\| \leq K\}$. Then S is a closed convex subset of $C^n[t_0, t_f]$ and it is easy to prove, from the Arzela–Ascoli theorem (Chapter 1), that P maps S into a compact set. Hence, by Schauder's fixed point theorem, there exists x^* such that

$$x^* = P(x^*)$$

and the result is proved. □

Further results using the method of fixed point theory can be found in Lukes (1974) where disturbances are allowed in the system and Klamka (1976) for delay systems.

2.2.3 Observability

In this section we shall study the dual problem to controllability, namely observability. We shall follow the work of Hermann and Krener (1977) in which the duality between controllability and observability is brought out strongly. In order to emphasize this duality, we shall first consider their formulation of controllability.

Given a subset $U \subseteq M$ we say that x_1 is *U-accessible* from x_0 if there is a control $u(t)$ (belonging to some admissible control set) such that the trajectory using this control passes through x_0 and x_1 and remains in U throughout. We write $x_1 A_U x_0$ in this case and $x_1 A x_0$ if $U = M$. For any relation R on M let

$$R(x_0) = \{x_1 \in M : x_1 R x_0\}.$$

Then $A(x_0)$ is the set of points *accessible* from x_0 and the system is controllable from x_0 if $A(x_0) = M$ and is controllable if $A(x) = M$ for each $x \in M$.

Even if a system is controllable from some point x_0, reaching points 'close' to x_0 may require trajectories which do not remain 'close' to x_0 and so it is desirable to have a local concept of controllability. A system is *locally controllable at* x_0 if, for every neighbourhood U of x_0, $A_U(x_0)$ is also a neighbourhood of x_0, and *locally controllable* if it is locally controllable everywhere. If $U \subseteq M$ is an open set there is a unique smallest equivalence relation on U which contains all U-accessible pairs. This is just the intersection of all equivalence relations containing such pairs. The resulting equivalence relation is called *weak U-accessibility* and is written WA_U. (Clearly $x_0 WA_U x_k$ if and only if there exist $x_1, \ldots, x_{k-1}$ such that

$x_i A_U x_{i-1}$ or $x_{i-1} A_U x_i$ for $i = 1, \ldots, k$, i.e. x_0 can be joined to x_k by trajectories going forward or backward in time. If $WA(x_0) = M$ we say that the system is *weakly controllable at* x_0 and if this is true for all $x_0 \in M$ we say that the system is *weakly controllable*. Finally, there is a local version of weak controllability. A system is *locally weakly controllable* (at x_0) if, for every neighbourhood U of x_0, $WA_U(x_0)$ is also a neighbourhood of x_0. Clearly we have the following relations between these types of controllability:

$$\begin{array}{ccc} \text{local controllability} & \Rightarrow & \text{controllability} \\ \Downarrow & & \Downarrow \\ \text{local weak controllability} & \Rightarrow & \text{weak controllability.} \end{array}$$

(These notions are, of course, equivalent for linear systems.) Using similar ideas to those above we can prove

Theorem 2.2.19
Let

$$\Sigma : \dot{x} = f(x, u)$$

be a system (written, for convenience, in local coordinates), and let $\mathscr{F}$ denote the Lie algebra generated by the vector fields $f(\,.\,, u)$ for constant values of u. Moreover, let $\mathscr{F}(x)$ denote the set of tangent vectors of these vector fields at x. Then if $\dim \mathscr{F}(x) = n (= \dim M)$ then Σ is locally weakly controllable at x. If, in addition Σ is analytic, then Σ is weakly controllable if and only if it is locally weakly controllable and this is true if and only if $\dim \mathscr{F}(x) = n$. □

NOTE The condition $\dim \mathscr{F}(x) = n$ is called the *controllability rank condition*.

Now suppose we add an observation equation to Σ so that we have the system

$$\Sigma : \dot{x} = f(x, u) \qquad (2.2.16)$$
$$y = g(x)$$

Given a fixed initial condition $x(t_0) = x_0$, we can obtain (in principle) a solution $x(t)$ of (2.2.16) on some interval $[t_0, t_1]$ and hence obtain the output $y(t) = g(x(t))$. We shall write

$$\Sigma(x_0) : u(t) \to y(t)$$

for this 'input–output' map or $\Sigma(x_0; t_0, t_1)$ if we wish to emphasize the time interval. Then we say that points x_0 and x_1 are *indistinguishable* (written $x_0 I x_1$) if $\Sigma(x_0; t_0, t_1)(u(t)) = \Sigma(x_1; t_0, t_1)(u(t))$ for all admissible inputs. Clearly, I is an equivalence relation on M. The system Σ is called *observable at* x_0 if $I(x_0) = \{x_0\}$ and Σ is *observable* if it is observable at each

$x_0 \in M$. As in the case of controllability it is useful to have a local concept of observability. Thus, let U be a subset of M and $x_0, x_1 \in U$. Then x_0 is said to be *U-indistinguishable* from x_1 ($x_0 I_U x_1$) if, for each control $u(t)$ ($t \in [t_0, t_1]$) such that the resulting trajectories starting at x_0 and x_1 both lie in U, we have

$$\Sigma(x_0)u(t) = \Sigma(x_1)u(t),$$

i.e. we cannot distinguish between x_0 and x_1 using such controls. Σ is said to be *locally observable at* x_0 if for every open neighbourhood U of x_0, we have $I_U(x_0) = \{x_0\}$, and Σ is *locally observable* if it is locally observable at each $x_0 \in M$.

Very often, distinguishing between initial points which are 'far apart' is unnecessary and so we introduce a weak concept of observability where we consider only neighbouring points. Hence we say that Σ is *weakly observable at* x_0 if there exists a neighbourhood U of x_0 such that $I(x_0) \cap U = \{x_0\}$ and *weakly observable* if it is weakly observable everywhere. Finally we say that Σ is *locally weakly observable at* x_0 if there exists an open neighbourhood U of x_0 such that for every open neighbourhood V of x_0 with $V \subseteq U$, $I_V(x_0) = \{x_0\}$ and if this holds for all $x_0 \in M$, Σ is called *locally weakly observable*. As for controllability, we have the relations:

$$\begin{array}{ccc} \text{local observability} & \Rightarrow & \text{observability} \\ \Downarrow & & \Downarrow \\ \text{local weak observability} & \Rightarrow & \text{weak observability} \end{array}$$

In order to derive a condition for observability similar to the controllability rank condition, recall first that a vector field $X \in D(M)$ operates on $C^\infty(M)$ in the following way. First write X locally as $X = \Sigma h_i \partial/\partial x_i$. Then X operates on $C^\infty(M)$ by Lie differentiation; i.e.

$$L_h(\phi)(x) = \frac{\partial \phi}{\partial x}(x)h(x) = \langle \mathrm{d}\phi(x), h(x) \rangle$$

for any $\phi \in C^\infty(M)$, where $\mathrm{d}\phi = \partial\phi/\partial x = (\partial\phi/\partial x_1, \ldots, \partial\phi/\partial x_n)$ is the exterior derivative or gradient of ϕ. Let $\mathscr{G}_0 = \{g_1, \ldots, g_m\} \subseteq C^\infty(M)$, where the functions g_i appear in the observation equation of Σ (see (2.2.16)) and let $\mathscr{G}$ denote the smallest linear subspace of $C^\infty(M)$ containing $\mathscr{G}_0$ and which is closed under Lie differentiation by the vector fields $f(\,.\,, u)$ for constant values of u. Then $\mathscr{G}$ consists of all elements of the form

$$L_{f_1}(\ldots(L_{f_k}(g_i))\ldots)$$

where $f_j(x) = f(x, u_j)$ for some constant u_j in the admissible control set. We associate with $\mathscr{G}_0$ and $\mathscr{G}$ the spaces of 1-forms

$$\mathrm{d}\mathscr{G}_0 = \{\mathrm{d}\phi : \phi \in \mathscr{G}_0\}, \qquad \mathrm{d}\mathscr{G} = \{\mathrm{d}\phi : \phi \in \mathscr{G}\}.$$

Recall from Lemma 1.3.2(v) that for any 1-form ω and vector fields h_1, h_2 we have the Leibnitz formula

$$L_{h_1}\langle \omega, h_2\rangle = \langle L_{h_1}\omega, h_2\rangle + \langle \omega, [h_1, h_2]\rangle. \tag{2.2.17}$$

Writing this expression explicitly, it is easy to check that the Lie derivative of a 1-form ω with respect to a vector field h is given by

$$L_h(\omega)(x) = \left(\frac{\partial \omega^{\mathrm{T}}}{\partial x}(s)h(x)\right)^{\mathrm{T}} + \omega(x)\frac{\partial h}{\partial x}(x) \tag{2.2.18}$$

where T denotes transpose. Also, L_h and d commute and so we have

$$L_h(\mathrm{d}\phi) = \mathrm{d}(L_h(\phi)) \qquad \text{for any } \phi \in C^{\infty}(M).$$

Since, for any $h_1, h_2 \in D(M)$ and $\phi \in C_{\infty}(M)$ we have

$$L_{h_1}(L_{h_2}(\phi)) - L_{h_2}(L_{h_1}(\phi)) = L_{[h_1, h_2]}(\phi),$$

it follows that $\mathrm{d}\mathscr{G}$ is the space of all finite linear combinations of elements of the form

$$L_{f_1}(\ldots(L_{f_k}(\mathrm{d}g_i))\ldots)$$

where $f_j(x) = f(x, u_j)$ for some constant admissible control u_j. Evaluating the elements of $\mathrm{d}\mathscr{G}$ at x we obtain the space $\mathrm{d}\mathscr{G}(x)$.

We say that the system Σ satisfies the *observability rank condition* at x if $\dim \mathrm{d}\mathscr{G}(x) = n$. We then require the lemma

Lemma 2.2.20
Let $V \subseteq M$ be any open set. Then if $x_0, x_1 \in V$ and $x_0 I_V x_1$ we have $\phi(x_0) = \phi(x_1)$ for every $\phi \in \mathscr{G}$.

Proof
If $x_0 I_V x_1$ and k is any positive integer, then for any constant admissible controls $u_1, \ldots, u_k$, small $s_1, \ldots, \ldots s_k \geqslant 0$ and any $g_i, i = 1, \ldots, m$ we have

$$g_i(\gamma^k_{s_k} \circ \cdots \circ \gamma^2_{s_2} \circ \gamma^1_{s_1}(x_0)) = g_i(\gamma^k_{s_k} \circ \cdots \circ \gamma^2_{s_2} \circ \gamma^1_{s_1}(x_1)),$$

where $\gamma^i_{s_i}(x)$ is the flow of $f_i(x) = f(x, u_i)$. Since the Lie derivative equals the derivative along the corresponding flow at 0, differentiating with respect to $s_k, \ldots, s_1$ at 0 gives

$$L_{f_1}(\ldots(L_{f_k}(g_i))\ldots)(x_0) = L_{f_1}(\ldots(L_{f_k}(g_i))\ldots)(x_1).$$

The lemma now follows since $\mathscr{G}$ is spanned by functions of this form. □

We can now prove the following result on observability.

Theorem 2.2.21
If Σ satisfies the observability rank condition at x, then Σ is locally weakly observable at x.

Proof If dim $d\mathcal{G}(x) = n$, there exist n functions $\phi_1, \ldots, \phi_n \in \mathcal{G}$ such that $d\phi_1(x), \ldots, d\phi_n(x)$ are linearly independent. Hence the map

$$\Phi : x \mapsto \text{col}(\phi_1(x), \ldots, \phi_n(x))$$

has nonsingular Jacobian at x and so Φ is 1–1 on some open neighbourhood U of x. If $V \subseteq U$ is any open neighbourhood of x, then, by Lemma 2.2.20, $I_V(x) = \{x\}$ and so Σ is locally weakly observable at x. □

Example
Consider the system

$$\dot{x} = u, \qquad y_1 = \cos x, \qquad y_2 = \sin x$$

where $x \in M = \mathbb{R}$, $y \in \mathbb{R}^2$, $u \in \mathbb{R}$. Clearly, $d\mathcal{G}(x)$ is spanned by $\cos x\, dx$, $\sin x\, dx$ for any x and so if $x_0 = 0$, then this system is locally weakly observable at x_0. However, it is not observable since x_0 and $x_0 + 2k\pi$ are indistinguishable for any x_0 and any integer k.

2.3 REALIZATION THEORY

2.3.1 The realization problem

We shall now return to the input–output description of a system as discussed above and shown in Fig. 2.1. The *realization problem* for the input–output system $(\mathcal{U}, \mathcal{Y}, S)$ is to specify a manifold M, called the *state space*, with a distinguished point x_0 (the *initial condition*) and functions $f: M \times U \to TM$, $h: M \to \mathbb{R}^r$, where the control values belong to U, such that the system

$$\dot{x} = f(x, u), \qquad x(0) = x_0, \quad y = h(x)$$

realizes the input-output system $(\mathcal{U}, \mathcal{Y}, S)$, i.e.

$$S(u(\,.\,))(t) = h(\phi_t(x_0, u(\,.\,); f)),$$

where $\phi_t(x_0, u(\,.\,); f)$ is the flow of $f(x, u(\,.\,))$, starting at x_0.

The realization theory of linear systems is well known (see Kalman *et al.* (1969)) and Brockett (1972) obtained a realization theory for systems defined on groups. Sussmann (1977) and Hermann and Krener (1977) have made fundamental contributions to the general theory of nonlinear realization. In this section we shall follow Jakubczyk (1980) and derive necessary and sufficient conditions for the existence of realizations of a system $(\mathscr{U}, \mathscr{Y}, S)$. We shall also discuss the notion of minimality of realizations.

We begin by defining a semigroup structure on the space of control functions as given by Lobry (1973). The controls will be assumed to be piecewise constant and we shall denote such a control $u(t)$, defined by

$$u(t) = u_i \in U, \qquad t \in [\sigma_{i-1}, \sigma_i), \quad 1 \leqslant i \leqslant k,$$

by $\tilde{u} = (t_k u_k)\dots(t_2 u_2)(t_1 u_1)$, where $t_i = \sigma_i - \sigma_{i-1}$ and $\sigma_0 = 0$ (i.e. the control u_i is applied for a time t_i). Let $\mathscr{U}_{\mathrm{pc}}$ denote the set of all such functions. Then $\mathscr{U}_{\mathrm{pc}}$ has a natural semigroup structure given by the multiplication

$$\tilde{v}\tilde{u} = (\tau_m v_m)\dots(\tau_1 v_1)(t_k u_k)\dots(t_1 u_1)$$

and with identity e equal to the empty sequence. A natural action of $\mathbb{R}_+$ on $\mathscr{U}_{\mathrm{pc}}$ is defined by

$$t\tilde{u} = ((tt_k)u_k)\dots((tt_1)u_1) \tag{2.3.1}$$

and $\tilde{u}$ is identified with $1\tilde{u}$. We can also extend $\mathscr{U}_{\mathrm{pc}}$ to a group $G\mathscr{U}_{\mathrm{pc}}$ consisting of formal sequences as before with $t_i \in \mathbb{R}$ and multiplication defined as above. We identify $(t_1 u)(t_2 u)$ with $(t_1 + t_2)u$ and $0u$ with e. $t\tilde{u}$ is defined by (2.3.1) for $t < 0$ as well as for $t \geqslant 0$.

For any sequences $\tilde{\mathbf{v}} = (\tilde{v}_1, \dots, \tilde{v}_m)$, $m \geqslant 1$, $\tilde{\mathbf{u}} = (\tilde{u}_1 \dots, \tilde{u}_p)$, $p \geqslant 1$ and sequences $\mathbf{t} = (t_1, \dots, t_p)$, $t_i \in \mathbb{R}_+$ of times we define $\psi_{\tilde{\mathbf{u}}}^{\tilde{\mathbf{v}}} : \mathbb{R}_+^p \to \mathbb{R}^{mr}$ as $\psi^{\tilde{\mathbf{v}}}{}_{\tilde{\mathbf{u}}} = (\psi_{\tilde{\mathbf{u}}}^{\tilde{v}_1}, \dots, \psi_{\tilde{\mathbf{u}}}^{\tilde{v}_m})$, where

$$\psi_{\tilde{\mathbf{u}}}^{\tilde{v}_i}(\mathbf{t}) = S(\tilde{v}_i(t_p\tilde{u}_p)\dots(t_1\tilde{u}_1)).$$

Note that if we define $\psi_{\tilde{\mathbf{u}}} : \mathbb{R}_+^p \to M$ and $\psi^{\tilde{\mathbf{v}}} : M \to \mathbb{R}^{mr}$ by

$$\psi_{\tilde{\mathbf{u}}}(\mathbf{t}) = \phi(x_0, (t_p\tilde{u}_P)\dots(t_1\tilde{u}_1); f)$$

and

$$\psi^{\tilde{\mathbf{v}}} = (\psi^{\tilde{v}_1}, \dots, \psi^{\tilde{v}_m}), \; \psi^{\tilde{v}_i}(x) = h(\phi(x, \tilde{v}_i; f))$$

then

$$\psi^{\tilde{\mathbf{v}}}{}_{\tilde{\mathbf{u}}} = \psi^{\tilde{\mathbf{v}}} \circ \psi_{\tilde{\mathbf{u}}}. \tag{2.3.2}$$

A system $(\mathscr{U}, \mathscr{Y}, S)$ is said to be of *class* C^k (or C^k-*smooth*) for $k = 0, 1, \dots, \infty, \omega$ if the functions $\psi^{\tilde{\mathbf{v}}}{}_{\tilde{\mathbf{u}}}$ are of class C^k for all sequences $\tilde{\mathbf{u}}, \tilde{\mathbf{v}}$. When $k = \omega$ this means that the functions $\psi_{\tilde{\mathbf{u}}}^{\tilde{\mathbf{v}}}$ have (real) analytic extensions to $\mathbb{R}^p$.

We define the *rank* of the system S by

$$\operatorname{rank} S = \sup_{\tilde{\mathbf{u}}, \tilde{\mathbf{v}}, \mathbf{t}} \operatorname{rank} D\psi_{\tilde{\mathbf{u}}}^{\tilde{\mathbf{v}}}(\mathbf{t}),$$

where the sup is taken over all sequences $\tilde{\mathbf{u}}, \tilde{\mathbf{v}}, \mathbf{t}$, for any $m \geqslant 1$, $p \geqslant 1$. Since for any realization on a manifold M we must have

$$\operatorname{rank} S \leqslant \sup_{\tilde{\mathbf{u}}, \tilde{\mathbf{v}}, \mathbf{t}} \max\{\operatorname{rank} D\psi_{\tilde{\mathbf{u}}}(\mathbf{t}), \operatorname{rank} D\psi^{\tilde{\mathbf{v}}}(\psi_{\tilde{\mathbf{u}}}(\mathbf{t}))\} \leqslant \dim M$$

(by (2.3.2)), to obtain finite-dimensional realizations we must assume that

$$\operatorname{rank} S \triangleq n < \infty. \tag{2.3.3}$$

We shall be interested in *symmetric realizations*, i.e. those for which, given any $u \in U$, there exists $v \in U$ such that $f(\,.\,, u) = -f(\,.\,, v)$. Then we can effectively reverse time in the realization. Formally, we say that a system $(\mathscr{U}, \mathscr{Y}, S)$ is *time-invertible* if for any $u \in U$ there exists $v \in U$ such that for all $\tilde{u}, \tilde{v} \in \mathscr{U}_{\text{pc}}$, and $t > 0$ we have

$$S(\tilde{v}(tv)(tu)\tilde{u}) = S(\tilde{v}\tilde{u}) = S(\tilde{v}(tu)(tv)\tilde{u}).$$

Note that we can define time-invertibility in the equivalent form: there exists a function $\alpha : U \to U$ such that

$$S(\tilde{v}\tilde{w}_\alpha\tilde{w}\tilde{u}) = S(\tilde{v}\tilde{u}) = S(\tilde{v}\tilde{w}\tilde{w}_\alpha\tilde{u})$$

for any $\tilde{u}, \tilde{v}, \tilde{w} \in \mathscr{U}_{\text{pc}}$, where

$$\tilde{w}_\alpha(\tau) = \alpha(\tilde{w}(t-\tau)), \qquad \text{for almost all } \tau \in [0, t)$$

if $\tilde{w}$ is defined on $[0, t)$.

We can now define precisely the concept of realization. For any $k = 2, \ldots, \infty, \omega$ we define a C^k *realization* of the system $(\mathscr{U}, \mathscr{Y}, S)$ to be a quadruple (M, f, h, x_0), where M is a C^k Hausdorff manifold, $x_0 \in M$, $f\colon M \times U \to TM$ is a map such that for any $u \in U$, $\phi(x_0, tu; f)$ exists and is of class C^k for $(t, x) \in R \times M$, where ϕ is the flow of f, $h : M \to \mathbb{R}^r$ is of class C^k, and we have

$$S(\tilde{u}) = h(\phi(x_0, \tilde{u}; f)) \qquad \text{for all } \tilde{u} \in \mathscr{U}_{\text{pc}}.$$

A realization (M, f, h, x_0) is *reachable* (or *weakly reachable*) if for any $x_1 \in M$, there exists $\tilde{u} \in \mathscr{U}_{\text{pc}}$ (or $\tilde{u} \in G\mathscr{U}_{\text{pc}}$) such that $\phi(x_0, \tilde{u}; f) = x_1$. Of course, we define

$$\phi(x_0, \tilde{u}; f) = \phi(\,.\,, t_k u_k; f) \circ \cdots \circ \phi(x_0, t_1 u_1; f)$$
$$\phi(x, tu; f) = (\phi(x, -tu; f))^{-1}$$

for $\tilde{u} = (t_k u_k) \cdots (t_1 u_1) \in G\mathscr{U}_{\text{pc}}$.

A realization is *observable* if for any $x_1, x_2 \in M$, $x_1 \neq x_2$, there exists $\tilde{u} \in \mathscr{U}_{\text{pc}}$ such that $h(\phi(x_1, \tilde{u}; f)) \neq h(\phi(x_2, \tilde{u}; f))$. A reachable and observable realization is called *minimal*. A weakly reachable and observable C^ω realization is called C^ω*-minimal*. As for linear systems it is important to have a notion of 'equivalence' of two realizations (M, f, h, x_0) and (M', f', h', x_0'). We say that these realizations are C^k*-diffeomorphic* if there is a C^k-diffeomorphism $\chi : M \to M'$ such that $f' = (D\chi \,.\, f) \circ \chi^{-1}$, $h' = h \circ \chi^{-1}$, $x_0' = \chi(x_0)$. Note that, in the case of linear systems $(\mathbb{R}^n, A, B, C, x_0)$, $(\mathbb{R}^m, A', B', C', x_0')$, this reduces to the usual equivalence, since then $\chi : \mathbb{R}^n \to \mathbb{R}^m$ is a linear map which is an isomorphism with matrix representation P, so $n = m$ and $D\chi = P$, so that

$$(D\chi \cdot f) \circ \chi^{-1}(x) = PAP^{-1}x + PBu$$

and

$$C' = h' = h \circ \chi^{-1} = CP^{-1}.$$

Hence

$$A' = PAP^{-1}, \quad B' = PB, \quad C' = CP^{-1}, \quad x_0' = Px_0.$$

In order to prove the existence and uniqueness of realizations of a system it is convenient to define the notion of an abstract system. If G is a group, let $\mathbb{R} \times G \to G$ be a surjective map which we write $(t, g) \to tg$. Then if $S : G \to \mathbb{R}^r$ is a map we call $(G, \mathbb{R}^r, S)$ an *abstract system*. A C^k *representation* $(k = 1, \ldots, \infty, \omega)$ of the system $(G, \mathbb{R}^k, S)$ is a quadruple $(M, \{\phi_g\}_{g \in G}, h, x_0)$, where

(a) M is a C^k Hausdorff manifold with a distinguished point $x_0 \in M$.
(b) ϕ_g, $g \in G$ are C^k diffeomorphisms of M, such that

$\phi_{g_1 g_2} = \phi_{g_1}\phi_{g_2}$, $g_1, g_2 \in G$ and the map $\psi_{\mathbf{g}} : \mathbb{R}^p \to M$ defined by

$$\psi_{\mathbf{g}}(\mathbf{t}) = \phi_{(t_p g_p) \ldots (t_1 g_1)}(x_0)$$

is of class C^k for any sequence $\mathbf{g} = (t_p g_p), \ldots, (t_1 g_1)$.
(c) $h : M \to \mathbb{R}^r$ is of class C^k.
(d) $S(g) = h(\phi_g(x_0))$ for all $g \in G$.

The representation is *transitive* if for any $x_1, x_2 \in M$ there exists $g \in G$ such that $\phi_g(x_1) = x_2$ and *distinguishable* if for any $x_1, x_2 \in M$, $x_1 \neq x_2$, there exists $g \in G$ such that $h(\phi_g(x_1)) \neq h(\phi_g(x_2))$. A representation is *minimal* if it is transitive and distinguishable. Two representations $(M, \{\phi_g\}, h, x_0)$ and $(M', \{\phi_g'\}, h', x_0')$ are C^k*-diffeomorphic* if there is a C^k diffeomorphism $\chi : M \to M'$ such that $\phi_g' \circ \chi = \chi \circ \phi_g$ for all $g \in G$, $h = h' \circ \chi$ and $x_0' = \chi(x_0)$.

2.3.2 Solution of the realization problem

We can now prove

Theorem 2.3.1
Every C^k-smooth $(k = 1, 2, \ldots, \infty, \omega)$ abstract system $(G, \mathbb{R}^r, S)$ with finite rank has a minimal C^k representation $(M, \{\phi_g\}, h, x_0)$ with dim $M =$ rank S, and any two minimal representations are C^k-diffeomorphic.

Proof
First we factor out from G all control sequences in G which are indistin-

guishable at the output, i.e. we introduce the equivalence relation

$$g_1 \sim g_2 \Leftrightarrow S(\gamma g_1) = S(\gamma g_2) \qquad \text{for all } g \in G$$

on G and define (set theoretically) the state space

$$M = G/\sim.$$

Writing $[g]$ for the equivalence class of $g \in G$ we define

$$\phi_g(x) = [g\gamma], \qquad h(x) = S(\gamma), \qquad x_0 = [e],$$

where $x = [\gamma]$. Then, if $x = [\gamma]$,

$$\phi_{g_1 g_2}(x) = [g_1 g_2 \gamma] = \phi_{g_1}[g_2\gamma] = \phi_{g_1}\phi_{g_2}[\gamma] = \phi_{g_1}\phi_{g_2}(x)$$

and

$$S(g) = h([g]) = h([ge]) = h(\phi_g(x_0)),$$

so that conditions (b) and (d) above hold. Moreover, the representation $(M, \{\phi_g\}, h, x_0)$ is minimal since if $x_1 = [\gamma_1]$, $x_2 = [\gamma_2]$ and we define $g = \gamma_2\gamma_1^{-1}$, then we have

$$\phi_g(x_1) = [g\gamma_1] = [\gamma_2\gamma_1^{-1}\gamma_1] = [\gamma_2] = x_2$$

and if $x_1 = [\gamma_1] \neq x_2 = [\gamma_2]$, then $S(g\gamma_1) \neq S(g\gamma_2)$ for some $g \in G$ (by definition of $\sim$) and so $h([g\gamma_1]) \neq h([g\gamma_2])$, i.e. $h(\phi_g(x_1)) \neq h(\phi_g(x_2))$. Note also that the mappings ϕ_g are invertible, since

$$\phi_g \circ \phi_{g^{-1}}(x) = [gg^{-1}\gamma] = x = [g^{-1}g\gamma] = \phi_{g^{-1}} \circ \phi_g(x)$$

where $x = [\gamma]$.

We next define a topology on M so that ϕ_g is a homeomorphism for all $g \in G$ and h is continuous. Define, as before, the functions $\psi_{\mathbf{g}} : \mathbb{R}^p \to M$ and $\psi^{\boldsymbol{\gamma}} : M \to \mathbb{R}^{mr}$ by

$$\psi_{\mathbf{g}}(\mathbf{t}) = \phi_{(t_p g_p)\dots(t_1 g_1)}(x_0)$$

and

$$\psi^{\boldsymbol{\gamma}}(\mathbf{t}) = (\psi^{\gamma_1}, \dots, \gamma^{\gamma_m}),$$

where

$$\psi^{\gamma_i}([g]) = h(\phi_{\gamma_i}([g])) = S(\gamma_i g).$$

Clearly, $\psi^{\boldsymbol{\gamma}} \circ \psi_{\mathbf{g}} = \psi_{\mathbf{g}}^{\boldsymbol{\gamma}}$, where $\psi_{\mathbf{g}}^{\boldsymbol{\gamma}}$ is as defined above. The topology which we define on M should make the functions $\psi_{\mathbf{g}}$ continuous for any sequence $\mathbf{g}$ and so we put on M the finest topology such that all these functions are continuous, so that $V \subseteq M$ is open if and only if $\psi_{\mathbf{g}}^{-1}(V)$ is open for all $\mathbf{g}$. Since $\psi_{\mathbf{g}}^{\boldsymbol{\gamma}} = \psi^{\boldsymbol{\gamma}} \circ \psi_{\mathbf{g}}$ is continuous for any $\boldsymbol{\gamma}$, $\mathbf{g}$ by definition of a C^k system, it follows that the functions $\psi^{\boldsymbol{\gamma}} : M \to \mathbb{R}^{mr}$ are continuous. To show that the functions $\phi_g : M \to M$ are continuous, note that if $\mathbf{g} = (g_1, \dots, g_p)$,

$\mathbf{t}=(t_1,\ldots,t_p)$, $\mathbf{g}'=(g_1,\ldots,g_p,g')$, $\mathbf{t}=(t_1,\ldots,t_p,t')$ and $g=(t'g')$, then

$$\phi_g\circ\psi_{\mathbf{g}}(\mathbf{t}')=\phi_g\circ\phi_{(t_pg_p)\ldots(t_1g_1)}(x_0)=\phi_{(t'g')(t_pg_p)\ldots(t_1g_1)}(x_0)$$
$$=\psi_{\mathbf{g}'}(t')$$

and so if $V\subseteq M$ is open in M and $W=\phi_g^{-1}(V)$, then $\psi_{\mathbf{g}}^{-1}(W)\subseteq\mathbb{R}^p$ may be identified with $\psi_{\mathbf{g}'}^{-1}(V)\cap(\mathbb{R}^p\times\{t'\})$, which is open in $\mathbb{R}^p\times\{t'\}$. However, if $\psi_{\mathbf{g}}^{-1}(W)$ is open then W is open by definition of the topology on M, i.e. ϕ_g is continuous for any g. The inverse ϕ_g^{-1} is continuous since it equals ϕ_{g-1}.

M is a Hausdorff space since if $x_1,x_2\in M$ with $x_1\neq x_2$, then there exists $g\in G$ such that $\psi^g(x_1)\neq\psi^g(x_2)$ and we can use the continuity of ψ^g.

The next task is to show that M has an n-dimensional manifold structure, by demonstrating that each point $x\in M$ has a neighbourhood homeomorphic to an open subset of $\mathbb{R}^n$. We can clearly do this for a single point $x\in M$ by the transitivity and homeomorphic nature of the maps $\{\phi_g\}_{g\in G}$. By assumption, we can find $\mathbf{g},\boldsymbol{\gamma},\mathbf{t}_0$ such that

$$\text{rank } D\psi_{\mathbf{g}}^{\gamma}(\mathbf{t}_0)=n=\text{rank } S.$$

It follows that $\mathbf{t}_0$ has at least n independent variables. Hence, if $\mathbf{t}=(t_1,\ldots,t_p)$ we shall fix $p-n$ of these variables equal to the corresponding values of $\mathbf{t}_0$ and denote the remaining variables by $\boldsymbol{\tau}=(t_{i_1},\ldots,t_{i_n})$. For such a sequence $\boldsymbol{\tau}$, $\mathbf{t}(\boldsymbol{\tau})$ denotes the sequence $(t_1,\ldots,t_p)$ with the $p-n$ variables not in $\boldsymbol{\tau}$ set to the corresponding $p-n$ variables of $\mathbf{t}_0$ and we choose $\boldsymbol{\tau}$ so that $D_{\tau}\psi_{\mathbf{g}}^{\gamma}(\mathbf{t}_0)$ has full rank. Then we can define the functions $\tilde{\psi}_{\mathbf{g}}:\mathbb{R}^n\to M$ and $\tilde{\psi}_{\mathbf{g}}^{\gamma}=\psi^{\gamma}\circ\tilde{\psi}_{\mathbf{g}}:\mathbb{R}^n\to\mathbb{R}^{rm}$, where

$$\tilde{\psi}_{\mathbf{g}}(\boldsymbol{\tau})=\psi_{\mathbf{g}}(\mathbf{t}(\boldsymbol{\tau})).$$

Thus, $\tilde{\psi}_{\mathbf{g}}^{\gamma}$ is of class C^k and rank $D\tilde{\psi}_{\mathbf{g}}^{\gamma}(\boldsymbol{\tau}_0)=n$, and so there is a neighbourhood $V\subseteq\mathbb{R}^n$ of $\boldsymbol{\tau}_0$ such that $D\tilde{\psi}_{\mathbf{g}}^{\gamma}$ has full rank on V. Hence the set $N=\tilde{\psi}_{\mathbf{g}}^{\gamma}(V)$ is an n-dimensional submanifold of $\mathbb{R}^{rm}$ of class C^k and $\tilde{\psi}_{\mathbf{g}}^{\gamma}\mid V: V\to N$ is a C^k-diffeomorphism from V onto N.

A chart on M will be defined as the inverse of $\psi=\tilde{\psi}_{\mathbf{g}}\mid V: V\to M$. It is injective because $\tilde{\psi}_{\mathbf{g}}^{\gamma}\mid V=\psi^{\gamma}\circ\psi$ is injective and it is continuous since $\psi_{\mathbf{g}}$ is continuous (by definition). We show that ψ is a homeomorphism. To do this we prove that $\psi(V)=W$ is open in M. If W is not open, then there exist $q\geqslant 1$ and $\mathbf{c}=(c_1,\ldots,c_q)\in G\times\cdots\times G$ such that $\psi_{\mathbf{c}}^{-1}(W)$ is not open in $\mathbb{R}^q$, i.e. there exists $\mathbf{s}'\in V_1=\psi_{\mathbf{c}}^{-1}(W)$ such that $\mathbf{s}'\notin \text{int } V_1$. Let $c=(s_q'c_q)\cdots(s_1'c_1)$. Then $c^{-1}=(s'c')$ for some $s'\in\mathbb{R}$, $c'\in G$ since the map $\mathbb{R}\times G\to G$ is assumed to be surjective. If $\mathbf{g}c'\mathbf{c}=(g_1,\ldots,g_p,c',c_1,\ldots,c_q)$, then we define the map

$$\tilde{\psi}_{\mathbf{g}c'\mathbf{c}}(\boldsymbol{\tau},\mathbf{s})=\psi_{\mathbf{g}c'\mathbf{c}}(\mathbf{t}(\boldsymbol{\tau}),s',\mathbf{s}).$$

Since $\tilde{\psi}_{\mathrm{g}}(V) = W$ and $\mathbf{s}' \in \psi_{\mathrm{c}}^{-1}(W)$, there exists a point $\boldsymbol{\tau}' \in \mathbb{R}^n$ such that

$$\tilde{\psi}_{\mathrm{g}}(\boldsymbol{\tau}') = \psi_{\mathrm{c}}(\mathbf{s}') = \phi_c(x_0)$$

(the latter by definition) and so

$$\phi_{(s'c')}(\tilde{\psi}_{\mathrm{g}}(\boldsymbol{\tau}')) = \phi_{c^{-1}}(\tilde{\psi}_{\mathrm{g}}(\boldsymbol{\tau}')) = x_0.$$

Hence

$$\begin{aligned}\psi_{\mathrm{c}}(\mathbf{s}) &= \phi_{(s_q c_q)\dots(s_1 c_1)}(x_0) = \phi_{(s_q c_q)\dots(s_1 c_1)}\phi_{(s'c')}(\tilde{\psi}_{\mathrm{g}}(\boldsymbol{\tau}'))\\ &= \phi_{(s_q c_q)\dots(s_1 c_1)}\phi_{(s'c')}\psi_{\mathrm{g}}(\mathbf{t}(\boldsymbol{\tau}'))\\ &= \psi_{\mathrm{g}c'\mathrm{c}}(\mathbf{t}(\boldsymbol{\tau}'), s', \mathbf{s})\\ &= \tilde{\psi}_{\mathrm{g}c'\mathrm{c}}(\boldsymbol{\tau}', \mathbf{s}).\end{aligned} \tag{2.3.4}$$

and similarly,

$$\tilde{\psi}_{\mathrm{g}c'\mathrm{c}}(\boldsymbol{\tau}, \mathbf{s}') = \tilde{\psi}_{\mathrm{g}}(\boldsymbol{\tau}), \tag{2.3.5}$$

since $(s_q'c_q)\cdots(s_1'c_1)(s'c') = e$. Thus, since

$$\operatorname{rank} D\tilde{\psi}_{\mathrm{g}}^{\gamma}(\boldsymbol{\tau}') = n \tag{2.3.6}$$

and

$$\tilde{\psi}_{\mathrm{g}}^{\gamma} = \psi^{\gamma} \circ \tilde{\psi}_{\mathrm{g}}, \qquad \tilde{\psi}_{\mathrm{g}c'(\mathrm{c})}^{\gamma} = \psi^{\gamma} \circ \tilde{\psi}_{\mathrm{g}c'\mathrm{c}}, \tag{2.3.7}$$

it follows that $\operatorname{rank} D\tilde{\psi}_{\mathrm{g}c'\mathrm{c}}^{\gamma}(\boldsymbol{\tau}, \mathbf{s}) = n$ on a neighbourhood $Q \subseteq \mathbb{R}^{n+q}$ of $(\boldsymbol{\tau}', \mathbf{s}')$. From the general theory of submanifolds in Chapter 1, it follows that the level sets

$$\{(\boldsymbol{\tau}, \mathbf{s}) \in Q : \tilde{\psi}_{\mathrm{g}c'\mathrm{c}}(\boldsymbol{\tau}, \mathbf{s}) = \text{constant}\}$$

are submanifolds of $Q \subseteq \mathbb{R}^{n+q}$ of dimension q. By the implicit function theorem, (2.3.6) and (2.3.7) it can easily be seen that all the level submanifolds close to the one through $(\boldsymbol{\tau}', \mathbf{s}')$ intersect $V \times \{\mathbf{s}'\}$ transversally, i.e. the tangent space of each submanifold and that of $V \times \{\mathbf{s}'\}$ span $\mathbb{R}^{n+q}$.

Since $\mathbf{s}' \notin \operatorname{int} V_1$ and $V_1 = \psi_{\mathrm{c}}^{-1}(W)$ it follows that we can choose $\mathbf{s}'' \in \mathbb{R}^q$ close to $\mathbf{s}'$ so that $\psi_{\mathrm{c}}(\mathbf{s}'') \in W$ and the level submanifold through $(\boldsymbol{\tau}', \mathbf{s}'')$ intersects $V \times \{\mathbf{s}'\}$ at $(\boldsymbol{\tau}'', \mathbf{s}')$ (see Fig. 2.7). We can also ensure that $(\boldsymbol{\tau}'', \mathbf{s}')$ and $(\boldsymbol{\tau}', \mathbf{s}'')$ lie on the same connected component of a level submanifold. Hence, using (2.3.4),

$$x_1 \triangleq \tilde{\psi}_{\mathrm{g}c'\mathrm{c}}(\boldsymbol{\tau}', \mathbf{s}'') = \psi_{\mathrm{c}}(\mathbf{s}'') \notin W$$

and, using (2.3.5),

$$x_2 \triangleq \tilde{\psi}_{\mathrm{g}c'\mathrm{c}}(\boldsymbol{\tau}'', \mathbf{s}') = \psi_{\mathrm{g}}(\boldsymbol{\tau}'') \in W$$

and so $x_1 \neq x_2$.

Since the representation $(M, \{\phi_g\}, h, x_0)$ is distinguishable, there exists

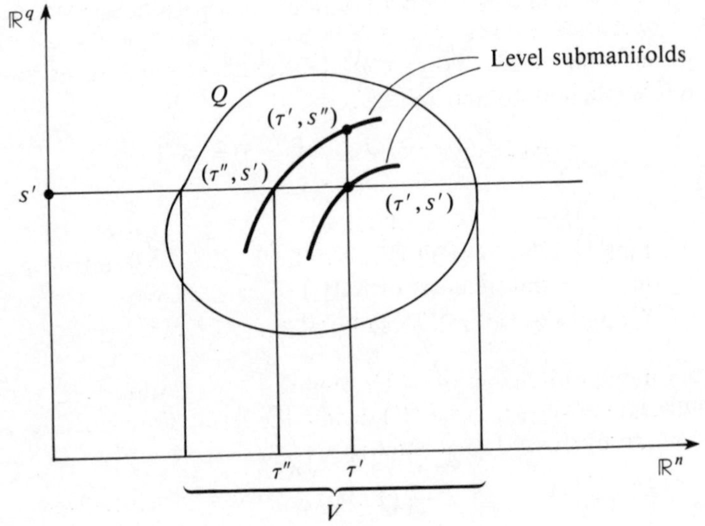

Fig. 2.7 Level submanifolds of $\tilde{\psi}$

$\gamma \in G$ such that

$$h \circ \phi_\gamma(x_1) \neq h \circ \phi_\gamma(x_2), \qquad \text{i.e.} \qquad \psi^\gamma(x_1) \neq \psi^\gamma(x_2). \tag{2.3.8}$$

Now, $(\boldsymbol{\tau}'', \mathbf{s}')$ and $(\boldsymbol{\tau}', \mathbf{s}'')$ lie on a level submanifold, so

$$\tilde{\psi}^\gamma_{\mathrm{gc'c}}(\boldsymbol{\tau}'', \mathbf{s}') = \tilde{\psi}^\gamma_{\mathrm{gc'c}}(\boldsymbol{\tau}', \mathbf{s}''). \tag{2.3.9}$$

but, by (2.3.8),

$$\tilde{\psi}^{(\gamma,\gamma)}_{\mathrm{gc'c}}(\boldsymbol{\tau}'', \mathbf{s}') \neq \tilde{\psi}^{(\gamma,\gamma)}_{\mathrm{gc'c}}(\boldsymbol{\tau}', \mathbf{s}''). \tag{2.3.10}$$

To complete the proof that W is open (and hence that ψ^{-1} is a chart) we apply the following lemma to show that (2.3.9) and (2.3.10) are contradictory.

Lemma 2.3.2
If f: $W \to \mathbb{R}^k$ and $\tilde{f}$: $W \to \mathbb{R}^{k+l}$ are C^1 maps, where $W \subseteq \mathbb{R}^m$ is open and $\tilde{f} = (f, f')$ for some function f' and if

$$\operatorname{rank} D\tilde{f}(x) = \operatorname{rank} Df(x) = \text{const.} = n$$

on W, then the connected components of the level submanifolds of f and $\tilde{f}$ coincide, i.e. if $f(x') = f(x'')$ with x', x'' in the same connected component of $N \triangleq \{x \in W: f(x) = f(x')\}$, then $\tilde{f}(x') = \tilde{f}(x'')$.

Proof
It is clearly sufficient to show that the set

$$\tilde{N} = \{x \in N: \tilde{f}(x) = \tilde{f}(x')\}$$

is open and closed in N. The latter property follows from the continuity of $\tilde{f}$. To show that $\tilde{N}$ is open let $x'' \in N$ be such that $\tilde{f}(x'') = \tilde{f}(x')$. Since Df has constant rank we can choose coordinates in a neighbourhood of x'' so that, in this neighbourhood, $Df = (f_{x_1}, \ldots, f_{x_n}, 0, \ldots, 0)$,

$$D\tilde{f} = \begin{bmatrix} f_{x_1} & \cdots & f_{x_n} & 0 & \cdots & 0 \\ f'_{x_1} & & f'_{x_n} & f'_{x_{n+1}} & \cdots & f'_{x_m} \end{bmatrix}$$

and locally, $N = \{x : x_i = x''_i, i = 1, \ldots, n\}$.

Since rank $D\tilde{f}(x) =$ rank $Df(x)$, it follows that $f'_{x_i} = 0$ for $i > n$ so that the functions $\tilde{f}$ are independent of $(x_{n+1}, \ldots, x_m)$. However, the equation $\tilde{f}(x'') = \tilde{f}(x')$ and the form of N imply that $\tilde{f}(x) = \tilde{f}(x')$ for $x \in N$. □

Continuing with the proof of Theorem 2.3.1 we next put a C^k-differentiable structure $\mathcal{F}$ on M. Let $F(V)$ denote the set of real-valued functions defined on an open set $V \subseteq M$ and let

$$\mathcal{F} = \{\alpha \in F(V_\alpha) : \alpha \circ \psi_{\mathbf{g}} \text{ is of class } C^k \text{ for any sequence } \mathbf{g}\}.$$

To prove that $\mathcal{F}$ is a C^k differentiable structure on M we must show

(a) if $\alpha \in \mathcal{F}$ and $V \subseteq V_\alpha$, then $\alpha \mid V \in \mathcal{F}$
(b) if $V_\alpha = \bigcup_\beta V_\beta$ and $\alpha \mid V_\beta \in \mathcal{F}$ for all β, then $\alpha \in \mathcal{F}$
(c) for any $x \in M$ there are open sets $x \in W \subseteq M$ and $V \subseteq \mathbb{R}^n$ and a homeomorphism $p : W \to V$ such that for each open subset $V_1 \subseteq W$ and any function $\alpha : V_1 \to \mathbb{R}$ we have $\alpha \in \mathcal{F}$ if and only if $\alpha \circ p^{-1}$ is of class C^k.

The only nontrivial part to prove is the implication $\alpha \circ p^{-1} \in C^k \Rightarrow \alpha \in \mathcal{F}$. Since local charts for M have been defined by the maps $\tilde{\psi}_{\mathbf{g}}^{-1}$ for any sequence $\mathbf{g}$, we must show that if $\alpha \circ \tilde{\psi}_{\mathbf{g}}^{-1} \in C^k$ then $\alpha \in \mathcal{F}$. Now

$$\alpha \circ \psi_{\mathbf{g}} = \alpha \circ \tilde{\psi_{\mathbf{g}}} \circ \tilde{\psi}_{\mathbf{g}}^{-1} \circ \psi_{\mathbf{g}}$$

and $\alpha \circ \tilde{\psi}_{\mathbf{g}} \in C^k$. Also $\tilde{\psi}_{\mathbf{g}}^{-1} \circ \psi_{\mathbf{g}} \in C^k$ since $\tilde{\psi}_{\mathbf{g}}^{-1} \circ \psi_{\mathbf{g}} = \tilde{\psi}_{\mathbf{g}}^{-1} \circ (\psi^\gamma)^{-1} \circ \psi^\gamma \circ \psi_{\mathbf{g}} = (\tilde{\psi}_{\mathbf{g}}^\gamma)^{-1} \circ \psi_{\mathbf{g}}^\gamma$ and the latter function is in C^k by definition of a C^k system. Hence $\alpha \circ \psi_{\mathbf{g}} \in C^k$ and so $\alpha \in \mathcal{F}$.

We must also show that ϕ_g, $g \in G$ and h are C^k-smooth. For ϕ_g, let $\alpha : M \to \mathbb{R}$ be C^k; then we show $\alpha \circ \phi_g \in C^k$. Now $\alpha \circ \psi_{\mathbf{g}} \in C^k$ for any $\mathbf{g}$ and we can write $\phi_g \circ \psi_{\mathbf{g}}(t) = \psi_{\mathbf{g}'}(\mathbf{t}')$ where $\mathbf{g}' = (g_1, \ldots, g_p, g')$, $\mathbf{t}' = (t_1, \ldots, t_p, t')$ and $g = (t' g')$ (since $\mathbb{R} \times G \to G$ is surjective). Hence, $\alpha \circ \phi_g \circ \psi_{\mathbf{g}} = \alpha \circ \psi_{\mathbf{g}'} \in C^k$ and so $\alpha \circ \phi_g \in C^k$. Since $D\phi_g(x) \circ D\phi_{g^{-1}}(\phi_g x) = Id$, we have the required rank condition rank $D\phi_g(x) = n$.

Since $h \circ \psi_{\mathbf{g}} = \psi_{\mathbf{g}}^e$ and the maps $\psi_{\mathbf{g}}$ (by definition of the C^k structure) and $\psi_{\mathbf{g}}^e$ (by assumption) are C^k it follows that h is C^k.

To complete the proof of Theorem 2.3.1 we show that any minimal C^k representation $(M', \{\phi'_g\}, h', x'_0)$ of the system $(G, \mathbb{R}^r, S)$ is C^k-diffeomorphic to the 'canonical' representation $(M, \{\phi_g\}, h, x_0)$ defined in the first part of the proof.

Define a function $\chi: M \to M'$ by

$$\chi(\phi_g(x_0)) = \phi'_g(x'_0).$$

To show that χ is well-defined note that if $\phi_{g_1}(x_0) = \phi_{g_2}(x_0)$ and $\phi'_{g_1}(x'_0) \neq \phi'_{g_2}(x'_0)$ then since the representation of M' is distinguishable, there is a $g \in G$ such that $h'(\phi'_g \circ \phi'_{g_1}(x'_0)) \neq h'(\phi'_g \circ \phi'_{g_2}(x'_0))$. But $h'(\phi_\gamma(x'_0)) = h(\phi_\gamma(x_0)) = S(\gamma)$ for any $\gamma \in G$ and so $h(\phi_g \circ \phi_{g_1}(x_0)) \neq h(\phi_g \circ \phi_{g_2}(x_0))$, and $\phi_{g_1}(x_0) \neq \phi_{g_2}(x_0)$, which is a contradiction. Similarly, if $\chi(\phi_{g_1}(x_0)) = \chi(\phi_{g_2}(x_0))$, then

$$h(\phi_g \circ \phi_{g_1}(x_0)) = h'(\phi'_g \circ \phi'_{g_1}(x'_0)) = h'(\phi'_g \circ \phi'_{g_2}(x'_0)) = h(\phi_g \circ \phi_{g_2}(x_0))$$

for any $g \in G$ and by distinguishability, $\phi_{g_1}(x_0) = \phi_{g_2}(x_0)$, i.e. χ is an injection. If $x' \in M'$ then, by transitivity, there exists g such that $\phi'_g(x'_0) = x'$. Hence $\chi(\phi_g(x_0)) = \phi'_g(x'_0) = x'$ and so χ is surjective.

To show the 'equivalence' of the M and M' representations, note that

$$\phi'_{g_1} \circ \chi(\phi_{g_2}(x_0)) = \phi'_{g_1} \circ \phi'_{g_2}(x'_0) = \phi'_{g_1 g_2}(x'_0) = \chi(\phi_{g_1 g_2}(x_0)) = \chi \circ \phi_{g_1}(\phi_{g_2}(x_0)),$$

and

$$h' \circ \chi(\phi_g(x_0)) = h' \circ \phi'_g(x'_0) = S(g) = h(\phi_g(x_0)),$$

so that

$$\phi'_g \circ \chi = \chi \circ \phi_g, \qquad h = h' \circ \chi, \chi(x_0) = \chi(\phi_e(x_0)) = \phi'_e(x'_0) = x'_0.$$

To prove that $\chi \in C^k$, let $\alpha: M' \to \mathbb{R}$ be C^k. Then we must show that $\alpha \circ \chi \in C^k$ or, by definition, $\alpha \circ \chi \circ \psi_{\mathbf{g}} \in C^k$ for any sequence $\mathbf{g}$. However,

$$\begin{aligned} \alpha \circ \chi(\psi_{\mathbf{g}}(\mathbf{t})) &= \alpha \circ \chi(\phi_{(t_p g_p) \dots (t_1 g_1)}(x_0)) \\ &= \alpha \circ (\phi'_{(t_p g_p) \dots (t_1 g_1)}(x'_0)) \end{aligned}$$

and the last function is C^k since $(M', \{\phi'_g\}, h', x'_0)$ is C^k.

Finally, $D\chi$ is nonsingular since $\psi^{\gamma}_{\mathbf{g}} = \psi'^{\gamma} \circ \chi \circ \psi_{\mathbf{g}}$ and since for any $x \in M$ there exist $\mathbf{g}, \boldsymbol{\gamma}, \mathbf{t}$ such that rank $D\psi^{\gamma}_{\mathbf{g}}(\mathbf{t}) = n$ and $\psi_{\mathbf{g}}(\mathbf{t}) = x$. Then use the C^k smoothness of $\psi'^{\gamma}, \chi, \psi_{\mathbf{g}}$ to show that rank $D_\chi(x) = n$. The proof of Theorem 2.3.1 is complete. □

Note that we may replace the output space $\mathbb{R}^r$ by any C^k (finite-dimensional) manifold, since we do not use the linear structure of $\mathbb{R}^r$ in any essential way.

Now return to a system $(\mathscr{U}, \mathscr{Y}, S)$ of the form defined above. By extending the system from $\mathscr{U}_{\text{pc}}$ to the group $G\mathscr{U}_{\text{pc}}$ and using methods similar to those in Theorem 2.3.1, we can prove

Theorem 2.3.3

(a) If $(\mathscr{U}, \mathscr{Y}, S)$ is a C^k-smooth ($k = 2, \dots, \infty$) time-invertible system with finite rank, then this system has a minimal, symmetric, C^k realization (M, f, h, x_0) with dim $M = \text{rank } S$. Moreover, any two minimal C^k realizations of the same input–output system are C^k-diffeomorphic.

(b) Every C^ω system $(\mathcal{U}, \mathcal{Y}, S)$ with finite rank has a C^ω-minimal realization such that dim $M = \text{rank } S$, and any two such realizations are C^ω-diffeomorphic.

(Details can be found in Jakubczyk (1980)) □

2.3.3 Systems with state-dependent input spaces

Consider the smooth nonlinear system

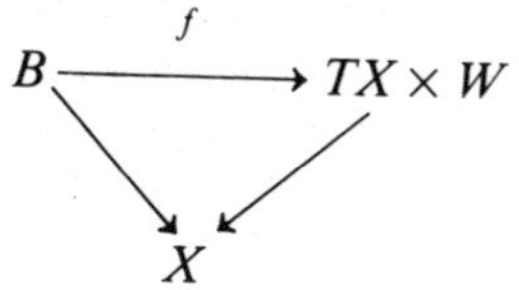

(2.3.11)

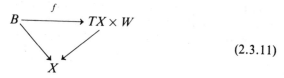

which we denote by $\Sigma(X, W, B, f)$. In defining minimality for linear systems we can partition the 'A, B, C' matrices of the system into controllable and observable parts and project the dynamics onto a smaller space so that the uncontrollable and unobservable parts are factored out. To generalize this idea to systems of the form (2.3.11) we require that the projections be submersions and so we say that the systems $\Sigma = \Sigma(X, W, B, f)$ and $\Sigma' = \Sigma'(X', W, B', f')$ satisfy the relation $\Sigma' \leqslant \Sigma$ if there exist surjective submersions $\phi: X \to X'$, $\Phi: B \to B'$ such that the diagram (2.3.12) commutes.

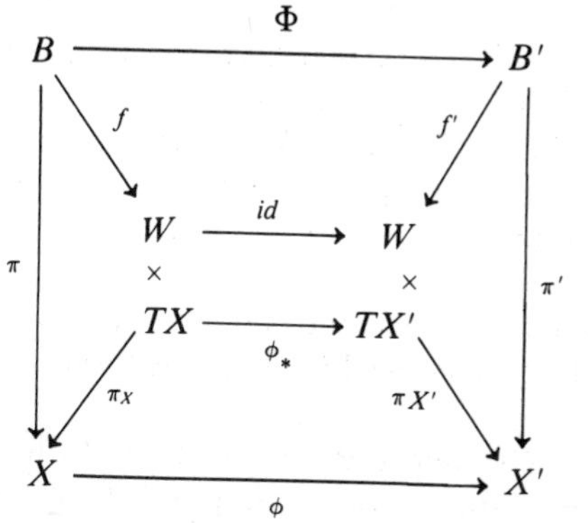

(2.3.12)

We say that Σ and Σ' are *equivalent* and we write $\Sigma \sim \Sigma'$ if ϕ and Φ are diffeomorphisms. Σ is called *minimal* if $\Sigma' \leqslant \Sigma$ implies that $\Sigma' \sim \Sigma$.

To describe the diagram (2.3.12) locally note that the submersion

Φ generates an involutive distribution D on B given by

$$D = \{ Z \in TB : \Phi_* Z = 0).$$

The vector fields in D are just the vertical vector fields to the submanifolds $\Phi^{-1}(c)$ of B, where c is constant. Similarly, ϕ induces an involutive distribution E on X. Then we have

Lemma 2.3.4
$\Sigma \sim \Sigma'$ is locally equivalent to the conditions

(a) $D \subseteq \ker dh$
(b) $\pi_* D = E$
(c) $g_* D \subseteq TE = T\pi_*(D)$

where $f = (g, h)$ and $f' = (g', h')$.

Proof
(a) and (b) follow directly from the subdiagrams

$$\begin{array}{ccc} B & \xrightarrow{\Phi} & B' \\ {\scriptstyle h}\downarrow & & \downarrow{\scriptstyle h'} \\ W & \xrightarrow{id} & W \end{array} \qquad \begin{array}{ccc} B & \xrightarrow{\Phi} & B' \\ {\scriptstyle \pi}\downarrow & & \downarrow{\scriptstyle \pi'} \\ X & \xrightarrow{\phi} & X' \end{array}$$

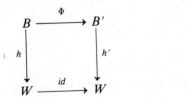

of (2.3.12). For (c) we have the subdiagram

$$\begin{array}{ccc} B & \xrightarrow{\Phi} & B' \\ {\scriptstyle g}\downarrow & & \downarrow{\scriptstyle g'} \\ TX & \xrightarrow{\phi_*} & TX' \end{array}$$

and ϕ_* induces the distribution TE on TX. □

Now suppose that

$$\dot{x} = f(x) + \sum_{i=1}^{m} u_i g_i(x)$$

is a local representation of a system on a manifold M. We can regard this as an affine distribution on M, i.e. for each $x \in M$ the right-hand side defines an affine subspace of $T_x M$ as $u = (u_1, \ldots, u_m)$ varies in the control space. Let

$$\Delta = \left\{ \mathscr{A} : \mathscr{A}(x) = f(x) + \sum_{i=1}^{m} u_i g_i(x) \in T_x M,\ x \in M,\ u_i \in U \right\}$$

and define

$$\Delta_0 = \Delta - \Delta = \{ Y - Z : Y, Z \in \Delta \}.$$

Clearly, Δ_0 is a distribution on M such that

$$\Delta_0(x) = \text{span}\{ g_1(x), \ldots, g_m(x) \}.$$

Moreover we introduce the set $\mathscr{D}$ of involutive distributions on M and define

$$A(D_0) = \{ D \in \mathscr{D} : D + \Delta_0 \in \mathscr{D} \}.$$

We shall prove the next result in Section 2.4.

Theorem 2.3.5
Let $D \in A(D_0)$ and suppose that $D \cap \Delta_0$ has constant dimension. Then the condition

$$[\Delta, D] \subseteq D + \Delta_0$$

(i.e. D is $\Delta (\text{mod}\, \Delta_0)$ invariant) is equivalent to the following two conditions:

(a) there exists $F \in \Delta$ such that $[F, D] \subseteq D$;
(b) there exist $B_i \in \Delta_0$ such that $\text{span}\{B_i\} = \Delta_0$ and $[B_i, D] \subseteq D$. □

We now define an 'infinitesimal version' of a system $\Sigma(X, W, B, f)$ in order to relate minimality of systems with invariant distributions. First let Δ_0 be the involutive distribution defined by the vertical vectors of B, i.e.

$$\Delta_0 = \{ Z \in TB : \pi_* Z = 0 \}.$$

(In the case where B is the trivial bundle $X \times U$ we have $\Delta_0 = TU$, since $T(X \times U) \cong TX \oplus TU$.) If $(x', v') \in B$ we define

$$\Delta(x', v') = \{ Z \in T_{(x', v')}B : \pi_* Z = g(x', v') \}$$

where $f = (g, h)$. Note that $\Delta(x', v')$ consists of all lifts of $g(x', v')$ in $T_{x'}X$ to $T_{(x', v')}B$. Clearly the map $\Delta : (x', v') \to \Delta(x', v')$ defines a distribution on B. If $Z \in \Delta$ is any element such that $\pi_* Z = g$, then we have $\Delta = Z + \Delta_0$ so that Δ is an affine distribution such that $\Delta - \Delta = \Delta_0$. The affine system (Δ, Δ_0) on B together with the output function $h : B \to W$ is called the *extended system and is denoted by* $\Sigma^e(X, W, B, f)$. In the trivial case with $B = X \times U$ the extended system has the form

$$\dot{x} = g(x, u)$$
$$\dot{u} = v$$

so that we specify the dynamics of the control u as well as x locally.

Lemma 2.3.6

(a) If D is an involutive distribution on B such that $D \cap \Delta_0$ has constant dimension, then $\pi_* D$ is an involutive distribution on X if and only if $D + \Delta_0$ is an involutive distribution.

(b) If D is an involutive distribution on B such that $D \cap \Delta_0$ has constant dimension then the following two conditions are equivalent:

(i) $\pi_* D$ is an involutive distribution on X, and $g_* D \subseteq T(\pi_* D)$

(ii) $[\Delta, D] \subseteq D + \Delta_0$.

Proof

(a) Since D and Δ_0 are involutive, it follows that $D + \Delta_0$ is involutive if and only if $[D, \Delta_0] \subseteq D + \Delta_0$. Hence by Theorem 2.3.5 we can find a basis $\{Z_1, \ldots, Z_k\}$ of D such that $[Z_i, \Delta_0) \subseteq \Delta_0$. If we choose local coordinates (x, u) in B then, since Δ_0 contains only derivatives in u, we can write $Z_i(x, u) = (Z_{ix}(x), Z_{iu}(x, u))$ where Z_{ix}, Z_{iu} are the x and u components of Z_i, respectively. Since π projects B onto X, π_* projects TB onto TX and so $\pi_* D = \text{span}\{Z_{1x}, \ldots, Z_{kx}\}$ which is a well-defined distribution, independent of u. Clearly $\pi_* D$ is involutive. The converse statement follows similarly.

(b) If (i) holds then $D = \{\partial/\partial x_1, \ldots, \partial/\partial x_k\}$ for some coordinates (x, u) in B. Recalling the definition of the differential it follows that the condition $g_* D \subseteq T(\pi_* D)$ is equivalent to the condition

$$\left(\frac{\partial g}{\partial x_i}\right)_j = 0, \qquad i = 1, \ldots, k, \quad j = k+1, \ldots, n,$$

where $n = \dim X$. Since Δ consists of lifts of g into TB and $\partial g/\partial x_i$ has zero components along $x_{k+1}, \ldots, x_n$ it follows that $[\Delta, D] \subseteq D + \Delta_0$. The converse follows from (a) by a similar argument. □

We can now define a concept of minimality for systems with state-dependent input spaces. The smooth system $\Sigma(X, W, B, f)$ is *locally minimal* at x if, given distributions D and E, defined in a neighbourhood of x, which satisfy the conditions of Lemma 2.3.4, we have $D = 0$ and $E = 0$. Note that we have defined minimality above by: $\Sigma' \leqslant \Sigma$ implies that $\Sigma' \sim \Sigma$. However, if this holds locally, then we must have $D = 0$ and $E = 0$, for otherwise we could factor D out locally and obtain a system Σ' with $\Sigma' < \Sigma$. Hence minimality of Σ locally implies local minimality of Σ. Lemmas 2.3.4 and 2.3.6 now imply

Theorem 2.3.7

$\Sigma(X, W, B, f = (g, h))$ is locally minimal if and only if the extended system $\Sigma^e(X, W, B, f)$ is such that there is no nonzero involutive distribution D on

B for which

(a) $[\Delta, D] \subseteq D + \Delta_0$
(b) $D \subseteq \ker\ dh.$ □

Nijmeijer (1981) gives a method for checking the local minimality of a system. Define

$\Delta^{-1}(\Delta_0 + D) = \{Z\colon Z$ is a vector field on B such that $[\Delta, Z] \subseteq \Delta_0 + D\}$.

Then we define a sequence $\{D^i\}$ by

$$D^0 = \ker\ dh \qquad D^i = D^{i-1} \cap \Delta^{-1}(\Delta_0 + D^{i-1}), \qquad i = 1, 2, \ldots \tag{2.3.13}$$

Clearly $\{D^i\}$ is a decreasing sequence of involutive distributions. Hence

$$D^k = D^i \qquad \text{for } i \geqslant k \tag{2.3.14}$$

where $k \leqslant \dim(\ker dh)$.

Corollary 2.3.8
$\Sigma(X, W, B, f)$ is locally minimal if and only if $D^k = 0$.

Proof
It follows from (2.3.13) and (2.3.14) that D^k is the maximal involutive distribution for which

$$D^k \subseteq \ker\ dh$$
$$[\Delta, D^k] \subseteq D^k + \Delta_0$$

and so by Theorem 2.3.7, $\Sigma(X, W, B, f)$ is locally minimal if and only if $D^k = 0$. □

We can now relate observability to minimality in systems for which $\partial h/\partial u$ is injective – as we saw in Section 2.1.2 this condition implies that there is a local input–output description of the system in the form

$$\dot{x} = g(x, u), \ y = h_1(x, u) \tag{2.3.15}$$

which is required to even consider observability. We shall say that the system with local description (2.3.15) is *locally distinguishable* if the existence of an involutive distribution E on X for which

(a) $[g(.\,, u), E] \subseteq E$ for all u
(b) $E \subseteq \ker\ d_x h_1(.\,, u)$ for all u

implies that $E = 0$. If the system (2.3.15) is locally minimal and such a distribution E exists, then we can lift E to B by defining a distribution D whose integral manifolds are contained in the sections $u =$ constant, locally.

Then $[\Delta, D] \subseteq D + \Delta_0$ and $D \subseteq \ker dh$, so that $D = 0$, $E = 0$, by definition. Hence local minimality implies local distinguishability. It is not difficult to prove that local minimality implies that the system (2.3.15) with h_1 independent of u (i.e. $y = h_1(x)$) is locally weakly observable in the sense defined in Section 2.2 (see Van der Schaft (1982a)).

In the case of controllability we can state the following result:

Theorem 2.3.9
If the fibres of B are connected, then the system $\Sigma(X, W, B, f = (g, h))$ is strongly accessible if and only if the extended system $\Sigma^{e}(X, W, B, f)$ is strongly accessible.

Proof
In local coordinates, we can write Σ and Σ^{e} in the forms

$$\Sigma : \dot{x} = g(x, u)$$
$$\Sigma^{e} : \dot{x} = g(x, v)$$
$$\dot{v} = u$$

If Σ^{e} is strongly accessible, then clearly so is Σ since the control does not have to be smooth.

For the converse note that if the fibres of B are connected, then the set of controls generated by the equation $\dot{v} = u$ is dense in the set of functions $L^2[0, T]$ and the result follows. □

2.4 INVERTIBILITY, INVARIANCE AND DECOUPLING OF NONLINEAR SYSTEMS

2.4.1 Invertibility of nonlinear control systems

Given a control system with an input–output map $S: \mathcal{U} \rightarrow \mathcal{Y}$, we say that the system is invertible if S is injective, so that given any output function we can, in principle, obtain the input which produced this output. In the case of linear systems of the form

$$\begin{aligned} \dot{x} &= Ax + bu, \qquad x(0) = x_0 \in \mathbb{R}^n, \\ y &= cx \end{aligned} \tag{2.4.1}$$

with a scalar control, we have the input–output map given by

$$y(t) = c\,\mathrm{e}^{At}x_0 + \int_0^t c\,\mathrm{e}^{A(t-s)}bu(s)\,\mathrm{d}s.$$

If two distinct controls $u_1 \neq u_2$ produce the same output function, then

$$\int_0^t c\,e^{A(t-s)}b(u_1(s)-u_2(s))\,ds = 0$$

for all $t \geqslant 0$. This can only hold if $c\,e^{At}b = 0$ and so the linear system (2.4.2) is invertible if and only if $cb, cAB, cA^2b, \ldots, cA^{n-1}b$ are not all zero (by the Cayley–Hamilton theorem). The *relative order* α is the least positive integer k such that $cA^{k-1}b \neq 0$ or ∞ if $cA^kb = 0$ for all $k \geqslant 0$. Hence, we have seen that $\alpha < \infty$ if and only if the system is invertible. Note that invertibility is independent of the initial value x_0 for linear systems, whereas it is not for general nonlinear systems as shown by Hirschorn (1979a). In this section we shall generalize the linear result to systems of the form

$$\begin{aligned} \dot{x}(t) &= A(x(t)) + u(t)B(x(t)), \qquad x(0) = x_0 \in M \\ y(t) &= c(x(t)) \end{aligned} \tag{2.4.2}$$

where M is a connected real analytic manifold, A and B are real analytic vector fields on M and $c: M \to \mathbb{R}$ is a real analytic map. Moreover, we shall assume that the controls $u(.)$ belong to the class $C^\omega([0,\infty))$ of real analytic functions from $[0,\infty)$ into $\mathbb{R}$. More generally, $C^\omega(M)$ will denote the ring of real analytic functions on M. We shall denote the solution of (2.4.2) by $x(t; u, x_0)$ and write $y(t; u, x_0) = c(x(t; u, x_0))$.

Since invertibility cannot be expected to be independent of x_0, in general, we shall say that the system (2.4.2) is *invertible* at $x_0 \in M$ if whenever $u_1, u_2 \in C^\omega$ are distinct controls, we have

$$y(t; u_1, x_0) \not\equiv y(t; u_2, x_0).$$

The system is called *strongly invertible* at x_0 if there is an open neighbourhood V of x_0 such that the system is invertible at x, for all $x \in V$. Finally, we shall say that the system is *strongly invertible* if it is strongly invertible on an open and dense submanifold of M.

Generalizing the above notion of relative order we say that the system (2.4.2) has *relative order* α if

(a) for $\alpha < \infty$, α is the least positive integer such that $(\mathrm{ad}_A^{\alpha-1}B)(c_i) \neq 0$
(b) for $\alpha = \infty$, $(\mathrm{ad}_A^kB)(c_i) = 0$ for all $k \geqslant 0, 0 \leqslant i \leqslant l$.

If $\alpha < \infty$, we denote the least integer j for which $(\mathrm{ad}_A^{\alpha-1}B)(c_j) \neq 0$ by i_α. In this definition, 0 denotes the zero function of $C^\omega(M)$. (Recall that the vector field ad_A^kB operates on c_i by

$$(\mathrm{ad}_A^kB)(c_i) = \sum_{j=1}^n a_j \frac{\partial c_i}{\partial x_j}$$

where $(\mathrm{ad}_A^kB) = \sum_{j=1}^n a_j \partial/\partial x_j$, locally.)

To derive invertibility results we need the notion of an inverse system.

For a linear system, since we have

$$Y(s) = G(s)U(s)$$

where G is the transfer function (in the Laplace domain), it follows that

$$U(s) = G^{-1}(s)Y(s)$$

and so, if the system is invertible, the inverse system is just $G^{-1}(s)$. However, since (assuming G is proper) G^{-1} is not proper, we must add derivatives of y to the inputs of the inverse system to ensure that the overall transfer function is proper.

Since we expect to add derivatives of y to the input of an inverse system, we shall say that the system

$$\begin{aligned} \dot{z} &= F(z) + vG(z), \qquad z(0) = z_0 \in N \\ w &= h(z) + vk(z) \end{aligned} \tag{2.4.3}$$

where N is a real analytic manifold, $F, G \in D(N)$ (analytic vector fields) and $h, k \in C^\omega(N)$, $v \in C^\omega([0, \infty))$, is a *left-inverse* system for (2.4.2) if for some $0 \leqslant i \leqslant l$,

$$w(\,.\,;\, y_i^{(\alpha)}, z_0) = u(\,.\,),$$

where $y_i^{(\alpha)}$ is the αth derivative of y_i. Hence, the left-inverse system produces the control $u(\,.\,)$ if the input v is chosen to be a derivative of some component of the output of (2.4.2).

Proposition 2.4.1
If a nonlinear system has relative order $\alpha < \infty$, then the set

$$M_\alpha = \{ x \in M\colon (\mathrm{ad}_A^{\alpha-1} B c_{i_\alpha})(x) \neq 0 \}$$

is open and dense (and hence a submanifold) in M.

Proof
Since, by definition, $(\mathrm{ad}_A^{\alpha-1} B)(c_{i_\alpha}) \neq 0$ and is an analytic function it cannot be zero on any open set and since this function is continuous, the result follows. □

M_α is called the *inverse submanifold* for the system (2.4.2).

Lemma 2.4.2
For $0 \leqslant k \leqslant \alpha - 2$ and $1 \leqslant i \leqslant l$ we have $(BA^k)c_i = 0$ and for all $x \in M_\alpha$ we have $(BA^{\alpha-1} c_{i_\alpha})(x) \neq 0$.

Proof
By definition,

$$(\mathrm{ad}_A^k B)c_i = 0 \qquad \text{for } 1 \leqslant i \leqslant l, \quad 0 \leqslant k \leqslant \alpha - 2 \tag{2.4.4}$$

We assert that

$$\mathrm{ad}_A^k Bc_i = (-1)^k \sum_{0 \leqslant \mu \leqslant k} (-1)^\mu \binom{k}{\mu} A^\mu BA^{k-\mu} c_i. \tag{2.4.5}$$

To prove this result let l_A (respectively r_A) denote the endomorphism $B \to AB$ (respectively $B \to BA$) of $D(M)$. Then l_A commutes with r_A and so

$$\mathrm{ad}_A^k = (l_A - r_A)^k = (-1)^k \sum_{0 \leqslant \mu \leqslant k} (-1)^\mu \binom{k}{\mu} l_A^\mu r_A^{k-\mu}$$

(2.4.5) now follows by applying this result to B.

It follows easily from (2.4.4) and (2.4.5) that $(BA^k)c_i = 0$ for $0 \leqslant k \leqslant \alpha - 2$ and $1 \leqslant i \leqslant l$. Hence by (2.4.5),

$$0 \neq (\mathrm{ad}_A^{\alpha-1} B)(c_{i_\alpha}) = (-1)^{\alpha-1}(BA^{\alpha-1})c_{i_\alpha}. \qquad \square$$

We can now prove

Theorem 2.4.3

The nonlinear system (2.4.2) is strongly invertible if and only if $\alpha < \infty$.

Proof

If $\alpha = \infty$ let $u_1, u_2 \in C^\omega([0, \infty))$ be distinct controls and let $x_0 \in M$. If $t \to A_t \, . \, x_0$ denotes the integral curve of the vector field A through x_0, then, by Corollary 2.2.15, $x(t; u_i, x_0) \in I(\mathscr{L}_0, A_t \, . \, X_0)$ for $i = 1, 2$ (provided the trajectories are defined), where $\mathscr{L}_0 = \{\mathrm{ad}_A^k B : k = 0, 1, \ldots\}_{LA}$ and $I(\mathscr{L}_0, A_t \, . \, x_0)$ is the maximal integral submanifold of M through $A_t \, . \, x_0$ determined by $\mathscr{L}_0$. Since $\alpha = \infty$, $(\mathrm{ad}_A^k B)(c_i) = 0$ for $i = 1, 2, \ldots, l$ and $k \geqslant 0$ and so the trajectory $t \to c((\mathrm{ad}_A^k B)_t \, . \, p)$ is constant (in $\mathbb{R}^l$) for any $p \in M$. However, $S = \{\mathrm{ad}_A^k B : k \geqslant 0\}$ is a set of generators of $\mathscr{L}_0$ and so by Chow's theorem (Theorem 2.2.1) any $q \in I(\mathscr{L}_{0,p})$ can be written as

$$q = X_{t_1}^1 \circ X_{t_2}^2 \circ \cdots \circ X_{t_k}^k \, . \, p, \; X^i \in S, \; t_i \in \mathbb{R}$$

and so

$$c(q) = c(X_{t_1}^1 \circ X_{t_2}^2 \circ \cdots \circ X_{t_k}^k \, . \, p) = c(X_{t_2}^2 \circ \cdots \circ X_{t_k}^k \, . \, p) = c(p).$$

Hence, the output map c is constant on the submanifolds $I(\mathscr{L}_0, p)$ for any $p \in M$ and so

$$y(t; u_1, x_0) = c(x(t; u_1, x_0)) = c(x(t; u_2, x_0)) = y(t; u_2, x_0)$$

for small t. Since the curves are analytic we must have

$$y(t; u_1, x_0) = y(t; u_2, x_0)$$

for all t and so the system is not strongly invertible.

To prove the converse, suppose that $\alpha < \infty$ and let M_α be the inverse submanifold, which is open and dense in M as noted above. Clearly, the system is strongly invertible if it is invertible for all $x_0 \in M_\alpha$. Thus, if $x_0 \in M_\alpha$, $u \in C^\omega([0,\infty))$ and $y(t) \triangleq y(t; u, x_0)$, then

$$y_i^{(1)}(t) \triangleq \frac{\mathrm{d}y_i}{\mathrm{d}t}(t) = (Ac_i)(x(t)) + u(t)(Bc_i)(x(t))$$

and so if $\alpha > 1$, $(Bc_i) = 0$ for $i = 1, \ldots, l$, and then

$$y_i^{(1)}(t) = (Ac_i)(x(t)).$$

Hence,

$$y_i^{(2)}(t) = (A^2c_i)(x(t)) + u(t)(BAc_i)(x(t))$$

and if $\alpha > 2$ then $BAc_i = 0$ for $i = 1, \ldots, l$, by Lemma 2.4.2, so that

$$y_i^{(2)}(t) = (A^2c_i)(x(t)).$$

By induction, we have

$$y_{i_\alpha}^{(\alpha)}(t) = (A^\alpha c_{i_\alpha})(x(t)) + u(t)(BA^{\alpha-1}c_{i_\alpha})(x(t)). \qquad (2.4.6)$$

Lemma 2.4.2 implies that $(BA^{\alpha-1}c_{i_\alpha})(p) \neq 0$ for all $p \in M_\alpha$. Hence we can define the system

$$\begin{aligned} \dot{z} &= (A(z) + h(z)B(z)) + vG(z), \qquad z(0) = x_0 \\ w &= h(z) + vk(z) \end{aligned} \qquad (2.4.7)$$

where $z \in M_\alpha$, $v \in C^\omega([0,\infty))$, $k(z) = 1/(BA^{\alpha-1}c_{i_\alpha})(z)$, $h(z) = -k(z)(A^\alpha c_{i_\alpha})(z)$, and $G(z) = k(z)B(z)$, and it is clear that this system is defined by real analytic vector fields on M_α. Let $v(t) = y_{i_\alpha}^{(\alpha)}(t)$ be the input to system (2.4.7) so that

$$\begin{aligned} \dot{z} &= A(z) + h(z)B(z) + y_{i_\alpha}^{(\alpha)}(t)G(z(t)) \\ &= A(z) + h(z)B(z) + [(A^\alpha c_{i_\alpha})(x(t)) + u(t)(BA^{\alpha-1}c_{i_\alpha})(x(t))]\,k(z)B(z) \\ &= A(z) + u(t)B(z) \end{aligned}$$

which is the same system as (2.4.2). Moreover,

$$\begin{aligned} w(t) &= h(z(t)) + y_{i_\alpha}^{(\alpha)}(t)k(z) \\ &= -k(z)(A^\alpha c_{i_\alpha})(z) + y_{i_\alpha}^{(\alpha)}(t)k(z) \\ &= u(t), \end{aligned}$$

by (2.4.6) and the definition of k. Hence the system (2.4.7), when driven with $v = y_{i_\alpha}^{(\alpha)}(t)$ produces the control u for system (2.4.2) and so is a left-inverse system. □

REMARK The proof above is constructive in that it produces a left-inverse system (2.4.7) for the control system (2.4.2).

Using similar techniques to those above, it can also be shown (Hirschorn,

1979a) that the system (2.4.7) acts as a right-inverse system to (2.4.2) in the sense that if $f(t) \in C^\omega(\mathbb{R})$ can be realized as the output of (2.4.2) for some control $u \in C^\omega([0, \infty))$, then the control which produces this output is given by the output w of the system (2.4.7) when the input v is chosen to be $f^{(\alpha)}$; i.e.

$$f(t) = y(t; u_f, x_0), \qquad \text{where } u_f(t) = w(t; f^{(\alpha)}, x_0).$$

The invertibility of multivariable nonlinear systems is discussed by Hirschorn (1979b), Singh (1982a, b) and Rebhuhn (1980).

2.4.2 Invariance in nonlinear systems

In order to introduce the notion of invariance we shall first consider the concept for the linear system

$$\Sigma : \dot{x}(t) = Ax(t) + Bu(t), \qquad x \in \mathbb{R}^n, \quad u \in \mathbb{R}^m. \tag{2.4.8}$$

A subspace $V \subseteq \mathbb{R}^n$ is *(A, B)-invariant* if

$$AV \subseteq V + \mathscr{B}$$

where $\mathscr{B}$ = range B. We can put this definition in a form suitable for generalization to nonlinear systems by defining the distributions

$$\mathscr{D}_V(x) = V \subseteq T_x(\mathbb{R}^n) \cong \mathbb{R}^n,$$

and

$$\mathscr{B}(x) = \mathscr{B} \subseteq T_x(\mathbb{R}^n).$$

Now, (A, B)-invariance of V implies that, for any basis $\{v_1, \dots, v_k\}$ of V,

$$Av_i \in V + \mathscr{B}, \qquad 1 \leqslant i \leqslant k.$$

Thus if $X_i(x)$ denotes the constant vector field $x \to v_i$, $x \in \mathbb{R}^n$, $1 \leqslant i \leqslant k$, we have

$$\mathscr{D}_V(x) = \text{span}\{X_1(x), \dots, X_k(x)\}$$

and

$$\text{ad}_A X_i(x) = (\text{d}X_i)_x Ax - (\text{d}A)_x X_i = 0 - Av_i \in \mathscr{D}_V + \mathscr{B}.$$

For any vector field $X \in \mathscr{D}_V$ with $X(x) = \sum_{i=1}^k a_i(x) X_i(x)$, we have

$$\begin{aligned}\text{ad}_A X(x) &= \left[A, \sum_{i=1}^k a_i X_i\right](x) \\ &= \sum_{i=1}^k (Aa_i)(x) X_i(x) + \sum_{i=1}^k a_i(x)(\text{ad}_A X_i(x)) \\ &\in (\mathscr{D}_V + \mathscr{B})(x).\end{aligned}$$

Hence,

$$\mathrm{ad}_A X \subseteq \mathscr{D}_V + \mathscr{B}, \qquad X \in \mathscr{D}_V$$

and so

$$\mathrm{ad}_A \mathscr{D}_V \subseteq \mathscr{D}_V + \mathscr{B}. \tag{2.4.9}$$

Moreover, we have

$$\mathrm{ad}_{b_i} X(x) = \sum_{j=1}^{m} [(b_i a_j)X_j + a_j \mathrm{ad}_{b_i} X_j],$$

where $X = \Sigma_{j=1}^{m} a_j X_j \in \mathscr{D}_V$, $X_j(x) = v_j$ and b_i is the ith column of the B matrix in (2.4.8). Since X_j and b_i are constant, it follows that $\mathrm{ad}_{b_i} X_j = 0$ and so $\mathrm{ad}_{b_i} X \in \mathscr{D}_V$. Hence

$$\mathrm{ad}_{b_i} \mathscr{D}_V \subseteq \mathscr{D}_V. \tag{2.4.10}$$

Similarly, it follows that $[\mathscr{D}_V, \mathscr{D}_V] \subseteq \mathscr{D}_V$, so that $\mathscr{D}_V$ is an involutive distribution.

The above consideration of linear systems leads us to define a generalization of (A, B)-invariance in the following way. Let

$$\dot{x}(t) = A(x(t)) + \sum_{i=1}^{m} u_i(t) B_i(x(t)), \qquad x \in M \tag{2.4.11}$$

be a nonlinear system with $A, B_1, \ldots, B_m \in D(M)$, and let $\mathscr{B}$ be the distribution defined on M by

$$\mathscr{B}(x) = \mathrm{span}\{B_1(x), \ldots, B_m(x)\}.$$

Then we say that an involutive distribution $\mathscr{D}$ on M is *$(A, \mathscr{B})$-invariant*‡ if the conditions (2.4.9) and (2.4.10) hold for $\mathscr{D}$, i.e.

$$\mathrm{ad}_A \mathscr{D} \subseteq \mathscr{D} + \mathscr{B}, \qquad \mathrm{ad}_{B_i} \mathscr{D} \subseteq \mathscr{D}. \tag{2.4.12}$$

Returning to the linear system (2.4.8), we say that a linear subspace $V \subseteq \mathbb{R}^n$ is *controlled invariant* if there exists a linear feedback $F: \mathbb{R}^n \to \mathbb{R}^m$ such that $A_F V \subseteq V$, where $A_F = A + BF$. This is easily seen to be equivalent to the condition

$$AV \subseteq V + \mathscr{B}.$$

In terms of distributions, the condition $A_F V \subseteq V$ is equivalent to $\mathrm{ad}\, A_F \mathscr{D}_V \subseteq \mathscr{D}_V$, as can be seen by using the above arguments. Hence from (2.4.12) we have

$$\mathrm{ad}\, A_F \mathscr{D} \subseteq \mathscr{D}, \qquad \mathrm{ad}_{B_i} \mathscr{D} \subseteq \mathscr{D}. \tag{2.4.13}$$

‡Note that Hirschorn (1981) requires (2.4.12) to hold only on an open dense submanifold $M_0 \subseteq M$.

To generalize the notion of controlled invariance to nonlinear systems of the form (2.4.11) we first recall from Section 2.3 that we may regard such a system as an *affine control system* on M; i.e. the system can be defined as an affine distribution $\triangle$ on M such that for all $x \in M$ there is a neighbourhood U of x and vector fields $X(=A), X_1(=B_1), \dots, X_m(=B_m) \in D(M)$ such that

$$\triangle(y) = X_0(y) + \text{span}\{X_1(y), \dots, X_m(y)\}, \text{ for all } y \in U(x).$$

The affine distribution $\triangle$ induces a distribution $\triangle_0$ given by

$$\triangle_0 = \{X - Y : X, Y \in \triangle\}$$

and in the local coordinates $y \in U$ used above we have

$$\triangle_0 = \text{span}\{X_1, \dots, X_m\}.$$

Hence we say that an involutive distribution $\mathscr{D}$ of fixed dimension is (locally) *controlled invariant* for an affine control system $\triangle$ if (locally) there exist vector fields $X_0, X_1, \dots, X_m \in D(M)$ such that

$$\triangle(x) = X_0(x) + \text{span}\{X_1(x), \dots, X_m(x)\}$$

and

$$[X_0, \mathscr{D}] \subseteq \mathscr{D}, \quad [X_i, \mathscr{D}] \subseteq \mathscr{D}, \qquad i = 1, \dots, m, \tag{2.4.14}$$

which generalize (2.4.13).

To derive a condition for an involutive distribution $\mathscr{D}$ to be controlled invariant we first prove Theorem 2.3.5 (stated in the last section) which we state in the following form:

Theorem 2.4.4 (Isidori *et al.* (1981), Nijmeijer (1981))
If an involutive distribution $\mathscr{D}$ of fixed dimension satisfies

$$[\triangle_0, \mathscr{D}] \subseteq \mathscr{D} + \triangle_0 \tag{2.4.15}$$

and $\mathscr{D} \cap \triangle_0$ has fixed dimension, then there exists a basis $\{X_1, \dots, X_m\}$ (locally) of $\triangle_0$ such that

$$[X_i, \mathscr{D}] \subseteq \mathscr{D}, i = 1, \dots, m.$$

Proof
Suppose that $[\triangle_0, \mathscr{D}] \subseteq \mathscr{D} + \triangle_0$ and let $p \in M$. Choose a coordinate neighbourhood U of p such that

$$\mathscr{D} = \text{span}\left\{\frac{\partial}{\partial x_1}, \dots, \frac{\partial}{\partial x_k}\right\}$$

(by Frobenius' theorem) and let $X_1, \dots, X_m$ be a basis of $\triangle_0$. Identifying X_i with its components we can regard

$$B(x) = (X_1(x), \dots, X_m(x))$$

as an $n \times n$ matrix. Then (2.4.15) implies that

$$\frac{\partial B(x)}{\partial x_i} = B(x)M_i(x) \pmod{\mathscr{D}} \qquad (1 \leqslant i \leqslant k) \tag{2.4.16}$$

for some $m \times m$ matrices $M_i(x)$ and we write $R = S \pmod{\mathscr{D}}$ for $n \times m$ matrices R and S if

$$R - S = \begin{bmatrix} * \\ 0_{n-k,m} \end{bmatrix}$$

for some matrix (*).

Next note that $[\mathscr{D}, \Delta_0] \subseteq \mathscr{D} + \Delta_0$ implies that the distribution $\mathscr{D} + \bar{\Delta}_0$ is involutive, where $\bar{\Delta}_0$ is the smallest involutive distribution containing Δ_0 (i.e. the *involutive closure* of Δ_0). This follows easily from the Jacobi identity. Since any distribution can be embedded in a distribution of fixed dimension (Sussmann (1973)), it follows by a simple extension of Frobenius' theorem that, for any $p \in M$, there is a coordinate neighbourhood U of p such that, in local coordinates,

$$\mathscr{D} = \operatorname{span}\left\{\frac{\partial}{\partial x_1}, \ldots, \frac{\partial}{\partial x_k}\right\}, \qquad \mathscr{D} + \bar{\Delta}_0 = \operatorname{span}\left\{\frac{\partial}{\partial x_1}, \ldots, \frac{\partial}{\partial x_k}, \ldots, \frac{\partial}{\partial x_q}\right\}.$$

We shall now prove the result for the special case in which $\mathscr{D} \cap \bar{\Delta}_0 = 0$. With this assumption it is clear that $\bar{\Delta}_0$ is spanned by $l \triangleq q - k$ vector fields of the form

$$Y_j = \frac{\partial}{\partial x_{k+j}} + \sum_{i=1}^{k} \alpha_i^j(x) \frac{\partial}{\partial x_i}, \qquad 1 \leqslant j \leqslant l.$$

Since $\bar{\Delta}_0$ is involutive and $\mathscr{D} \cap \bar{\Delta}_0 = 0$ it follows that $[Y_s, Y_t] = 0$, for $1 \leqslant s, t \leqslant l$. Moreover it is clear that $[\mathscr{D}, Y_s] \subseteq \mathscr{D}$ for $1 \leqslant s \leqslant l$.

Now let $X_1, \ldots, X_m$ be the basis of Δ_0 introduced above and let $X_{m+1}, \ldots, X_l$ be chosen so that

$$\bar{\Delta}_0 = \operatorname{span}\{X_1, \ldots, X_l\} \qquad (= \operatorname{span}\{Y_1, \ldots, Y_l\}).$$

Define the $n \times l$ matrices $B_1(x)$ and $\tilde{B}(x)$ by

$$B_1(x) = (X_1(x), \ldots, X_l(x)), \qquad \tilde{B}(x) = (Y_1(x), \ldots, Y_l(x)).$$

Then

$$B_1(x) = \tilde{B}(x)P(x)$$

for some nonsingular matrix P and so

$$\frac{\partial B_1(x)}{\partial x_i} = \frac{\partial \tilde{B}(x)}{\partial x_i} P(x) + \tilde{B}(x) \frac{\partial P(x)}{\partial x_i}$$

$$= B_1(x)P^{-1}(x) \frac{\partial P(x)}{\partial x_i} \pmod{\mathscr{D}}, \quad 1 \leqslant i \leqslant k \tag{2.4.17}$$

since $[\mathscr{D}, Y_s] \subseteq \mathscr{D}$, $1 \leqslant s \leqslant l$.

Since $\mathscr{D} + \bar{\Delta}_0$ is involutive it follows from $[\mathscr{D}, \Delta_0] \subseteq \mathscr{D} + \Delta_0$ that $[\mathscr{D}, \bar{\Delta}_0] \subseteq \mathscr{D} + \bar{\Delta}_0$ and so

$$\frac{\partial B_1(x)}{\partial x_i} = B_1(x)\bar{M}_i(x) \pmod{\mathscr{D}} \tag{2.4.18}$$

where $\bar{M}_i(x)$ is an $l \times l$ matrix of the form

$$\begin{bmatrix} M_i(x) & * \\ 0_{l-m,\, m} & * \end{bmatrix}. \tag{2.4.19}$$

(This follows from (2.4.16).) Thus (2.4.17) and (2.4.18) imply

$$B(x)\left[P^{-1}(x)\frac{\partial P(x)}{\partial x_i} - \bar{M}_i(x)\right] = 0 \pmod{\mathscr{D}}, \qquad 1 \leqslant i \leqslant k$$

and since $\mathscr{D} \cap \bar{\Delta}_0 = 0$ and $B(x)$ is a basis of $\bar{\Delta}_0$ it follows that

$$P^{-1}(x)\frac{\partial P(x)}{\partial x_i} - \bar{M}_i(x) = 0, \qquad 1 \leqslant i \leqslant k.$$

If we partition

$$P(x) = \begin{bmatrix} P_1(x) & P_2(x) \\ P_3(x) & P_4(x) \end{bmatrix}$$

as in (2.4.19) we have

$$\frac{\partial P_1}{\partial x_i} = P_1(x)M_i(x), \qquad 1 \leqslant i \leqslant k$$

and it is easy to see that the columns of the matrix $(X_1(x), \ldots, X_k(x))^{-1}P(x)$ form the required basis of Δ_0.

To prove the general case we apply the above method to the distributions $\mathscr{D}$ and $\Delta_0 \pmod{\Delta_0 \cap \mathscr{D}}$. □

Returning to the system (2.4.11) we next prove a result of Hirschorn (1981) which is useful in the disturbance decoupling problem.

Theorem 2.4.5

Let $\mathscr{D}$ be an involutive distribution on M of fixed dimension. Then $\mathscr{D}$ is $(A, \mathscr{B})$-invariant if and only if

(a) $\mathrm{ad}_{B_i}\mathscr{D} \subseteq \mathscr{D}$, $1 \leqslant i \leqslant m$

(b) for all $x \in M$, there exists an open set $U \subseteq M$ containing x and functions $k_1, \ldots, k_m \in \mathscr{F}(U)$ such that

$$\mathrm{ad}_{(A + \sum_{i=1}^{m} k_i B_i)}\mathscr{D} \subseteq \mathscr{D}.$$

Proof

If conditions (a) and (b) hold, then if $X \in \mathscr{D}$ we have

$$\mathrm{ad}_{(A+\sum_{i=1}^{m} k_i B_i)} X = \mathrm{ad}_A X + \sum_{i=1}^{m} (k_i \,\mathrm{ad}_{B_i} X - (Xk_i)B_i)$$

and since $\mathrm{ad}_{B_i}\mathscr{D} \subseteq \mathscr{D}$ it follows that $\mathrm{ad}_A X(x) \in \mathscr{D}(x) + \mathscr{B}(x)$ for all $x \in U$. Since x is arbitrary, we have $\mathrm{ad}_A \mathscr{D} \subseteq \mathscr{D} + \mathscr{B}$, so $\mathscr{D}$ is $(A, \mathscr{B})$-invariant.

To prove that the conditions (a) and (b) are necessary, suppose that $\mathscr{D}$ is $(A, \mathscr{B})$-invariant. Choose a minimal subset of $\{B_1, \dots, B_m\}$, say $\{B_1, \dots, B_q\}$ (by renumbering, if necessary), so that $\mathscr{D}(x) + \mathrm{span}\{B_1(x), \dots, B_q(x)\} = \mathscr{D}(x) + \mathscr{B}(x)$ for $x \in M$. For any fixed $x \in M$ let $U \subseteq M$ be an open set containing x and suppose that $\{\partial/\partial x_1, \dots, \partial/\partial x_\alpha\}$ spans $\mathscr{D}$ on U (by Frobenius' theorem). Since $\mathrm{ad}_A \mathscr{D} \subseteq \mathscr{D} + \mathscr{B}$ and $X_i = \partial/\partial x_i \in \mathscr{D}$ on U we have

$$\mathrm{ad}_A X_i = \sum_{j=1}^{\alpha} h_{ij} X_j + \sum_{j=1}^{q} g_{ij} B_j, \qquad 1 \leqslant i \leqslant \alpha \tag{2.4.20}$$

for some functions h_{ij}, $g_{ij} \in \mathscr{F}(U)$. We must prove that

$$\mathrm{ad}_{(A+\sum_{j=1}^{m} k_j B_j)} X_i \in \mathscr{D}, \ 1 \leqslant i \leqslant \alpha$$

i.e.

$$\mathrm{ad}_A X_i + \sum_{j=1}^{m} k_j \,\mathrm{ad}_{B_j} X_i - \sum_{j=1}^{m} (X_i k_j) B_j \in \mathscr{D}.$$

However, $\mathrm{ad}_{B_j} X_i \in \mathscr{D}$ and $\mathrm{ad}_A X_i - \sum_{j=1}^{q} g_{ij} B_j \in \mathscr{D}$ by (2.4.20) and so we must prove that

$$\sum_{j=1}^{m} (X_i k_j) B_j - \sum_{j=1}^{q} g_{ij} B_j \in \mathscr{D}, \qquad 1 \leqslant i \leqslant \alpha$$

If $m > q$ then let $k_{q+1}, \dots, k_m = 0$; then we prove that

$$\sum_{j=1}^{q} (X_i k_j - g_{ij}) B_j \in \mathscr{D}, \qquad 1 \leqslant i \leqslant \alpha.$$

Since we have chosen the set $\{B_1, \dots, B_q\}$ to be minimal, $B_j \notin \mathscr{D}$, $1 \leqslant j \leqslant q$ and so we must find functions k_j such that

$$X_i k_j = g_{ij}, \qquad 1 \leqslant j \leqslant q, \ \ 1 \leqslant i \leqslant \alpha. \tag{2.4.21}$$

This is a set of partial differential equations and for a solution to exist it is necessary that the g_{ij} satisfy the consistency or integrability condition

$$X_r g_{sj} = X_s g_{rj}, \qquad 1 \leqslant j \leqslant q, \ \ 1 \leqslant r, \ \ s \leqslant \alpha. \tag{2.4.22}$$

To verify this condition we apply ad_{X_r} to (2.4.20). We obtain

$$\mathrm{ad}_{X_r}\,\mathrm{ad}_A X_s = \sum_{j=1}^{\alpha} (X_r h_{sj}) X_j + \sum_{j=1}^{q} (X_r g_{sj}) B_j + \sum_{j=1}^{q} g_{sj}\,\mathrm{ad}_{X_r} B_j$$

$$= \sum_{j=1}^{\alpha} p_{sj} X_j + \sum_{j=1}^{q} (X_r g_{sj}) B_j$$

for some $p_{sj} \in \mathscr{F}(U)$, since, by assumption, $\mathrm{ad}_{B_j} X_r \in \mathscr{D}$, i.e. $-\mathrm{ad}_{X_r} B_j \in \mathscr{D}$. Similarly,

$$\mathrm{ad}_{X_s}\,\mathrm{ad}_A X_r = \sum_{j=1}^{\alpha} t_{sj} X_j + \sum_{j=1}^{q} (X_s g_{rj}) B_j, \qquad t_{sj} \in \mathscr{F}(U)$$

and since

$$\mathrm{ad}_{X_s}\,\mathrm{ad}_A X_r = \mathrm{ad}_{X_r}\,\mathrm{ad}_A X_s$$

by the Jacobi identity, the compatibility condition (2.4.22) holds. The existence of the functions k_j satisfying (2.4.21) now follows from the elementary theory of partial differential equations (see Hirschorn (1981) for details). □

REMARK For analytic distributions the result can be generalized to the case where dim $\mathscr{D}(x)$ is not constant. Then the theorem is true on an open dense submanifold of M. As an example, consider the system

$$\begin{bmatrix} \dot{x}_1 \\ \dot{x}_2 \\ \dot{x}_3 \end{bmatrix} = \begin{bmatrix} x_2 \\ 0 \\ 0 \end{bmatrix} + u \begin{bmatrix} x_1 \\ 0 \\ 0 \end{bmatrix}, \qquad y = x_1.$$

The distribution

$$\mathscr{D}(x) = \begin{cases} \{0\} \times \mathbb{R}^2 & \text{if } x_1 \neq 0 \\ \{(0,0)\} \times \mathbb{R} & \text{if } x_1 = 0 \end{cases}$$

has dimension 2 if $x \in M_0 \triangleq \mathbb{R}^3 \backslash \{x : x_1 = 0\}$ and dimension 1 if $x \notin M_0$. However, it is easy to check that $\mathscr{D}$ is $(A, \mathscr{B})$-invariant on M_0.

Corollary 2.4.6

An involutive distribution $\mathscr{D}$ of fixed dimension satisfying

$$[X_0, \mathscr{D}] \subseteq \mathscr{D} + \Delta_0$$
$$[X_i, \mathscr{D}] \subseteq \mathscr{D}, \qquad 1 \leqslant i \leqslant m$$

for an affine system $X_0 + \mathrm{span}\{X_1, \ldots, X_m\}$ is locally a controlled invariant distribution. □

Corollary 2.4.7
If $\mathscr{D}$ is an involutive distribution of fixed dimension and $\mathscr{D} \cap \Delta_0$ also has fixed dimension, then $\mathscr{D}$ is locally controlled invariant if and only if $[\Delta, \mathscr{D}] \subseteq \mathscr{D} + \Delta_0$.

This follows from Corollary 2.4.6 and Theorem 2.4.4. □

2.4.3 Decoupling of nonlinear systems using feedback

In this section we shall consider the problem of decoupling a nonlinear system by the use of feedback. By this we mean that the control is to be chosen as a function of the output so that in the resulting system certain variables do not depend on certain other variables in the system. For example, we may wish to isolate (part of) the output from noise which may be present in the system or we may require a noninteracting system (with m inputs and outputs) in which the output y_i depends only on the input u_i, $1 \leqslant i \leqslant m$. These problems have been considered in the linear case by Wonham (1979) and the disturbance decoupling problem has been solved for analytic systems by Hirschorn (1981). Here we shall follow Isidori *et al.* (1981) where the general input–output decoupling problem is considered.

First we return to the concept of invariant distributions. A distribution Δ defined on a Hausdorff manifold M is called *regular* if it is involutive, of constant rank, and partitions M into regularly embedded maximal integral submanifolds such that the quotient space M' of M consisting of these submanifolds has a C^∞ structure for which the canonical projection $\pi : M \to M'$ is a submersion. A distribution Δ is *invariant* under the dynamics defined by the nonlinear system‡

$$\dot{x} = f(x) + \sum_{i=1}^{m} u_i g_i(x), \qquad x(0) = x^0$$
$$y = h(x) \in \mathbb{R}^k \tag{2.4.23}$$

if

$$[f, \Delta] \subseteq \Delta$$
$$[g_i, \Delta] \subseteq \Delta, \qquad 1 \leqslant i \leqslant m. \tag{2.4.24}$$

Lemma 2.4.8
If Δ is an involutive distribution, then its involutive closure $\bar{\Delta}$ is also invariant.

Proof
If X and Y are vector fields in Δ, then

$$[f, X] \in \Delta, \qquad [f, Y] \in \Delta$$

‡ As usual we use a local representation of a globally defined system.

by (2.4.24) and so

$$[Y, [f, X]] \in [\Delta, \Delta] \subseteq \bar{\Delta}$$

and similarly

$$[X, [f, Y]] \in \bar{\Delta}.$$

Hence by the Jacobi identity,

$$[f, [X, Y]] = [X, [f, Y]] - [Y, [f, X]] \in [\Delta, \Delta] \subseteq \bar{\Delta},$$

i.e. $[f, \bar{\Delta}] \subseteq \bar{\Delta}$. A similar argument works for each g_i. □

To see the local consequences of the existence of an invariant distribution let Δ be a regular invariant distribution of rank k and choose local coordinates $x = (x_1, x_2)$, $x_1 \in \mathbb{R}^{n-k}$, $x_2 \in \mathbb{R}^k$ such that

$$\Delta = \text{span}\left\{\frac{\partial}{\partial x_{2,1}}, \ldots, \frac{\partial}{\partial x_{2,k}}\right\}.$$

Then we can write the system (2.4.23) in the form

$$\begin{aligned}\dot{x}_1 &= f_1(x_1, x_2) + g_1(x_1, x_2)u\\ \dot{x}_2 &= f_2(x_1, x_2) + g_2(x_1, x_2)u.\end{aligned}$$

By invariance,

$$[f, \Delta] \subseteq \Delta = \text{span}\left\{\frac{\partial}{\partial x_2}\right\}.$$

However,

$$\left[f, \frac{\partial}{\partial x_2}\right] = -\left(\frac{\partial f_1}{\partial x_2}\frac{\partial}{\partial x_1} + \frac{\partial f_2}{\partial x_2}\frac{\partial}{\partial x_2}\right)$$

and the only way the right-hand side can belong to Δ is if $\partial f_1/\partial x_2 = 0$, i.e. f_1 is independent of x_2. Since Δ is also invariant under each g_i, the same conclusion holds for g_1, and the system becomes

$$\begin{aligned}\dot{x}_1 &= f_1(x_1) + g_1(x_1)u\\ \dot{x}_2 &= f_2(x_1, x_2) + g_2(x_1, x_2)u.\end{aligned}$$

This means that we can locally 'factor out', the x_2 variable and reduce the system to one evolving on a manifold of dimension $n - k$. In the case of regular invariant distribution Δ we can obtain the global version of this result; namely, since Δ partitions M into maximal integral submanifolds (of the same dimension), Δ induces a dynamical system on the quotient space $M' = M/\Delta$ consisting of these submanifolds. If the system is controllable, then M' is Hausdorff.

We have considered above the generalization of (A, B)-invariance for a linear system to a nonlinear system. Now suppose we have an output equation in addition;

$$\dot{x} = Ax + Bu$$
$$y = Cu.$$

Then we say that a subspace $V \subseteq \mathbb{R}^n$ is *(C, A)-invariant* if there exists an $n \times m$ matrix F such that

$$(A + FC)V \subseteq V.$$

This is equivalent to the condition

$$A(V \cap \ker C) \subseteq V. \tag{2.4.25}$$

Combining the two concepts, we say that a subspace $V \subseteq \mathbb{R}^n$ is *(C, A, B)-invariant* if there exists a matrix F such that

$$(A + BFC)V \subseteq V.$$

The dynamics can then be made invariant under the output feedback $u = Fy + v$, since we then obtain the system

$$\dot{x} = (A + BFC)x + Bv.$$

Returning to the nonlinear system (2.4.23), we say that a distribution Δ is *(f, g)-invariant* (where g is the matrix with columns g_i) if, locally, there exist smooth functions $\alpha : U \subseteq M \to \mathbb{R}^k$, $\beta : U \subseteq M \to \mathbb{R}^{k\times r}$ such that

$$[\bar{f}, \Delta] \subseteq \Delta, \qquad [\bar{g}, \Delta] \subseteq \Delta \tag{2.4.26}$$

where

$$\begin{aligned} \bar{f}(x) &= f(x) + g(x)\alpha(x) \\ \bar{g}(x) &= g(x)\beta(x). \end{aligned} \tag{2.4.27}$$

Then Δ is invariant under the dynamics

$$\dot{x} = \bar{f}(x) + \bar{g}(x)v.$$

If $k = r$ and β is invertible we say that Δ is (f, g)-invariant with *full control* and otherwise Δ is (f, g)-invariant with *partial control*. In the former case this is just controlled invariance as introduced above.

To generalize (C, A)-invariance and (C, A, B)-invariance to nonlinear systems recall that $\mathrm{d}h_i \in T^*(M)$ $(1 \leqslant i \leqslant k)$ and operates on $T(M)$ by

$$\mathrm{d}h_i\left(\frac{\partial}{\partial x_j}\right) = \left(\frac{\partial h_i}{\partial x_i}\,\mathrm{d}x_1 + \cdots + \frac{\partial h_i}{\partial x_n}\,\mathrm{d}x_n, \frac{\partial}{\partial x_j}\right) = \frac{\partial h_i}{\partial x_j}$$

Hence we can consider the distribution Δ^0 defined by

$$\Delta^0(x) = \ker \mathrm{d}h(x) = \bigcap_{i=1}^{k} \ker \mathrm{d}h_i(x).$$

A simple application of the inverse function theorem shows that a function $\alpha(x)$ depends only on $y=h(x)$ (i.e. $\alpha(x)=\bar{\alpha}(h(x))=\bar{\alpha}(y)$) if

$$\Delta^0(x) \subseteq \ker \mathrm{d}\alpha(x).$$

Hence we say that a distribution Δ is *(h, f, g)-invariant* if there exist functions α, β which satisfy (2.4.26), (2.4.27) and

$$\Delta^0(x) \subseteq \ker \mathrm{d}\alpha(x) \cap \ker \mathrm{d}\beta(x). \tag{2.4.28}$$

Moreover, generalizing (2.4.24) we say that Δ is *(h, f)-invariant* if

$$[f, \Delta \cap \Delta^0] \subseteq \Delta, \qquad [g, \Delta \cap \Delta^0] \subseteq \Delta.$$

It follows from Lemma 2.4.8 that if Δ is (h, f, g)-invariant then so is $\bar{\Delta}$, but this is false in general for (h, f)-invariance.

Consider now the disturbance decoupling problem. Suppose that the system (2.4.23) is written in the form

$$\begin{aligned} \dot{x} &= f(x) + g(x)u + p(x)w \\ y_1 &= h_1(x) \\ y_2 &= h_2(x) \end{aligned}$$

where the input is composed of a control input u and a noise or disturbance input w (which is not controllable), and the output is split into two parts y_1, y_2. Then we wish to choose a control so that the output y_2 is independent of the input w. We say that the *static, state feedback, noise decoupling problem* is solvable (with full control) if there exist feedback functions $\alpha(x)$ and $\beta(x)$ such that y_2 is independent of w (and β is invertible). If we can find such functions α, β which depend only on y, then we say that the *static, output feedback, noise decoupling problem* is solvable. Note that we allow the noise $w(t)$ to be bounded and measurable whereas the control $u(t)$ must be piecewise smooth.

Theorem 2.4.9

The static, output feedback, noise decoupling problem is solvable in a regular way (with full control) if and only if there exists a regular distribution Δ such that

(a) Δ is (h_1, f, g) invariant (with full control)
(b) $p \subseteq \Delta \subseteq \ker \mathrm{d}h_2$.

Proof

If such a regular distribution Δ exists, then there exist α and β such that

$$\ker \mathrm{d}h_1 \subseteq \ker \mathrm{d}\alpha \cap \ker \mathrm{d}\beta. \tag{2.4.29}$$

Define $\bar{f}$ and $\bar{g}$ by

$$\bar{f}(x) = f(x) + g(x)\alpha(x)$$
$$\bar{g}(x) = g(x)\beta(x).$$

Then by the above remarks, we can write the system in the form

$$\dot{x}_1 = \bar{f}_1(x_1) + \bar{g}_1(x_1)v + p_1(x_1, x_2)w$$
$$\dot{x}_2 = \bar{f}_2(x_1, x_2) + \bar{g}_2(x_1, x_2)v + p_2(x_1, x_2)w$$
$$y_2 = h_2(x_1, x_2)$$

By (b) we have $p_1 = 0$ and $\partial k/\partial x_2 = 0$ so that k depends only on x_1, which is independent of w.

To prove the converse, note that if the problem is solvable in a regular way then α and β exist satisfying (2.4.29) and there exists a manifold M' and a projection $\pi : M \to M'$ given locally by

$$\pi(x_1, x_2) = x_1.$$

It is easy to check that $\Delta = \ker d\pi$ is a regular distribution which satisfies (a) and (b). □

In the case of state feedback, we therefore have

Corollary 2.4.10
The static, state feedback, noise decoupling problem is solvable in a regular way (with full control) if an only if there exists a regular distribution Δ such that

(a) Δ is (f, g) invariant (with full control)
(b) $p \subseteq \Delta \subseteq \ker dh_2$. □

It can also be shown (Isidori *et al.*, 1981) that for analytic systems the static, state feedback, noise decoupling problem is solvable (with full control) if and only if there exists a distribution Δ satisfying (a) and (b) above. Moreover, the dynamic feedback problem is also studied in this reference. In fact, the following result can be proved:

Theorem 2.4.11
Consider the system

$$\dot{x} = f(x) + g(x)u + p(x)w,\ x(0) = x^0$$
$$y_1 = h_1(x)$$
$$y_2 = h_2(x)$$

and suppose that

(a) there exists a regular (f, g) invariant distribution Δ^1 (with full control)
(b) there exists a regular (h, f) invariant distribution Δ^2
(c) $\Delta^2 \cap \Delta^0$ is a regular distribution
(d) $p \subseteq \Delta^2 \subseteq \Delta^1 \subseteq \ker dh_2$.

Then there exists a system

$$\dot{\xi} = \phi(\xi, \mu), \qquad \xi(0) = \theta(x^0)$$

for some function θ and feedback functions $\alpha(y_1, \xi)$, $\beta(y_1, \xi)$ such that y_2 is independent of w. □

In the final part of this section we shall study the problem of noninteracting control. To do this we first introduce, the concept of codistribution which is dual to that of distribution. A *codistribution* (or Pfaffian system) θ on a manifold M is a map which assigns to each $x \in M$ a subspace $\theta(x)$ of T_x^*M in a smooth way. For any distribution Δ on M we can define the codistribution $\Delta^\perp$ by

$$\Delta^\perp(x) = \{\omega \in T_x^*M : \langle \omega, \Delta(x) \rangle = 0\}.$$

Dually, if θ is a codistribution, we define

$$\theta^\perp(x) = \{\tau \in T_xM : \langle \theta(x), \tau \rangle = 0\}$$

Clearly, $(\Delta^\perp)^\perp = \Delta$ and $(\theta^\perp)^\perp = \theta$. Also, $\Delta^1 \subseteq \Delta^2 \Leftrightarrow (\Delta^1)^\perp \supseteq (\Delta^2)^\perp$.

A one-form ω is said to belong to a codistribution θ if $\omega(x) \in \theta(x)$ for all x, and we say that θ is *regular* if $\theta^\perp$ is regular. Since a regular distribution partitions M into integral submanifolds of constant dimension, θ is regular if and only if, for each $x \in M$, there is a neighbourhood U of x and smooth functions $\phi_1, \ldots, \phi_k$ such that $d\phi_1, \ldots, d\phi_k$ span θ. The integral manifolds of θ are then the intersections of the level sets of $\phi_1, \ldots, \phi_k$.

We may extend the notion of invariance to a codistribution θ: we say that θ is *invariant* under the dynamics (f, g) if

$$L_f(\theta) \subseteq \theta, \qquad L_g(\theta) \subseteq \theta.$$

Recall that, for any one-form ω and any vector field τ, we have

$$L_f\langle \omega, \tau \rangle(x) = \langle \omega, L_f\tau \rangle(x) + \langle L_f\omega, \tau \rangle(x).$$

If θ is invariant, then for any $\omega \in \theta$, $\tau \in \theta^\perp$ we have

$$\langle \omega, \tau \rangle = 0, \quad \langle L_f\omega, \tau \rangle = \langle \lambda, \tau \rangle = 0, \qquad \text{for some } \lambda \in \theta$$

and so $\langle \omega, L_f\tau \rangle(x) = 0$, for each $\omega \in \theta$ and each $x \in M$. Hence $L_f\tau \in \theta^\perp$ and similarly $L_g\tau \in \theta^\perp$; i.e. $\theta^\perp$ is invariant. By duality, $\theta^\perp$ invariant implies θ invariant and so θ is invariant if and only if $\theta^\perp$ is invariant. In a similar way we can prove the statements

(a) $\Delta^1 \subseteq \Delta^2, \quad [f, \Delta^1] \subseteq \Delta^2, \quad [g, \Delta^1] \subseteq \Delta^2$

(b) $(\Delta^2)^\perp \subseteq (\Delta^1)^\perp, \quad L_f((\Delta^2)^\perp) \subseteq (\Delta^1)^\perp, \quad L_g((\Delta^2)^\perp) \subseteq (\Delta^1)^\perp.$

Now recall that Theorem 2.4.5 implies that (f, g) invariance is equivalent, locally, to the conditions

$$[f, \Delta](x) \subseteq \Delta(x) \oplus \text{span } g(x)$$
$$[g, \Delta](x) \subseteq \Delta(x) \oplus \text{span } g(x).$$

However,

$$(\Delta + \text{span } g(x))^\perp = \Delta^\perp \cap (\text{span } g(x))^\perp$$

and by (a) and (b), Δ is locally (f, g) invariant at x if and only if

$$\begin{aligned} L_f(\Delta^\perp \cap (\text{span } g(x))^\perp) &\subseteq \Delta^\perp \\ L_g(\Delta^\perp \cap (\text{span } g(x))^\perp) &\subseteq \Delta^\perp \end{aligned} \tag{2.4.30}$$

on some neighbourhood of x.

Similarly, Δ is (h, f) invariant if and only if

$$\begin{aligned} L_f(\Delta^\perp) &\subseteq \Delta^\perp + \text{span } dh \\ L_g(\Delta^\perp) &\subseteq \Delta^\perp + \text{span } dh. \end{aligned} \tag{2.4.31}$$

Consider now the problem of noninteracting control; i.e. for the system

$$\dot{x} = f(x) + g(x)u$$
$$y = h(x)$$

we wish to find a (static) state feedback control of the form $u = \alpha(x) + \beta(x)v$ such that, in some coordinate system, the equations take the form

$$\begin{aligned} \dot{x}_1 &= \bar{f}_1(x_1) + \bar{g}_1(x_1)v_1 \\ &\vdots \\ \dot{x}_m &= \bar{f}_m(x_m) + \bar{g}_m(x_m)v_m \\ \dot{x}_{m+1} &= \bar{f}_{m+1}(x) + \bar{g}_{m+1}(x)v \\ y_1 &= h_1(x_1) \\ &\vdots \\ y_m &= h_m(x_m) \end{aligned} \tag{2.4.32}$$

where $x = (x_1, \ldots, x_{m+1})$, $v = (v_1, \ldots, v_m)$, and $y = (y_1, \ldots, y_m)$. (Each x_i, v_i, y_i may be a vector.) Then we can find manifolds M^i, $1 \leqslant i \leqslant m$, and projections $\pi_i: M \to M^i$, given by $\pi_i(x) = x_i$, locally, and on M^i the system becomes

$$\dot{x}_i = \bar{f}_i(x_i) + \bar{g}_i(x_i)v_i$$
$$y_i = h_i(x_i),$$

in local coordinates. We say that the *static, state feedback noninteracting control problem is solvable in a regular way* (with full control) if such functions α, β, π_i exist (and β is invertible).

We say that a family of (f, g) invariant distributions $\Delta_1, \ldots, \Delta_m$ are *compatible* if a single pair of functions α, β leaves each Δ_i invariant, i.e.

$$[f + g\alpha, \Delta_i] \subseteq \Delta_i, \quad [g\beta, \Delta_i] \subseteq \Delta_i, \quad 1 \leqslant i \leqslant m.$$

Theorem 2.4.12

The static, state feedback, noninteracting control problem is solvable in a regular way (with full control) if and only if there exists a family $\Delta_1, \ldots, \Delta_m$ of compatible (f, g) invariant distributions (with full control) such that

(a) each Δ_i is regular
(b) $\bar{g}_j \subseteq \Delta_i \subseteq \ker \mathrm{d}h_j$, for all $i \neq j$.
(c) if I and J are disjoint nonempty subsets of $\{1, \ldots, m)$, then

$$\left(\bigcap_{i \in I} \Delta_i\right) \oplus \left(\bigcap_{j \in J} \Delta_j\right) = TM.$$

Proof

If the problem is solvable, just take $\Delta_i = \ker \mathrm{d}\pi_i$.

To prove the converse, let $\Delta_1, \ldots, \Delta_m$ be compatible (f, g) invariant distributions (with full control) satisfying (a)–(c). Then there exist α and β leaving all the Δ_i invariant. Choose these functions and let π_i be the canonical projection $\pi_i: M \to M/\Delta_i$. Let $\theta_i = \Delta_i^{\perp}$. Then θ_i is regular so

$$\theta_i(x) = \text{span } \mathrm{d}\xi_i(x)$$
$$\Delta_i(x) = \ker \mathrm{d}\xi_i(x)$$

for some functions $\xi_i(x)$. If I and J are disjoint, nonempty subsets of $\{1, \ldots, m\}$, then, by (c),

$$\left(\sum_{i \in I} \theta_i\right) \cap \left(\sum_{j \in J} \theta_j\right) = 0.$$

and so the gradients $\mathrm{d}\xi_1(x), \ldots, \mathrm{d}\xi_m(x)$ are linearly independent. Hence there exists a vector of functions $\xi_{m+1}(x)$ such that the one-forms

$$\mathrm{d}\xi_1, \ldots, \mathrm{d}\xi_m, \mathrm{d}\xi_{m+1}$$

are linearly independent and span T^*M at x. Thus the map $x \to \xi(x)$ is a diffeomorphic coordinate change and so (b) and the invariance of each θ_i imply that the system is locally of the desired form. □

Using minimal (h, f) and maximal (f, g) invariant distributions, Isidori *et al.* (1981) prove:

Theorem 2.4.13
If the dimensions of u and y are equal to m, so that each subsystem in (2.4.32) is scalar, then the static, state feedback, noninteracting control problem is solvable in a regular way if

(a) the $m \times m$ matrix $A(x)$ with components

$$a_{ij}(x) = L_{g_i} L_f^{\rho_i} h_i(x)$$

is nonsingular for each x, where ρ_i is the largest integer such that

$$L_f L_f^r h_i(x) = 0 \qquad \text{for all } r < \rho_i, \quad x \in M.$$

(*b*) for each i the codistribution

$$\theta_i(x) = \text{span}\{ \mathrm{d}h_i(x), \mathrm{d}L_f h_i(x), \ldots, \mathrm{d}L_f^{\rho_i} h_i(x)\}$$

has constant dimension for all x.

(c) $\left(\sum_{i \in I} \theta_i\right) \cap \left(\sum_{j \in J} \theta_j\right) = \emptyset$, for $I, J \subseteq \{1, \ldots, m\}$, $I, J \neq \emptyset$, $I \cap J = \emptyset$. □

See also Falb and Wolovich (1967) for a linear version of the condition (a) in the last theorem. For a discussion of decoupling in systems with state-dependent input spaces, see Nijmeijer and Van der Schaft (1983).

2.5 SYSTEM SYMMETRIES AND DECOMPOSITION

2.5.1 Cascade decomposition

In this section we shall describe Krener's extension of the Krohn–Rhodes theory of finite automata (Krohn and Rhodes (1965)) to differentiable systems (Krener (1977)). Consider, therefore, a system defined on a manifold M with local representation

$$\Sigma : \dot{x} = f(x, u), \qquad u \in U \subseteq \mathbb{R}^k \tag{2.5.1}$$

where f is analytic in x, continuous in u and locally Lipschitz in x, uniformly in u; i.e. for any compact set $C \subseteq U$ and any $x_0 \in M$ there is a neighbourhood V of x_0 such that

$$| f(x_1, u) - f(x_2, u) | \leqslant K(x_1 - x_2), \qquad x_1, x_2 \in V, \quad u \in C,$$

for some constant K (depending on x_0).

We shall consider two classes of controls: the set $\mathscr{U}_{\text{pc}}$ of piecewise constant functions $u(.) : [0, T] \to U$ and the set $\mathscr{U}_{\text{m}}$ of all bounded measurable functions $u(.) : [0, T] \to U$ for any $T \geqslant 0$. The sets $\mathscr{U}_{\text{m}}$ and

$\mathcal{U}_{pc}$ are semigroups under the operation of concatenation (*) defined by

$$u_1 * u_2(.): [0, T_1 + T_2] \to U,$$

$$u_1 * u_2(t) = \begin{cases} u_1(t), & t \in [0, T_1] \\ u_2(t - T_1), & t \in [T_1, T_1 + T_2] \end{cases}$$

for any $u_1(.): [0, T_1] \to U$, $u_2(.): [0, T_2] \to U$, $u_1, u_2 \in \mathcal{U}_{pc}$ or $u_1, u_2 \in \mathcal{U}_m$. The set $\mathcal{U}_{pc}$ is a subsemigroup of $\mathcal{U}_m$ and we shall allow the control functions for the system (2.5.1) to belong to any subsemigroup $\mathcal{U}$ of $\mathcal{U}_m$ containing $\mathcal{U}_{pc}$.

As usual we let L denote the Lie subalgebra of $D(M)$ generated by the vector fields $f(.,u)$ for all $u \in U$. We shall assume that Σ is complete, i.e. each vector field in L is complete. A result of Palais (1957) shows that Σ is complete if L is finite dimensional and $f(., u)$ is complete for each $u \in U$. If $X \in L$ is complete and $\phi_t(.): M \to M$ is the flow of X, then ϕ_t, is an analytic diffeomorphism. The set of all analytic diffeomorphisms $\phi: M \to M$ will be denoted by Diff(M).

Define a map $\Phi: \mathcal{U}_{pc} \to \text{Diff}(M)$ as follows. If $u(t) = u$, $t \in [0, T]$, is a constant control and $\phi_t(.,u)$ is the flow of $f(.,u)$, then let $\Phi(u(.)) = \phi_T(.,u) \in \text{Diff}(M)$. For a concatenation $u_1 * u_2(.)$ of two controls $u_1(t) = u_1$, $u_2(t) = u_2$, we define $\Phi(u_1 * u_2(.)) = \Phi(u_1) \circ \Phi(u_2)$. By the semigroup property of flows it can be seen that this is well-defined if $u_1 = u_2$. Since any control in $\mathcal{U}_{pc}$ is a finite concatenation of constant controls, this defines Φ uniquely. Note that by construction, Φ is a semigroup homomorphism and we call $S \triangleq \text{Range } \Phi$ the *semigroup of the system* Σ. If G is the subgroup of Diff(M) generated by S, then G is called the *group of* Σ.

For any $x_0 \in M$ define

$$S(x_0) = \{\phi(x_0) : \phi \in S\}$$
$$G(x_0) = \{\phi(x_0) : \phi \in G\}$$

i.e. the orbits of x_0 under S and G, respectively. S is, of course, the reachable set of Σ from x_0 under the controls $\mathcal{U}_{pc}$ and we have defined, previously, a system to be controllable (or weakly controllable) if $M = S(x_0)$ $(M = G(x_0))$. If $G(x_0) \neq M$ we replace L by the smallest subalgebra of $D(G(x_0))$ containing the vector fields $f(.,u)$, $u \in U$. (Note that $G(x_0)$ is an analytic submanifold of M, since $f(x, u)$ is analytic in x; this follows from a result of Nagano (1966.)) It can also be shown that the trajectories arising from $\mathcal{U}_m$ can be approximated on compact sets by trajectories arising from controls in $\mathcal{U}_{pc}$ so that any control semigroup $\mathcal{U}$ with $\mathcal{U}_{pc} \subseteq \mathcal{U} \subseteq \mathcal{U}_m$ can be used in the following discussion.

In developing the theory of cascaded systems it is necessary to lift the dynamics of the system to G. If L is finite dimensional, then G can be given a Lie group structure such that L is the Lie algebra (of right invariant vector fields) of G (see Palais (1957)). By Ado's theorem (Varadarajan (1974)) L

is isomorphic to a subalgebra L of $gl(m; \mathbb{R})$ and so a neighbourhood of the identity in G is isomorphic as a Lie group to a neighbourhood of the identity in the Lie subgroup G of $GL(m; \mathbb{R})$ corresponding to L. If $F(u)$ denotes the matrix of L corresponding to $f(., u)$ under the above isomorphism, for each $u \in U$, then the dynamics

$$\dot{X} = F(u)X \tag{2.5.2}$$

locally represent the lift of the system (2.5.1) to G.

We shall exemplify the above lifting theory to linear and bilinear systems. Thus let

$$\dot{x} = Ax + \sum_{i=1}^{k} u_i b_i, \qquad x(0) = x_0, \tag{2.5.3}$$

be a linear system with $M = \mathbb{R}^n$, $U = \mathbb{R}^k$. As we have seen before, the Lie algebra L generated by Ax and b_i is finite dimensional since

$$\begin{aligned} [Ax, b_i] &= Ab_i, \qquad [b_i, b_j] &&= 0, \\ \mathrm{ad}^l(Ax)b_i &= A^l b_i, \qquad [b_i, \mathrm{ad}^l(Ax)b_j] &&= 0. \end{aligned}$$

The flow of (the vector field) (2.5.3) is given by

$$\phi_t(x; u) = e^{At}x + \int_0^t e^{A(t-s)} \sum_i u_i(s) b_i \, ds$$

and so, for $t, u \in \mathcal{U}_m$ fixed, $\phi_t(.;u)$ is an invertible affine map from $\mathbb{R}^n$ onto $\mathbb{R}^n$. Hence G is the group of affine transformations of $\mathbb{R}^n$ generated by all the functions $\phi_t(.;u)$ for $t \geqslant 0$, $u \in \mathcal{U}_m$ and so lifting the dynamics of (2.5.3) to G gives the matrix equation

$$\dot{X} = AX + \left(\sum_i u_i b_i \ldots \sum_i u_i b_i \right).$$

Note that this is not the form (2.5.2) since G consists of affine transformations here. However, we can put it in the form of (2.5.2) by noting that any linear system of the form (2.5.3) can be written as a bilinear system

$$\dot{x} = \left(A + \sum_{i=1}^{k} u_i B_i \right) x, \tag{2.5.4}$$

by adding to (2.5.3) the equation $\dot{x}_{n+1} = 0$ where $x_{n+1} \equiv 1$ and expressing (2.5.3) as $\dot{x} = Ax + \Sigma u_i b_i x_{n+1}$ which, together with $\dot{x}_{n+1} = 0$, is of the form (2.5.4).

Now, for a bilinear system (2.5.4), L is easily seen to be equal to the Lie subalgebra of $gl(n; \mathbb{R})$ generated by $A, B_1, \ldots, B_k$. The lifted dynamics are

$$\dot{X} = \left(A + \sum_i u_i B_i \right) X$$

which are defined on $GL(n; \mathbb{R})$ and are of the form (2.5.2).

Next we consider when two systems have the same state-space or input–output behaviour under changes in the input space and the state or output spaces. Denote by $\Sigma = \Sigma(M, U, f, \mathscr{U}, N, g, x_0)$ an 'initialized' system‡ with output space N and output map $g: M \to N (= \mathbb{R}^k$, for some $k)$. Introduce the sets

$\mathscr{X}_{x_1} = \{x(.): [0, T] \to M \mid T \geqslant 0$ and x is absolutely continuous with $x(0) = x_1\}$.

$\mathscr{Y}_{x_1} = \{y(.): [0, T] \to N \mid T \geqslant 0$ and y is absolutely continuous with $y(0) = g(x_1)\}$.

Then we can define maps $\mathscr{F}_{x_1}: \mathscr{U} \to \mathscr{X}_{x_1}$, $\mathscr{G}_{x_1}: \mathscr{U} \to \mathscr{Y}_{x_1}$ by

$$\mathscr{F}_{x_1}(u(t)) = \phi_t(x_1; u), \quad \mathscr{G}_{x_1}(u(t)) = g(\phi_t(x_1; u)).$$

Now, if $\Sigma^i = \Sigma^i(M^i, U^i, f^i, \mathscr{U}^i, N^i, g^i, x_0^i)$, $i = 1, 2$, is a pair of initialized systems, $\alpha: U^2 \to U^1$ is continuous and $\beta: N^1 \to N^2$ is analytic, then we say that Σ^1 *simulates* Σ^2 with encoder α and decoder β if the diagram

$$\begin{array}{ccc} \mathscr{U}^1 & \xrightarrow{\mathscr{G}^{1_1}_{x_0}} & \mathscr{Y}^{1_1}_{x_0} \\ \uparrow{\scriptstyle\alpha} & & \downarrow{\scriptstyle\beta} \\ \mathscr{U}^2 & \xrightarrow[\mathscr{G}^{2_2}_{x_0}]{} & \mathscr{Y}^{2_2}_{x_0} \end{array} \tag{2.5.5}$$

commutes, where we have extended α and β pointwise to $\mathscr{U}^2$ and $\mathscr{Y}^{1_1}_{x_0}$; i.e.

$$\alpha(u(.))(t) = \alpha(u(t)), \qquad \beta(y(.))(t) = \beta(y(t)).$$

We say that Σ^1 is *equivalent* to Σ^2 if Σ^1 simulates Σ^2, $U^1 = U^2$, $N^1 = N^2$ and α, β are identity maps.

In the state-space form, if $\alpha: U^2 \to U^1$ is continuous and $\gamma: G^1(x_0^1) \to M^2$ is analytic, then we say that Σ^1 is *homomorphic to* Σ^2 if the diagram

$$\begin{array}{ccc} \mathscr{U}^1 & \xrightarrow{\mathscr{F}^{1_1}_{x_0}} & \mathscr{X}^{1_1}_{x_0} \\ \uparrow{\scriptstyle\alpha} & & \downarrow{\scriptstyle\gamma} \\ \mathscr{U}^2 & \xrightarrow[\mathscr{F}^{2_2}_{x_0}]{} & \mathscr{X}^{2_2}_{x_0} \end{array} \tag{2.5.6}$$

commutes, (if Σ^1 is weakly controllable, $G^1(x_0^1) = M^1$. Otherwise the range of $\mathscr{Y}^{1_1}_{x_0}$ has elements which take values in $G^1(x^1)$.) *Isomorphic* systems are

‡ i.e. the system has a distinguished initial point x_0

defined as those for which $\alpha : U^1 = U^2$ is the identity map and $\gamma : M^1 \to M^2$ is a diffeomorphism. Local forms of the above definitions can be given by restricting the functions in $\mathscr{X}_{x*}$ and $\mathscr{Y}_{x*}$ to be defined only for small T.

The diagram (2.5.6) has the obvious local form

$$\begin{array}{ccc} U^1 & \xrightarrow{f^1(x^1,.)} & T_{x^1}M^1 \\ \uparrow \alpha & & \downarrow \gamma_{*x^1} \\ U^2 & \xrightarrow{f^2(\gamma(x^1),.)} & T_{x^2}M^2 \end{array} \tag{2.5.7}$$

It follows easily that if Σ^1 has the accessibility property and is homomorphic to Σ^2, then $\gamma_* : L^1 \to L^2$ is a Lie algebra homomorphism between the Lie algebras L^1 and L^2 of Σ^1 and Σ^2, respectively. This justifies the use of the term 'homomorphism' above.

Consider next the map $\mathscr{T}: f^1(.,\alpha(u)) \to f^2(.,u)$. If $u, v, w \in U^2$, then

$$f^1(x^1, \alpha(u)) = \lambda f^1(x^1, \alpha(v)) + \mu f^1(x^1, \alpha(w))$$

implies that

$$f^2(\gamma(x^1), u) = \lambda f^2(\gamma(x^1), v) + \mu f^2(\gamma(x^1), w)$$

by the linearity of γ_{*x^1}. It follows that $\mathscr{T}$ is well defined and has a linear extension from $\text{span}\{f^1(.,\alpha(u)) : u \in U^2\}$ to $\text{span}\{f^2(.,u) : u \in U^2\}$ and hence to a Lie algebra homomorphism from $\alpha(L^2)$ to L^2 where $\alpha(L^2)$ is the subalgebra of L^1 generated by $\{f^1(.,\alpha(u)) : u \in U^2\}$. Let L_0^i be the subalgebra of L^i given by

$$L_0^i = \{h(.) \in L^i : h(x_0^i) = 0\}.$$

L_0^i is the *isotropy subalgebra* of L^i at x_0^i and corresponds to the vector fields which have an equilibrium point at x_0^i. Finally put $\alpha(L^2)_0 = L_0^1 \cap \alpha(L^2)$.

If L^i, $i = 1, 2$, are the Lie algabras of systems Σ^i, then we say that

(a) L^2 *divides* L^1 if L^2 is homomorphic (as a Lie algebra) to a subalgebra of L^1.

(b) L^2 *G-divides* L^1 if there exists a map $\alpha : U^2 \to U^1$ such that $\mathscr{T}$ is a homomorphism of the Lie algebra $\alpha(L^2)$ onto L^2.

(c) L^2 *Σ-divides* l^1 if it G-divides L^1 and $\mathscr{T}(\alpha(L^2)_0) \subseteq L_0^2$.

Then we shall show in Chapter 3 on linearization of systems that Σ^1 is locally homomorphic to Σ^2 if L^2 Σ-divides L^1 (see Theorem 3.1.9). If $G^1(x_0^1)$ is simply connected then, since map γ is analytic, we can extend the local homomorphism to a global one by the theory of analytic continuation (see Forster 1981). Hence, if L^2 Σ-divides L^1 and $G^1(x_0^1)$ is simply connected, then Σ^1 is homomorphic to Σ^2.

If L^2 Σ-divides L^1, then suppose $M^1 = G^1(x_0^1)$, without loss of generality. If M^1 is not simply connected, then let $\pi: M \to M^1$ be the unique simply connected covering manifold of M^1 (see Spanier (1966)). Since π is a local diffeomorphism π_* is invertible and we can define the lift of f^1 to M by

$$f(x,u) = \pi_*^{-1}(x) f^1(\pi(x), u), \; x \in M, u \in U = U^1.$$

For any initial point $x_0 \in \pi^{-1}(x_0^1)$ it follows from Chow's theorem and the local equivalence of M and M^1 that $G(x_0) = M$. Moreover we have the diagram

$$\begin{array}{ccc} U = U^1 & \xrightarrow{\mathscr{F}_{x_0}} & \mathscr{X}_{x_0} \\ \uparrow{\scriptstyle id} & & \downarrow{\scriptstyle \pi} \\ U^1 & \xrightarrow{\mathscr{F}^1_{x_0}} & \mathscr{X}^1_{x_0} \end{array}$$

so that the system $\Sigma = \Sigma(M, U, f, x_0)$ is homomorphic to Σ^1 and is called the *simply connected cover* of Σ^1. For example, if Σ^1 consists of $M^1 = S^1$ (the unit circle), $U^1 = \mathbb{R}, \theta_0 = 0, \dot{\theta} = u$, then the simply connected cover of Σ^1 consists of $M = \mathbb{R}$, $U = \mathbb{R}$, $x_0 = 0, \dot{x} = u$.

If L^2 G-divides L^1, then lift L^1 to the group G^1 as in (2.5.2). Then define a system $\Sigma = \Sigma(M, U, f, \phi_0)$, where $M = G^1, U = U^1$, f is given locally by (2.5.2) and ϕ_0 is the identity map of G^1. Then Σ is homomorphic to Σ^1 if we define

$$\alpha = id: U^1 \to U, \qquad \gamma(\phi) = \phi(x_0), \quad \phi \in G^1.$$

Since the isotropy subalgebra L_0 of Σ is trivial (since ϕ_0 is the identity) and since the Lie algebra of Σ is isomorphic to that of Σ^1, by construction, L^2 Σ-divides L. Hence if L^2 G-divides L^1, then by lifting the dynamics of Σ^1 we can replace Σ^1 by the system Σ so that L^2 Σ-divides L, where L is the Lie algebra of Σ. Recall, however, from the construction of the local system (2.5.2) that L^1 must be finite dimensional. Σ is called the *group cover* of Σ^1.

Let Σ^1 be a system whose Lie algebra L^1 is finite dimensional and let $\Sigma = \Sigma(M, U, f, x_0)$ be the system defined by

$$M = M^1, \quad U = L^1, \quad x_0 = x_0^1 \;\text{ and }\; f(x, h(.)) = h(x),$$
$$\text{for } x \in M^1, \quad h(.) \in L^1.$$

Then Σ is homomorphic to Σ^1 if we define $\alpha: U^1 \to U$ by $\alpha(u) = f^1(.,u)$ and $\gamma = \text{id}: M \to M^1$. Σ is called the *fully controllable cover* of Σ^1 since Σ

is given locally by

$$\dot{x} = u.$$

Note, however, that u is not arbitrary here, but is a vector field on M. If Σ^2 and Σ^1 such that L^2 divides L^1, then if L is the fully controllable cover of Σ^1 it is clear that L^2 G-divides L.

Hence we have shown that if Σ^2 and Σ^1 are systems such that the Lie algebra L^2 of Σ^2 divides that of Σ^1, then we can lift Σ^1 to a system Σ such that $L^2 \Sigma$-divides L and the group $G(x_0)$ of Σ is simply connected, provided the Lie algebras are finite dimensional.

We can now consider the cascade decomposition of systems. Let $\Sigma^i = \Sigma^i(M^i, U^i, f^i, x^i)$ be systems for $i = 1, 2$ and let $v: M^1 \times U^1 \to U^2$ be a function which is analytic in $x^1 \in M^1$ and continuous in $u^1 \in U^1$. Then the *cascade* $\Sigma^1 \ominus_v \Sigma^2$ of Σ^1 and Σ^2 with *linking map* v is defined as the system

$$(M^1 \times M^2, U^1, f^1 \ominus_v f^2, (x_0^1, x_0^2))$$

with the local dynamics given by

$$f^1 \ominus_v f^2(x^1, x^2, u^1) = (f^1(x^1, u^1), f^2(x^2, v(x^1, u^1)).$$

i.e. the control input to Σ^2 is a function of the state x^1 and control input u^1 of the system Σ^1. If v is independent of x^1 we call $\Sigma^1 \ominus_v \Sigma^2$ a *parallel cascade* and if v is independent of u^1, it is called a *series cascade* (see Fig. 2.8). A system Σ is said to have a *nontrivial cascade decomposition* if there exist Σ^1, Σ^2 and v such that $\Sigma^1 \ominus_v \Sigma^2$ is homomorphic to Σ but neither of Σ^1, Σ^2 is homomorphic to Σ.

Suppose that Σ is homomorphic to Σ^1 and consider the system $\Sigma^1 \ominus_v \Sigma^2$. Locally this system has the form

$$(f^1(x^1, u^1), f^2(x^2, v(x^1, u^1)).$$

Since Σ is homomorphic to Σ^1 there exist maps $\alpha: U^1 \to U$, $\gamma: M \to M^1$ such that the diagram (2.5.6) commutes. Define

$$w(x, u) = v(\gamma(x), u^1), \qquad x \in M, \quad u \in \alpha(U)$$

for any $u^1 \in \alpha^{-1}(u)$ (and define $w(x, u)$ arbitrarily for $u \notin \alpha(U)$). Then it is easy to see that $\Sigma \ominus_w \Sigma^2$ is homomorphic to $\Sigma^1 \ominus_v \Sigma^2$. Similarly, $\Sigma^2 \ominus_s \Sigma$ is homomorphic to $\Sigma^2 \ominus_v \Sigma^1$ for some linking map s. It follows that if $\Sigma^1 \ominus_v \Sigma^2$ is homomorphic to Σ and $\Sigma^3 \ominus_w \Sigma^4$ is homomorphic to Σ^1, then $(\Sigma^3 \ominus_w \Sigma^4) \ominus_s \Sigma^2$ is homomorphic to Σ for some linking map s and so we have a kind of associative law for $\ominus$.

Theorem 2.5.1

If the Lie algebra L of a system $\Sigma = \Sigma(M, U, f, x_0)$ is the nontrivial direct sum of a finite-dimensional subalgebra L^1 and an ideal L^2 (as vector spaces) then the system has a nontrivial cascade decomposition. If it is the direct sum of two ideals, then the system has a parallel cascade decomposition.

Proof
Since $L = L^1 \oplus L^2$ (as vector spaces) we can write

$$f(.,u) = f^1(.,u) + g(.,u)$$

where $f^1 \in L^1$, $g \in L^2$. Let $\tilde{\Sigma}^1 = \Sigma(M, U, f^1, x_0)$ be the system associated with L^1 in the obvious way. Since L^1 is finite dimensional, then so is the group G^1 of $\tilde{\Sigma}^1$ and so G^1 can be embedded in $\mathbb{R}^k$ for some k, by Whitney's theorem. Hence we can define the control system

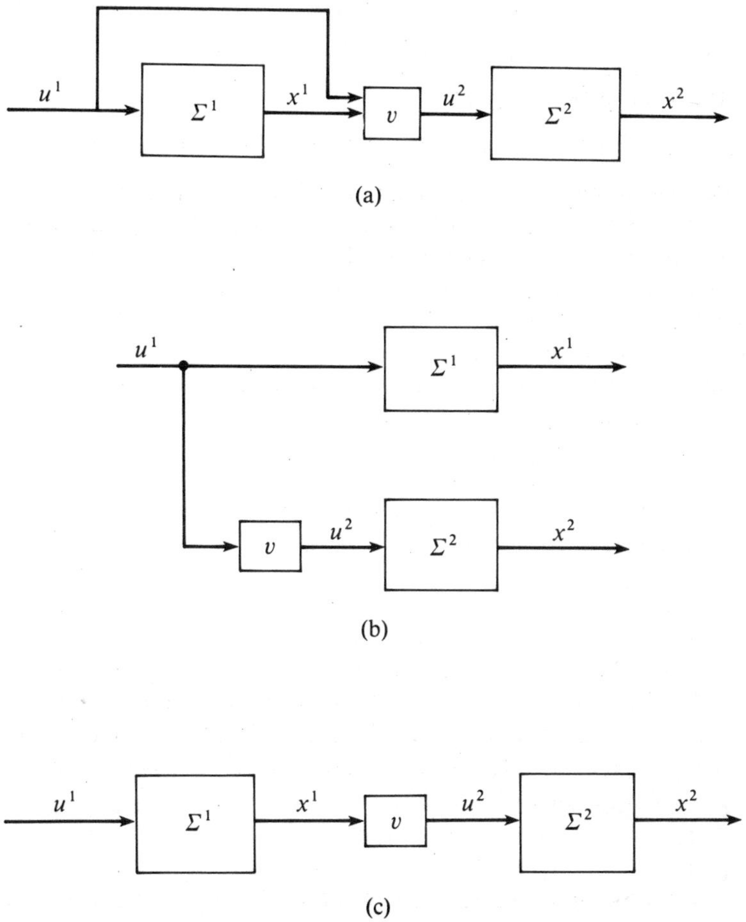

Fig. 2.8 Cascade decomposition of a system: (a) general cascade; (b) parallel cascade; (c) series cascade

$\Sigma^2 = \Sigma(M, G^1 \times U, f^2, x_0)$, where

$$f^2(x, \phi, u) = \phi_*^{-1}(x) g(\phi(x), u), \qquad x \in M, \quad \phi \in G^1, \quad u \in U.$$

Note that the Lie algebra of Σ^2 is L^2 since $\phi_*(x)$ is an isomorphism. Let Σ^1 be the group cover of $\tilde{\Sigma}^1$, so that G^1 is the state space of Σ^1. Then we can define the cascade $\Sigma^1 \ominus_v \Sigma^2$ where $v: G^1 \times U \to U^2 (= G^1 \times U)$ is the identity. To show that $\Sigma^1 \ominus_v \Sigma^2$ is homomorphic to Σ let $u(.) \in \mathscr{U}$ be a fixed control and let $x(t), \phi(t;.)$ and $x^2(t)$ be the trajectories arising from this control for Σ, Σ^1 and Σ^2, respectively. Then $x(0) = x_0 = \phi(0, x_0) = \phi(0, x_0^2)$ and

$$\begin{aligned} \frac{\mathrm{d}\phi}{\mathrm{d}t}(t, x^2(t)) &= f^1(\phi(t, x^2(t)), u(t)) + \phi_*(t, x^2(t)) \frac{\mathrm{d}x^2}{\mathrm{d}t}(t) \\ &= f^1(\phi(t, x^2(t)), u(t)) + g(\phi(t, x^2(t)), u(t)) \\ &= f(\phi(t, x^2(t)), u(t)). \end{aligned}$$

Hence $\Sigma^1 \ominus_v \Sigma^2$ is homomorphic to Σ if we choose $\alpha = id$ and $\gamma: G^1 \times M^2 \to M$ to be given by $\gamma(\phi, x) = \phi(x)$. Neither Σ^1 nor Σ^2 is homomorphic to Σ since L^1 and L^2 are nontrivial.

Finally, if L^1 and L^2 are ideals then $[L^1, L^2] \subseteq L^1 \cap L^2 = 0$ and so $f^2(x, \phi, u) = \phi_*^{-1}(x) g(\phi(x), u) = g(x, u)$, for any $\phi \in G^1$. Hence f^2 is independent of ϕ and so $\Sigma^1 \ominus_v \Sigma^2$ is a parallel cascade. □

Corollary 2.5.2

If the Lie algebra L of Σ is finite dimensional, then Σ can be decomposed into a parallel cascade of systems with simple Lie algebras followed by a cascade of one-dimensional systems.

Proof

Since L is finite dimensional it can be written as the vector space direct sum of a semisimple subalgebra L^1 and a maximal solvable ideal L^2. (This is called the Levi decomposition – see Sagle and Walde (1973).) Thus, Σ has a cascade decomposition $\Sigma^1 \ominus_v \Sigma^2$.

Write $L^1 = L_1^1 \oplus \ldots \oplus L_k^1$ as a direct sum of simple ideals. Theorem 2.5.1 now gives the first part of this result. As for Σ^2, since L^2 is solvable, $[L^2, L^2]$ is a proper ideal of L^2 and so we can find a subspace L_1^2 of codimension one in L^2 containing $[L^2, L^2]$, which is clearly an ideal, and so we can write L^2 as a vector space direct sum $L^2 = L_1^2 \oplus L_2^2$, where L_2^2 is one dimensional and L_1^2 is a solvable ideal of dimension one less than that of L^2. The second part now follows from Theorem 2.5.1. □

Krener (1977) gives an example to show that a system may admit a non-trivial cascade decomposition even when its Lie algebra is simple. However, this decomposition is based on a splitting of the controls which does not split the Lie algebra of the system. Hence we say that a finite-dimensional system is *indecomposable* if, for any finite-dimensional cascade decomposition $\Sigma^1 \ominus_v \Sigma^2$ of Σ into finite-dimensional systems Σ^1, Σ^2, L must divide L^1 or L^2. Then we have

Theorem 2.5.3
If Σ is finite dimensional, then Σ is indecomposable if and only if its Lie algebra Σ is one dimensional or simple. (See Krener (1977) for a proof.)□

A similar approach to the decomposition of bilinear systems is given by Wichmann (1961), again in terms of solvable and simple Lie algebras. However, in both this work and that described above, the sum of the dimensions of the state spaces of the decomposed subsystems is generally much greater than the dimension of the state space of the original system. In the next section we shall consider systems which possess symmetries and show that, under certain mild conditions, such systems have decompositions in which the dimensions of the subsystems add exactly to that of the main system. Before this let us note finally that Nijmeijer (1983) has given a solution of the parallel decomposition problem by feedback.

2.5.2 Systems with symmetries

In this section we shall consider the decomposition of systems with symmetry following Grizzle and Marcus (1985). We begin with a simple example to illustrate how the existence of a symmetry can lead to a decomposition of the system. Let $M = (\mathbb{R}^2 - \{0\}) \times \mathbb{R}^2$ and consider on M the two-dimensional motion of a particle of unit mass under a central (gravitational) force. If (q_1, q_2) and (p_1, p_2) are the usual rectangular coordinates and the corresponding momenta, respectively, then we have the dynamical equations

$$\dot{x} = f(x)$$

where $x = (q_1, q_2, p_1, p_2)^{\mathrm{T}}$ and $f(x) = (p_1, p_2 - q_1/(q_1^2 + q_2^2)^{3/2}, -q_2/(q_1^2 + q_2^2)^{3/2})^{\mathrm{T}}$. Since this is a radially symmetric problem it follows that f must be invariant under the transformation group $SO(2)$ acting on $\mathbb{R}^2$. Indeed we have

$$\begin{bmatrix} g & 0 \\ 0 & g \end{bmatrix} f(x) = f\left[\begin{bmatrix} g & 0 \\ 0 & g \end{bmatrix} x\right]$$

for any $g \in SO(2)$. (This can also be checked by direct computation, of course.) If we add acceleration controls along the q_1, q_2 directions, then we obtain the system

$$\dot{x} = \hat{f}(x, u)$$

where

$$\hat{f}(x, u) = f(x) = (0, 0, u_1, u_2)^{\mathrm{T}}. \tag{2.5.8}$$

Clearly, to obtain an invariance in this system we must also transform the controls, so that we have

$$\begin{bmatrix} g & 0 \\ 0 & g \end{bmatrix} \hat{f}(x, u) = \hat{f}\left\{ \begin{bmatrix} g & 0 & 0 \\ 0 & g & 0 \\ 0 & 0 & g \end{bmatrix} \begin{bmatrix} x \\ u \end{bmatrix} \right\} \tag{2.5.9}$$

However, if we express the controls in terms of radial and tangential components v_1, v_2, where

$$\begin{bmatrix} v_1 \\ v_2 \end{bmatrix} = \frac{1}{r} \begin{bmatrix} q_1 & -q_2 \\ q_2 & q_1 \end{bmatrix} \begin{bmatrix} u_1 \\ u_2 \end{bmatrix}, r = (q^2 + q^2)^{1/2},$$

then the system is transformed into

$$\dot{x} = f'(x, v) \triangleq f(x) + \frac{1}{r}(0 \quad 0 \quad q_1 \quad q_2)^{\mathrm{T}} v_1 + \frac{1}{r}(0 \quad 0 \quad -q_2 \quad q_1)^{\mathrm{T}} v_2$$

and we clearly have

$$\begin{bmatrix} g & 0 \\ 0 & g \end{bmatrix} f'(x, v) = f'\left[\begin{bmatrix} g & 0 \\ 0 & g \end{bmatrix} x, v \right], \, g \in SO(2). \tag{2.5.10}$$

In polar coordinates we can write the system as

$$\begin{bmatrix} \dot{r} \\ \dot{p}_r \\ \dot{p}_\theta \end{bmatrix} = \begin{bmatrix} p_r \\ p_\theta^2/r^3 - 1/r^2 \\ 0 \end{bmatrix} + \begin{bmatrix} 0 \\ v_1 \\ v_2 \end{bmatrix}$$

$$\dot{\theta} = p_\theta / r^2.$$

Clearly, the radial symmetry allows the decomposition of the system into the (r, p_r, p_θ) dynamics followed by a 'quadrature' for the evaluation of θ.

Consider now a general system $\Sigma(B, M, f)$ (without an output space) given by the commutative diagram

$$\begin{array}{ccc} B & \xrightarrow{\quad f \quad} & TM \\ & {\scriptstyle \pi} \searrow \quad \swarrow {\scriptstyle \pi_M} & \\ & M & \end{array}$$

where $\pi : B \to M$ is a smooth fibre bundle and f is a smooth map. As usual, we write this in local (fibre preserving) coordinates as

$$\dot{x} = f(x, u).$$

In the case of a state-independent input space U, we have $B = M \times U$.

Recall that a left action of a Lie group G on M is a smooth map $\Phi: G \times M \to M$ such that

(a) $\Phi(e, x) = x, x \in M$
(b) $\Phi(g, \Phi(h, x)) = \Phi(gh, x), x \in M, g, h \in G.$

Let $\Phi_g: M \to M, \Phi_x: G \to M$ be the maps given by

$$\Phi_g(x) = \Phi(g, x), \Phi_x(g) = \Phi(g, x).$$

Then Φ_g is a diffeomorphism, for each $g \in G$, with $(\Phi_g)^{-1} = \Phi_{g^{-1}}$. The orbit of x is $G.x = \{\Phi_g(x): g \in G\}$.

If θ and Φ are actions of G on B and M, respectively, then we say that Σ has the *symmetry* (G, θ, Φ) if the diagram

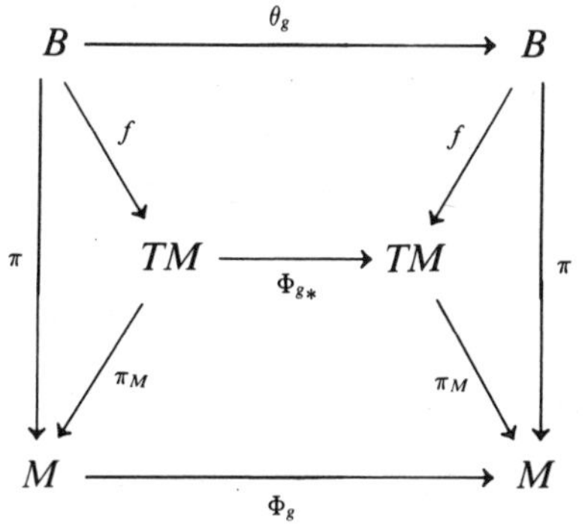

(2.5.11)

commutes for all $g \in G$. If $B = M \times U$ (i.e. B is trivial) then we say that the system has the *state-space symmetry* (G, Φ) if it has the symmetry (G, θ, Φ) where $\theta_g = (\Phi_g, Id_U): (x, u) \mapsto (\Phi_g(x), u)$. For a general system we can clearly define local state-space symmetries in a similar way.

As an example consider the system given by (2.5.8) on the manifolds $M = \mathbb{R}^2 \times (\mathbb{R}^2/\{0\})$ and $B = M \times \mathbb{R}^2$. If $G = SO(2)$ define the actions Φ, θ by

$$\Phi_g(x) = \begin{bmatrix} g & 0 \\ 0 & g \end{bmatrix} x, \theta_g(x, u) = \begin{bmatrix} g & 0 & 0 \\ 0 & g & 0 \\ 0 & 0 & g \end{bmatrix} \begin{pmatrix} x \\ u \end{pmatrix}. \qquad (2.5.12)$$

Then (2.5.9) shows that the system has the symmetry (G, θ, Φ) since $\Phi_{g*}: M \to M$ is given by $(x, \dot{x}) \to (\Phi_g(x), \Phi_g(\dot{x}))$. Similarly (2.5.10) shows that the system has the state-space symmetry (G, Φ).

We can also define a local version (in space and time) of the above symmetry concept. To do this we need to introduce the infinitesimal generator of a G action $\Phi: G \times M \to M$. Let T_eG be the Lie algebra of G and let $\xi \in T_eG$. If $\exp: T_eG \to G$ denotes the exponential map, then let $\Phi^\xi: \mathbb{R} \times M \to M$ be given by

$$\Phi^\xi(t, x) = \Phi(\exp t\xi, x).$$

Since Φ^{ξ} is a complete flow on M, this is an $\mathbb{R}$-action on M. The vector field

$$\xi_M^{\Phi}(x)(\text{or } \xi_M(x)) = \frac{\mathrm{d}}{\mathrm{d}t}\,\Phi(\exp t\xi, x)|_{t=0}$$

is called the *infinitesimal generator* of this action. We have

$$[\xi_M, \eta_M] = -[\xi, \eta]_M$$

for any $\xi, \eta \in T_eG$. (This can be checked in the case $G = SO(3)$, $T_eG = so(3)$ with the action of G on $\mathbb{R}^3$ given by $(g, x) \mapsto gx$.)

We say that (G, θ, Φ) is an *infinitesimal symmetry* of $\Sigma = \Sigma(B, M, f)$ if for each $x_0 \in M$ there exists an open neighbourhood V of x_0 and an $\varepsilon > 0$ such that

$$T(\xi_M)_t f(b) = f((\xi_B)_t(b)),$$

for all $b \in \pi^{-1}(V)$, $|t| < \varepsilon$ and $\xi \in T_eG$ with $\|\xi\| < 1$, where X_t denotes the flow of the vector field X and $\|\,.\,\|$ is any norm on T_eG. Since the diagram (2.5.11) implies the commutativity of the diagram

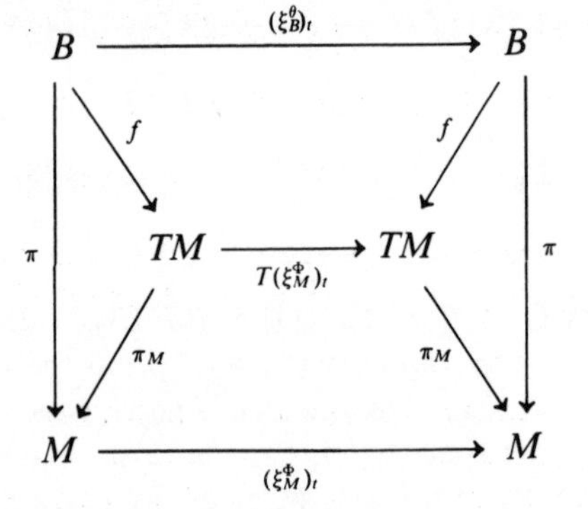

(2.5.13)

for all $t \in \mathbb{R}$ and $\xi \in T_eG$, a symmetry (G, θ, Φ) of Σ is also an infinitesimal symmetry. An infinitesimal symmetry (G, θ, Φ) is an *infinitesimal state-space symmetry* if B has an integrable connection H and the infinitesimal generators of θ are horizontal, i.e. ξ_B^{θ} is the horizontal lift of ξ_M^{Φ}. (See Chapter 1 for the concept of horizontal and vertical vectors of a bundle $\pi: B \to M$ with a connection H.) Since the horizontal lift is unique we can omit the reference to θ and write the symmetry in this case as (G, Φ).

Since H and V (where $H(b) \oplus V(b) = T_bB$) are integrable we can find coordinates (x, u) for B such that the vectors $\partial/\partial x_i$ are horizontal and the vectors $\partial/\partial u_j$ are vertical. These are called *connection respecting coordinates.* In such coordinates an infinitesimal state-space symmetry takes the form

$$(\xi_M)_{t*} f(x, u) = f((\xi_M)_t(x), u).$$

For example, it is easy to check that the system $\Sigma(B, M, \hat{f})$ has an infinitesimal symmetry (G, θ, Φ), where $\hat{f}$ is given by (2.5.8), $G = SO(2)$ and Φ, θ are given by (2.5.12). However, this is not a state-space symmetry under the obvious horizontal connection.

To see the connection between symmetries and invariant distributions (which we expect to exist since both give rise to system decompositions, as we shall see) let $\Sigma(B, M, f)$ have the infinitesimal symmetry (G, θ, Φ). Then the diagram (2.5.13) implies that

$$f_* \xi_B = \dot{\xi}_M, \; \pi_* \xi_B = \xi_M.$$

If $\triangle_B$ is the distribution $\text{span}\{\xi_B : \xi \in T_e G\}$ and $\triangle_M = \text{span}\{\xi_M : \xi \in TM\}$, then

$$f_* \triangle_B \subseteq \dot{\triangle}_M \qquad \text{and} \qquad \pi_* \triangle_B = \triangle_M. \tag{2.5.14}$$

Nijmeijer and Van der Schaft (1982) show that this is equivalent to local controlled invariance. Here $\dot{\triangle}_M$ is defined in the following way. First define the codistribution Γ by

$$\Gamma(x) = \{\theta \in T_x^* M : \theta(X) = 0 \text{ for all } X \in \triangle_M(x), x \in M\}$$

Let $\theta_1, \ldots, \theta_{n-k}$ be a basis for Γ and define $\dot{\theta}_i \in T^*TM$ by

$$\dot{\theta}_i(X) = X(\theta_i), \qquad X \in D(TM) \quad (= \text{vector fields on } TM).$$

(This makes sense since θ_i can be regarded as a real-valued function on TM.) Note that $\pi_M^* \theta_i \in T^*TM$, and so we can define the codistribution $\dot{\Gamma}$ on TM by

$$\dot{\Gamma} = \text{span}\{\pi^* \theta_1, \ldots, \pi^* \theta_{n-k}, \dot{\theta}_1, \ldots, \dot{\theta}_{n-k}\}.$$

Then $\dot{\triangle}_M$ is defined by

$$\dot{\triangle}_M(z) = \{X \in T_z TM : \Theta(X) = 0, \Theta \in \dot{\Gamma}(z), z \in TM\}.$$

(If $\triangle_M$ is involutive and is given locally by $\text{span}\{\partial/\partial x_1, \ldots, \partial/\partial x_k\}$, then $\dot{\triangle}_M$ is given by $\dot{\triangle}_M = \{\partial/\partial x_1, \ldots, \partial/\partial x_k, \partial/\partial \dot{x}_1, \ldots, \partial/\partial \dot{x}_k\}$, where $(x_1, \ldots, x_n, \dot{x}_1, \ldots, \dot{x}_n)$ are the coordinates of TM, with $\dot{x}_j(X) = \mathrm{d}x_j(X)$, $X \in TM$.)

Now we shall assume that M and G have dimensions n and k, respectively, with $k < n$. We say that the action $\Phi : G \times M \to M$ is *nondegenerate* at m if it is free, i.e. if $\Phi_m : G \to M$ is one-to-one. This condition is equivalent to Φ_{m*} having full rank $(= k)$ and since

$$\xi_m(x) = \frac{\mathrm{d}}{\mathrm{d}t} \Phi(\exp t\xi, x)|_{t=0}$$

it is also equivalent to $\xi_m^1(m), \ldots, \xi_m^k(m)$ being linearly independent whenever $\xi^1, \ldots, \xi^k \in T_e G$ are linearly independent.

Theorem 2.5.4
If the system $\Sigma(B, M, f)$ has an infinitesimal state-space symmetry (G, Φ) where G is abelian and Φ is nondegenerate at some point $m \in M$, then there exist connection respecting coordinates $(x_1, \ldots, x_n, u)$ on B for x in a neighbourhood of $x(m)$ such that Σ has the local form

$$\dot{x} = f(x_1, \ldots, x_{n-k}, u).$$

Proof
By assumption, we have

$$(\xi_m)_{t*} f(x, u) = f((\xi_m)_t(x), u) \tag{2.5.15}$$

for all $(x, u) \in \pi^{-1}(V)$, $|t| < \varepsilon$ and $\|\xi\| < 1$, for some neighbourhood V of m. Let $\xi^1, \ldots, \xi^k \in T_eG$ be linearly independent vectors with $\|\xi^i\| < 1$. Then $[\xi^i_m, \xi^j_m] = 0$, $1 \leqslant i, j \leqslant k$, since G is abelian. Moreover, $\xi^1_m(m), \ldots, \xi^k_m(m)$ are linearly independent by nondegeneracy. Hence, by Frobenius' theorem we can find a chart (V, x) so that

$$\xi^i_m \bigg|_V = \frac{\partial}{\partial x_{n-k+i}}\bigg|_V, \qquad 1 \leqslant i \leqslant k,$$

i.e. $(\xi^i_m)_t(x) = x + te_{n-k+i}$, where $\{e_i\}$ is the standard basis of $\mathbb{R}^n$. Hence $(\xi^i_m)_{t*} = I$ and so (2.5.15) gives

$$f(x, u) = f(x + te_{n-k+i}, u), \qquad 1 \leqslant i \leqslant k$$

for all $(x, u) \in \pi^{-1}(V)$ with $|t| < \varepsilon$. This implies that $\partial f/\partial x_{n-k+i} = 0$ and the result follows directly. □

This theorem shows that if a system $\dot{x} = f(x, u)$ has an infinitesimal state-space symmetry of the above form, then we can find coordinates so that the system becomes

$$\dot{\xi}_1 = F_1(\xi_1, u)$$

$$\xi_2 = \xi_2(0) + \int_0^t F_2(\xi_1, u)\, \mathrm{d}t$$

where $\xi_1 = (x_1, \ldots, x_{n-k})^{\mathrm{T}}$, $\xi_2 = (x_{n-k+1}, \ldots, x_n)^{\mathrm{T}}$, and

$$F_1(\xi_1, u) = (f_1(x_1, \ldots, x_{n-k}, u), \ldots, f_{n-k}(x_1, \ldots, x_{n-k}, u))^{\mathrm{T}}$$
$$F_2(\xi_1, u) = (f_{n-k+1}(x_1, \ldots, x_{n-k}, u), \ldots, f_n(x_1, \ldots, x_{n-k}, u))^{\mathrm{T}},$$

and so the symmetry reduces the system to one of dimension $n - k$ followed by a quadrature (i.e. a static nonlinearity followed by an integrator). In the case of a non-abelian Lie group G a similar result can be obtained by considering the centre of G, i.e. the set $C = \{h \in G : hg = gh \text{ for all } g \in G\}$; for details see Grizzle and Marcus (1985).

Next consider the following version of feedback equivalence (discussed in Section 2.4). A system $\Sigma(B, M, f)$ is said to be *feedback equivalent* to a system $\Sigma'(B, M, f')$ if there exists a bundle isomorphism $\gamma: B \to B$ such that the diagram

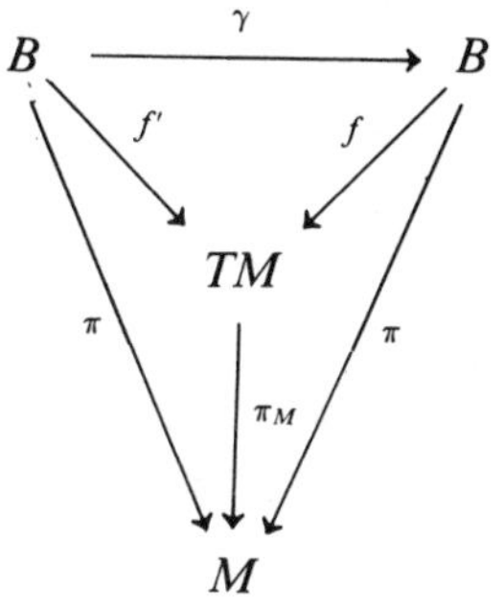

commutes. Moreover, we define the *restriction* $\Sigma \mid V$ of Σ to an open subset $V \subseteq M$ to be the system $\Sigma(\pi^{-1}(V), V, f \mid \pi^{-1}(V))$.

Theorem 2.5.5
If $\Sigma(B, M, f)$ has the infinitesimal symmetry (G, θ, Φ) and Φ is nondegenerate at m, then there exists a system $\Sigma'(B, M, f')$ with infinitesimal symmetry (G, Ψ, Φ) such that $\Sigma \mid V$ is feedback equivalent to $\Sigma' \mid V$ for some open set $V \subseteq M$.

Proof
We show that there exists an open set V containing m, $\varepsilon > 0$, a subset $S \subseteq B$ with $S \supseteq \pi^{-1}(V)$ and an isomorphism $\gamma: S \to S$ such that

$$(\xi_B^\theta)_t \circ \gamma = \gamma \circ (\xi_B^\Psi)_t$$

for all $\| \xi \| < \varepsilon$ and $| t | < 1$. This is sufficient, as follows from the diagrams

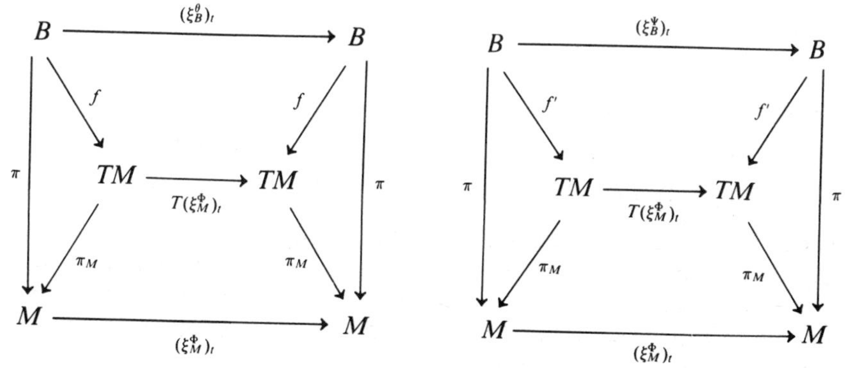

(2.5.16)

restricted to $S \subseteq B$. Let $\Delta = \{\xi_M^\Phi : \xi \in T_e G\}$. From the relation $[\xi_M, \eta_m] = -[\xi, \eta]_M$ it follows that Δ is an involutive distribution on M. By the nondegeneracy of Φ at m we have rank $(\Delta \mid \bar{V}) = k$ $(=\dim G)$ in some open set $\bar{V}$ containing m and so

$$\Delta \mid \bar{V} = \text{span}\left\{\frac{\partial}{\partial x_1}, \ldots, \frac{\partial}{\partial x_k}\right\},$$

in appropriate coordinates. We want to 'factor out' the first k coordinates from $\bar{V}$ and replace this part of $\bar{V}$ by a subset of $T_e G$ (which has dimension k). Let $\bar{V}/\Delta$ denote the set of equivalence classes of points in $\bar{V}$ with the same first k coordinates. By nondegeneracy of Φ and a simple application of the inverse function theorem it follows that there exists an open set $W \subseteq T_e G$ containing the origin such that the maps

$$F^\theta : \bar{V}/\Delta \times W \times U \to B, \qquad F^\theta(y, \xi, u) = (\xi_B^\theta)_1(\sigma(y), u)$$
$$F^\Psi : \bar{V}/\Delta \times W \times U \to B, \qquad F^\Psi(y, \xi, u) = (\xi_B)_1(\sigma(y), u)$$

where $\sigma : \bar{V}/\Delta \to V$ is given by $\sigma(y) = (0, \ldots, 0, y_1, \ldots, y_{n-k})$, are diffeomorphisms onto the same range. (Since we have effectively 'factored out' the action Φ the diagrams (2.5.16) imply that F^θ and F^Ψ have the same range.) Now exp is a local diffeomorphism and so there exists an open set $W_1 \subseteq W$ with $0 \in W_1$ and

$$\exp(W_1) \circ \exp(W_1) \subseteq \exp(W). \tag{2.5.17}$$

Restricting F^θ and F^Ψ to $\bar{V}/\Delta \times W_1 \times U$ and denoting their common range on this set by S we define the diffeomorphism $\gamma : S \to S$ by

$$\gamma(s) = F^\theta \circ (F^\Psi)^{-1}(s), \qquad s \in S$$

and put $V = \pi(S)$. γ is a bundle isomorphism because we have

$$\pi \circ (\xi_B^\theta)_t = (\xi_M)_t \circ \pi, \ \pi \circ (\xi_B^\Psi)_t = (\xi_M)_t \circ \pi.$$

Finally let $(x, u) \in \pi^{-1}(V)$ and let $(y, \xi, v) = (F^\Psi)^{-1}(x, u)$. Then if $\eta \in W_1$ with $\|\eta\| < \varepsilon$ and $|t| < 1$ we have

$$\begin{aligned}(\eta_B^\theta)_t \circ \gamma(x, u) &= \theta(\exp t\eta, \theta(\exp \xi, (\sigma(y), v))) \\ &= \theta(\exp t\eta \circ \exp \xi, (\sigma(y), v))\end{aligned}$$

and

$$\begin{aligned}\gamma \circ (\eta_B^\Psi)_t(x, u) &= \gamma \circ (\eta_B^\Psi)_t \circ (\xi_B^\Psi)_1(\sigma(y), v) \\ &= \gamma \circ \Psi(\exp t\eta \circ \exp \xi, (\sigma(y), v)) \\ &= \theta(\exp t\eta \circ \exp \xi, (\sigma(y), v)),\end{aligned}$$

using (2.5.17). □

It follows from this result that if $\Sigma(B, M, f)$ has the symmetry (G, θ, Φ)

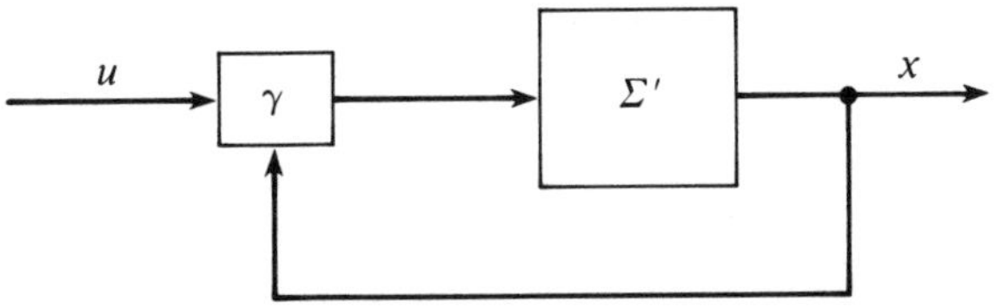

Fig. 2.9 Local decomposition of a symmetric system

where Φ is nondegenerate at $m \in M$, then Σ is locally feedback equivalent to a system Σ' with infinitesimal state-space symmetry (G, Φ) near m. Hence Σ can be decomposed locally as in Fig. 2.9.

The above results can be globalized by considering the global factorization of M by G. Then, under certain conditions, $p: M \to M/G$ is a principal fibre bundle and the local results globalize if there exists a cross section of p. (See Grizzle and Marcus (1985) for details.) Symmetry theory for Hamiltonian systems (see below) is discussed in detail by Abraham and Marsden (1978).

2.6 HAMILTONIAN SYSTEMS

2.6.1 Symplectic geometry

In the final section of this chapter we shall consider a particularly important class of nonlinear systems, namely those described by a Hamiltonian form. We shall begin by discussing symplectic geometry, which is the natural apparatus for the study of Hamiltonian systems. The exposition given here will follow closely that of Abraham and Marsden (1978).

We start with local symplectic theory on the finite-dimensional vector space $\mathbb{R}^n$. A bilinear form $\omega: \mathbb{R}^n \times \mathbb{R}^n \to \mathbb{R}$ is said to be *nondegenerate* if

$$\omega(x_1, x_2) = 0 \quad \text{for all } x_1 \in \mathbb{R}^n \qquad \text{implies that} \quad x_2 = 0$$

and is *symmetric* or *skew-symmetric* if $\omega^{\mathsf{T}} = \omega$ or $\omega^{\mathsf{T}} = -\omega$, respectively, where ω^{T} is the transpose of ω defined by

$$\omega^{\mathsf{T}}(x_1, x_2) = \omega(x_2, x_1).$$

If $\{e_i\}$ is a basis of $\mathbb{R}^n$ and $\{f^j\}$ is the dual basis, then we can write

$$\omega = \omega_{ij} f^i \otimes f^j, \qquad \omega_{ij} = \omega(e_i, e_j)$$

where ω_{ij} is the matrix representation of ω in this basis. If we change the basis to $e_i' = \sum_{j=1}^n a_{ij} e_j$, then, if ω_{ij}' is the matrix of ω in this basis and

$A = (a_{ij})$, we obtain the relation

$$\omega'_{ij} = \omega(e'_i, e'_j) = \omega\left(\sum_k a_{ik} e_k, \sum a_{jl} e_l\right)$$

$$= \sum_k \sum_l a_{ik}\omega(e_k, e_l) a_{jl}$$

that is

$$(\omega') = A(\omega)A^{\mathsf{T}},$$

where (ω) denotes the matrix of ω in the appropriate basis.

Given a bilinear form ω, we associate with it a linear map $L_\omega : \mathbb{R}^n \to (\mathbb{R}^n)^* = \mathbb{R}^n$ given by

$$\langle L_\omega(x), y \rangle = \omega(x, y), \qquad x, y \in \mathbb{R}^n. \tag{2.6.1}$$

Note that

$$\langle L_\omega(e_i), e_j \rangle = \omega(e_i, e_j) = \omega_{ij}$$

and so

$$L_\omega(e_i) = \sum_{j=1}^{n} \omega_{ij} f^j.$$

The *rank* of a bilinear form is defined as the rank of any matrix representation of the form. Clearly, ω is nondegenerate if and only if it has rank n, or equivalently, if L_ω is an isomorphism. If ω is symmetric and of rank s then, by classical linear algebra, (ω) has s nonzero eigenvalues and we can choose a basis in $\mathbb{R}^n$ so that (ω) is diagonal with the eigenvalues on the diagonal. Similarly, if ω is skew symmetric of rank r, then $r = 2m$, i.e. r is even, and we can write ω in the form

$$\omega = 2 \sum_{i=1}^{m} f^i \wedge f^{i+m} = \sum_{i=1}^{m} (f^i \otimes f^{i+m} - f^{i+m} \otimes f^i)$$

in a suitable (dual) basis $\{f^i\}$, so that, in this basis, ω has the matrix representation

$$(\omega) = \begin{bmatrix} 0 & I_{m\times m} & 0 \\ -I_{m\times m} & 0 & 0 \\ 0 & 0 & 0 \end{bmatrix}. \tag{2.6.2}$$

If $\{e_i\}$ is the dual basis to $\{f^i\}$, then

$$\omega(x, y) = \sum_{i=1}^{m} (x_i y_{i+m} - y_i x_{i+m})$$

where

$$x = \sum_{i=1}^{n} x_i e_i, \qquad y = \sum_{i=1}^{n} y_i e_i.$$

A *symplectic form* on $\mathbb{R}^n$ is a nondegenerate 2-form $\omega \in \Omega_2(\mathbb{R}^n)$ and the pair $(\mathbb{R}^n, \omega)$ is called a *symplectic (vector) space.* A map $f: (\mathbb{R}^n, \omega_1) \to (\mathbb{R}^n, \omega_2)$ is called *symplectic* if $f^*\omega_2 = \omega_1$, where f^* is the pull-back function defined by

$$(f^*\omega_2)(x, y) = \omega_2(f(x), f(y)). \tag{2.6.3}$$

The set of all symplectic maps $f: (\mathbb{R}^n, \omega) \to (\mathbb{R}^n, \omega)$ is called the *symplectic group*, $Sp(n)$. The condition (2.6.3) reduces to

$$A^{\mathsf{T}} JA = J$$

where

$$J = \begin{bmatrix} 0 & I \\ -I & 0 \end{bmatrix}$$

is the matrix of ω and A is the matrix of f in suitable coordinates. As we have seen in Chapter 1, the Lie algebra $sp(n)$ of $Sp(n)$ is the set of matrices A which satisfy

$$A^{\mathsf{T}} J + JA = 0. \tag{2.6.4}$$

We also say that a linear map $L: (\mathbb{R}^n, \omega) \to (\mathbb{R}^n, \omega)$ is *infinitesimally symplectic* if

$$\omega(Lx, y) + \omega(x, Ly) = 0 \tag{2.6.5}$$

for all $x, y \in \mathbb{R}^n$. Then if A is the matrix of L in the basis for which the matrix of ω is

$$J = \begin{bmatrix} 0 & I \\ -I & 0 \end{bmatrix},$$

this is equivalent to (2.6.4).

We now turn to global symplectic geometry on a manifold M. If $\omega \in \Omega_2(M)$, i.e. ω is a 2-form on M, then we define the vector bundle mapping $\Gamma_\omega: TM \to T^*M$ by

$$\Gamma_\omega(p): T_pM \to T_p^*M$$

where

$$\Gamma_\omega(p) = L_{\omega(p)}: T_pM \to T_p^*M \tag{2.6.6}$$

and $L_{\omega(p)}$ is given by (2.6.1). It is easy to check that Γ_ω *is* a vector bundle mapping. If $X \in D(M)$, we define $\Gamma_\omega X: M \to T^*M$ by $\Gamma_\omega X(p) = \Gamma_\omega(p)(X_p)$.

Nondegenerate closed 2-forms have a particularly simple local expression as the following theorem of Darboux shows.

Theorem 2.6.1

If ω is a nondegenerate‡ 2-form on a $2n$-manifold M, then ω is closed (i.e. $\mathrm{d}\omega = 0$) if and only if there is a chart (U, ϕ) at each $m \in M$ such that $\phi(m) = 0$ and

$$\omega \mid U = \sum_{i=1}^{n} \mathrm{d}x^i \wedge \mathrm{d}y^i$$

where $\phi(p) = (x^1(p), \ldots, x^n(p), y^1(p), \ldots, y^n(p))$.

Proof

The sufficiency of the condition is clear and so we suppose that $\mathrm{d}\omega = 0$. If we can find a chart in which ω is constant, then the result follows from the representation (2.6.2). Since the result is local we can assume that $M = \mathbb{R}^{2n}$ and $m = 0$.

Let $\tilde{\omega} = \omega(0) - \omega$, where $\omega(0)$ is regarded as a constant 2-form on $\mathbb{R}^{2n}$, and define $\omega_t = \omega + t\tilde{\omega}$ for $0 \leqslant t \leqslant 1$. Note that

$$\omega_t(0) = \omega(0), \qquad 0 \leqslant t \leqslant 1$$

and so $\omega_t(0)$ is nondegenerate for each t. Hence there is a spherical neighbourhood V of 0 in $\mathbb{R}^n$ on which ω_t is nondegenerate for each t.

Since $\mathrm{d}\tilde{\omega} = \mathrm{d}\omega(0) - \mathrm{d}\omega = 0$ on V we can write $\tilde{\omega} = \mathrm{d}\alpha$ for a 1-form α with $\alpha(0) = 0$, by the Poincaré lemma (Lang (1972)). Since ω_t is nondegenerate, we can find a smooth vector field $X_t (0 \leqslant t \leqslant 1)$ such that

$$X_t \lrcorner\, \omega_t = -\alpha$$

(see Section 1.3 for the definition of the inner product). Since $X_t(0) = 0$, the flow ϕ_t of X_t (a time-dependent vector field) exists in V, by shrinking V if necessary. Thus,

$$\begin{aligned}
\frac{\mathrm{d}}{\mathrm{d}t}(\phi_t^* \omega_t) &= \phi_t^*(L_{X_t}\omega_t) + \phi_t^* \frac{\mathrm{d}}{\mathrm{d}t}\omega_t \\
&= \phi_t^*(d \circ (X_t \lrcorner\, \omega_t) + X_t \lrcorner \circ \mathrm{d}\omega_t) + \phi_t^* \frac{\mathrm{d}}{\mathrm{d}t}\omega_t \\
&= \phi_t^*(-\mathrm{d}\alpha + \tilde{\omega}) \\
&= 0,
\end{aligned}$$

since $\mathrm{d}\omega_t = 0$. (Again, see Section 1.3 for the relation between the flow ϕ_t and the Lie derivative.) Hence

$$\phi_1^* \omega_1 = \phi_0^* \omega_0 = \omega_0 = \omega$$

since $\phi_0 = Id$. Changing coordinates by using $y = \phi_1(x)$ transforms ω into ω_1, which is constant. □

‡ That is, $\omega(m)$ is nondegenerate for each $m \in M$.

A *symplectic manifold* (M, ω) is a manifold M with a nondegenerate, closed 2-form ω on M (called a *symplectic form* or *symplectic structure*). The chart given in Darboux's theorem is called a *symplectic chart* and the functions x^i, y^i are called *canonical coordinates.* A *symplectic transformation* $f: (M, \omega) \to (N, \rho)$ is a C^∞ map between symplectic manifolds for which $f^*\rho = \omega$, where the pull-back f^* of f is defined pointwise as above.

Note that if Q is a manifold which is regarded as the space of 'generalized coordinates' representing a dynamical system and T^*Q is the canonical momentum space (see earlier in this chapter), then T^*Q carries a canonical symplectic structure ω given locally by

$$\omega = \sum_{i=1}^{n} \mathrm{d}q^i \wedge \mathrm{d}p_i \tag{2.6.7}$$

where $(q^1, \ldots, q^n)$ are coordinates on Q and $(q^1, \ldots, q^n, p_1, \ldots, p_n)$ are the corresponding coordinates on T^*Q. ω is given by $\omega = -\mathrm{d}\theta$, where θ is the 1-form

$$\theta = \sum_{i=1}^{n} p_i \, \mathrm{d}q^i.$$

By considering local charts it can be shown that θ has the following global description. let $\pi: T^*Q \to Q$ and $\pi_*: T(T^*Q) \to TQ$ be the canonical projection and its differential, respectively. Then if $v \in T^*Q$ we define $\theta_V: T_V(T^*Q) \to \mathbb{R}$ by

$$\theta_V(w) = \langle v, \pi_* w \rangle$$

and then $\theta: v \to \theta_V$ is the required 1-form.

2.6.2 Hamiltonian systems

If (M, ω) is a symplectic manifold and $H: M \to \mathbb{R}$ is differentiable then we consider the 1-form $\mathrm{d}H$ defined on M. Since ω is nondegenerate, there exists a vector field X_H such that

$$\omega(X_H, .) = \mathrm{d}H \tag{2.6.8}$$

(this is trivial pointwise and it is easy to see that X_H is smooth). This can also be written

$$X_H \lrcorner\, \omega = \mathrm{d}H \qquad \text{or} \qquad \Gamma_\omega^{-1}(\mathrm{d}H)$$

where Γ_ω is given by (2.6.6) and $\Gamma_\omega^{-1}(p) = (\Gamma_\omega(p))^{-1}$, which exists since ω is nondegenerate.

To verify that this global structure has the expected form, write ω in canonical coordinates as in (2.6.7) and then note that

$$X_H \lrcorner\, \mathrm{d}q^i = \frac{\partial X_H}{\partial q^i}, \qquad X_H \lrcorner\, \mathrm{d}p_i = \frac{\partial X_H}{\partial p_i}$$

where

$$X_H = \sum_{i=1}^{n} \left(\frac{\partial X_H}{\partial q^i} \, \mathrm{d}q^i + \frac{\partial X_H}{\partial p_i} \, \mathrm{d}p_i \right).$$

Hence,

$$\begin{aligned} X_H \,\lrcorner\, \omega &= \Sigma X_H(\mathrm{d}q^i \wedge \mathrm{d}p_i) = \Sigma (X_H \,\lrcorner\, \mathrm{d}q^i) \wedge \mathrm{d}p_i - \Sigma \mathrm{d}q^i \wedge (X_H \,\lrcorner\, \mathrm{d}p_i) \\ &= \Sigma \left(\frac{\partial X_H}{\partial q^i} \, \mathrm{d}p_i - \frac{\partial X_H}{\partial p_i} \right) \mathrm{d}q^i. \end{aligned}$$

Since we require $X_H \,\lrcorner\, \omega = \mathrm{d}H$ we must have

$$X_H = \left(\frac{\partial H}{\partial p_i}, \; - \frac{\partial H}{\partial q^i} \right) = J \, \mathrm{d}H$$

where

$$J = \begin{bmatrix} 0 & I \\ -I & 0 \end{bmatrix}.$$

We therefore see that the system (2.6.8) has the local form

$$q^i = \frac{\partial H}{\partial p_i}, \quad \dot{p}_i = - \frac{\partial H}{\partial q^i}, \qquad 1 \leqslant i \leqslant n$$

and so we call (2.6.8) (or the triple (M, ω, X_H)) a *Hamiltonian system*.

If (M, ω, X_H) is a Hamiltonian system and ϕ_t is the flow of X_H, then

$$\begin{aligned} \frac{\mathrm{d}}{\mathrm{d}t} \phi_t^* \omega = \phi_t^* L_{X_H} \omega &= \phi_t^* (X_H \,\lrcorner\, \mathrm{d}\omega + \mathrm{d}(X_H \,\lrcorner\, \omega)) \\ &= \phi_t^* (0 + \mathrm{dd}H) \\ &= 0 \end{aligned} \qquad (2.6.9)$$

since $\mathrm{d}\omega = 0$. Hence $\phi_t^* \omega$ is constant and equals ω; i.e. ϕ_t^* is a symplectic map for each t.

Let (M, ω) be a symplectic manifold and let X be a vector field on M. Then X is called *locally Hamiltonian* if for each $m \in M$ there exists a neighbourhood V of m such that $X \mid V$ is Hamiltonian. Note that X is locally Hamiltonian if and only if $X \,\lrcorner\, \omega$ is locally exact (i.e. equals $\mathrm{d}H$ locally, for some H) and this is true if and only if $d(X \,\lrcorner\, \omega) = 0$ by the Poincaré lemma. By (2.6.9) X is also locally Hamiltonian if and only if $L_X \omega = 0$. Hence it follows from the relation

$$L_{[X, Y]} \omega = L_X L_Y \omega - L_Y L_X \omega$$

that the set of locally Hamiltonian vector fields is a Lie subalgebra of $D(M)$.

We shall next introduce the global version of the classical Poisson bracket. If (M, ω) is a symplectic manifold, α is a 1-form and X is a vector field (on M), then define the maps $\# : \Omega_1(M) \to D(M)$ and $\flat : D(M) \to \Omega_1(M)$ by

$$(\#\alpha) \lrcorner \omega = \alpha, \qquad \flat X = X \lrcorner \omega.$$

Then we define the *Poisson bracket* $\{.,.\} : \Omega_1 \times \Omega_1 \to \Omega_1$ by

$$\{\alpha, \beta\} = -\flat[\#\alpha, \#\beta].$$

Since we have, in general,

$$\begin{aligned}(\mathrm{d}\omega)(X, Y, Z) &= L_X(\omega(Y, Z)) + L_Y(\omega(Z, X)) + L_Z(\omega(X, Y)) \\ &\quad - \omega([X, Y], Z) - \omega([Y, Z], X) - \omega([Z, X], Y)\end{aligned}$$

and since $\mathrm{d}\omega = 0$ if ω is a symplectic form, we have

$$\begin{aligned}0 &= L_{\#\alpha}(\beta(Z)) - L_{\#\beta}(\alpha(Z)) - L_Z(\#\alpha \lrcorner (\#\beta \lrcorner \omega)) \\ &\quad + \{\alpha, \beta\}(Z) + \alpha(L_{\#\beta} Z) - \beta(L_{\#\alpha} Z)\end{aligned}$$

if we put $X = \#\alpha$, $Y = \#\beta$. However,

$$(L_{\#\alpha}\beta)(Z) = \#\alpha(\beta(Z)) - \beta(L_{\#\alpha} Z) \qquad \text{(by (1.3.1))}$$

and

$$L_{\#}\alpha(\beta(Z)) = \mathrm{d}(\beta(Z))\#\alpha = \#\alpha(\beta(Z)) \qquad \text{(by (1.1.1))}$$

and so, using similar expressions for $(L_{\#\beta}\alpha)(Z)$ and $L_{\#\beta}(\alpha Z))$, we obtain

$$\{\alpha, \beta\} = -L_{\#\alpha}\beta + L_{\#\beta}\alpha + \mathrm{d}(\#\alpha \lrcorner \#\beta \lrcorner \omega).$$

Note that if α is closed then $L_{\#\beta}\alpha = \#\beta \lrcorner \mathrm{d}\alpha + \mathrm{d}\#\beta \lrcorner \alpha = \mathrm{d}\#\beta \lrcorner \alpha$ and similarly, if β is closed, $L_{\#\alpha}\beta = \mathrm{d}\#\alpha \lrcorner \beta$. Hence, if α and β are closed, $\{\alpha, \beta\}$ is exact.

We have defined the Poisson bracket of 1-forms above. We can also define the *Poisson bracket* of functions $f, g \in \mathscr{F}(M)$ by

$$\{f, g\} = -\#(\mathrm{d}f) \lrcorner \#(\mathrm{d}g) \lrcorner \omega = \omega(\#\ \mathrm{d}f, \#\ \mathrm{d}g).$$

We have

$$\{f, g\} = -\#(\mathrm{d}f) \lrcorner \mathrm{d}g = -L_{\#\ \mathrm{d}f}g$$

and similarly,

$$\{f, g\} = L_{\#\mathrm{d}g}f.$$

If ϕ_t is the flow of $\#\mathrm{d}f$, then

$$\frac{\mathrm{d}}{\mathrm{d}t}(g \circ \phi_t) = \phi_t^* L_{\#\mathrm{d}f}g = -\phi_t^*\{f, g\} = 0$$

if and only if $\{f, g\} = 0$. Hence g is constant on the orbits of $\#f$ if and only if $\{f, g\} = 0$, and a similar result holds for f. Moreover, we have the well-known equations of motion in the Poisson bracket formulation:

$$\frac{\mathrm{d}}{\mathrm{d}t}(f \circ \phi_t) = \frac{\mathrm{d}}{\mathrm{d}t}(\phi_t^* f) = \phi_t^* L_{X_H} f$$
$$= L_{X_H}(f \circ \phi_t) = \{f \circ \phi_t, H\}, \qquad \text{for all } f \in \mathscr{F}(M)$$

where X_H is a Hamiltonian vector field with Hamiltonian H and flow ϕ_t.

We can write the Poisson bracket in canonical symplectic coordinates $(q^1, \ldots, q^n, p_1, \ldots, p_n)$ as follows. Since $\{f, g\} = L_{\#\mathrm{d}g} f = \mathrm{d}f(\#\mathrm{d}g)$, and

$$\#\mathrm{d}g = \left(\frac{\partial g}{\partial p_i}, -\frac{\partial g}{\partial q^i}\right)$$

in canonical coordinates, we have

$$\{f, g\} = \sum_{i=1}^{n} \left(\frac{\partial f}{\partial q^i}\frac{\partial g}{\partial p_i} - \frac{\partial f}{\partial p_i}\frac{\partial g}{\partial q^i}\right).$$

Let us note finally that the vector space $\mathscr{F}(M)$ with the Poisson bracket is a Lie algebra. For, $\{f, g\}$ is clearly bilinear in f and g and $\{f, f\} = 0$. Moreover we have

$$\{f, \{g, h\}\} = -L_{\#\mathrm{d}f}\{g, h\} = L_{\#\mathrm{d}f} L_{\#\mathrm{d}g} h$$
$$\{g, \{h, f\}\} = -L_{\#\mathrm{d}g} L_{\#\mathrm{d}f} h$$

and

$$\{h, \{f, g\}\} = L_{\#\mathrm{d}\{f, g\}} h = L_{\#\{\mathrm{d}f, \mathrm{d}g\}} h = L_{-[\#\mathrm{d}f, \#\mathrm{d}g]} h,$$

since $\mathrm{d}\{f, g\} = \{\mathrm{d}f, \mathrm{d}g\}$. Hence the Jacobi identity holds. Since, for a Hamiltonian vector field X_H we have $X_H = \#\mathrm{d}H$, the above reasoning also shows that the globally Hamiltonian vector fields on a symplectic manifold (M, ω) form a Lie algebra.

2.6.3 Affine Hamiltonian systems

Consider first the linear system

$$\begin{aligned} \dot{x} &= Ax + Bu, \qquad x \in X = \mathbb{R}^{2n} \\ y &= Cx + Du, \qquad u \in \mathbb{R}^m, \quad y \in \mathbb{R}^m \end{aligned} \tag{2.6.10}$$

where

$$A^{\mathrm{T}} J + JA = 0, \qquad B^{\mathrm{T}} J = C, \qquad D = D^{\mathrm{T}} \tag{2.6.11}$$

and

$$J = \begin{bmatrix} 0 & -I \\ I & 0 \end{bmatrix},$$

as before. Then if $G(s)$ is the transfer function of this system it is easy to check that $G(s) = G^{\mathrm{T}}(-s)$. Conversely, Brockett and Rahimi (1972) have shown that a system with a transfer function which satisfies $G(s) = G^{\mathrm{T}}(-s)$ has a minimal realization of the form (2.6.10). The linear system (2.6.10) which satisfies (2.6.11) is called a *(linear) Hamiltonian system*. It can be seen that such a system is controllable if and only if it is observable.

If we now consider the affine system

$$\dot{x} = A(x) + \sum_{i=1}^{m} u_i B_i(x), \qquad y_i = C_i(x), \quad 1 \leqslant i \leqslant m, \qquad (2.6.12)$$

defined on a symplectic manifold (M, ω), the above discussion of linear systems leads us to define an *affine Hamiltonian system* to be a system of the form (2.6.12) for which A is locally Hamiltonian, i.e.

$$L_A \omega = 0$$

and such that

$$\omega(B_i, .) = \mathrm{d}C_i,$$

that is $\qquad B_i = X_{C_i} \quad \text{(by (2.6.8))} \qquad \text{for } 1 \leqslant i \leqslant m.$

We would like to relate controllability and observability for the affine Hamiltonian system (2.6.12). Let us first recall the definitions of strong accessibility and local weak observability. For the system (2.6.12), define

$$\Gamma_0 = \operatorname{span}\{B_1, \ldots, B_m\}, \qquad \Gamma = A + \Gamma_0.$$

(Γ is an affine subspace of $D(M)$.) By induction, define

$$\Gamma_k = [\Gamma, \Gamma_{k-1}] + \Gamma_{k-1}, \qquad k \geqslant 1$$

and let

$$K = \bigcup_{k \geqslant 0} \Gamma_k.$$

The system is *strongly accessible* if $K(x) = T_x M$, $x \in M$. Next define

$$F_0 = \operatorname{span}\{C_1, \ldots, C_m\}, \qquad F_k = L_\Gamma F_{k-1} + F_{k-1},$$

and

$$G = \bigcup_{k \geqslant 0} F_k.$$

Then the system is *locally weakly observable* if $\mathrm{d}G(x) = T_x^* M$, $x \in M$.

Since $L_A \omega = 0$ there exists a function $H: M \to \mathbb{R}$ such that, locally, $A = X_H$. Elements of F_k are sums of functions of the form

$$L_{f_1} L_{f_2} \cdots L_{f_k} C_j$$

where $f_i = A$ or $f_i = B_l$, which we can write as

$$(\pm)\{h_1, \{h_2, \{h_3, \ldots, \{h_k, C_j\}\ldots\}$$

where $\{.,.\}$ is the Poisson bracket and $h_i = H$ or $h_i = C_l$. This follows because $A = X_H$ and $B_i = X_{C_i}$ and we have

$$\{N_1, N_2\} = \omega(X_{N_1}, X_{N_2})$$

and so, for example,

$$\begin{aligned}\{H, C_j\} &= \omega(A, B_j)\\ &= -\mathrm{d}C_j(A)\\ &= -L_A C_j.\end{aligned}$$

Hence, if we define

$$F = H + \mathrm{span}\{C_1, \ldots, C_m\} \subseteq \mathscr{F}(M)$$

it follows that

$$F_k = \{F, F_{k-1}\} + F_{k-1}.$$

Theorem 2.6.2
The map $\alpha : N \to X_N$ is an isomorphism of $F_k/\mathbb{R}$ onto Γ_k.

Proof
Clearly, $X_N = 0$ if and only if N is constant. Hence we shall factor out $\mathbb{R}$ from F_k and write F_k in place of $F_k/\mathbb{R}$.

We shall prove that F_k is isomorphic to Γ_k by induction. For $k = 0$, $\Gamma_0 = \mathrm{span}\{B_1, \ldots, B_m\}$ and $F_0 = \mathrm{span}\{C_1, \ldots, C_m\}$ with $\omega(B_i,.) = \mathrm{d}C_i$ and so the result is true for $k = 0$, since ω is nondegenerate. Suppose the result is true for $k-1$. Since $F_k = \{F, F_{k-1}\} + F_{k-1}$ and F_{k-1} is isomorphic to Γ_{k-1}, by induction, we must prove that $\{F, F_{k-1}\}$ is mapped onto $\{\Gamma, \Gamma_{k-1}\}$ by α. Now,

$$\{F, F_{k-1}\} = \{H + F_0, F_{k-1}\} = \{H, F_{k-1}\} + \{F_0, F_{k-1}\}$$

and

$$\{\Gamma, \Gamma_{k-1}\} = [A, \Gamma_{k-1}] + [\Gamma_0, \Gamma_{k-1}] = [X_H, \Gamma_{k-1}] + [\Gamma_0, \Gamma_{k-1}].$$

Since the local Hamiltonian vector fields form a Lie algebra, we have

$$\alpha(\{N_1, N_2\}) = [X_{N1}, X_{N2}]$$

and the induction hypothesis implies that $\alpha(\{F, F_{k-1}\}) = [\Gamma, \Gamma_{k-1}]$. □

Corollary 2.6.3. An affine Hamiltonian system is locally weakly observable if and only if it is strongly accessible.

Proof
By Theorem 2.6.2, $\alpha: G \to K$ is an isomorphism, so that $\mathrm{d}G(x) = T_x^* M$ if and only if $K(x) = T_x M$. □

The minimality of affine Hamiltonian systems is discussed by Van der Schaft (1982b) and the theory of general nonlinear Hamiltonian systems is given by Van der Schaft (1982a).

3 LINEARIZATION THEORY

3.1 GLOBAL LINEARIZATION BY A DIFFEOMORPHIC CHANGE OF VARIABLES

3.1.1 Local theory

In this chapter we shall be interested in methods for reducing a nonlinear system to an equivalent linear one. As opposed to the familiar local linearization of a nonlinear system by retaining only the Jacobian of the nonlinearity in its Taylor series expansion, we shall be concerned here largely with *exact* linear representations. Given a linear control system

$$\dot{x} = Ax + Bu, \tag{3.1.1}$$

with $x(t) \in \mathbb{R}^n$, we can consider orthogonal linear transformations of the state space

$$y = Px \tag{3.1.2}$$

and we obtain an 'equivalent' linear system

$$\dot{y} = PAP^{-1}y + PBu. \tag{3.1.3}$$

If we apply a nonlinear transformation

$$y = g(x)$$

where g is a diffeomorphism of $\mathbb{R}^n$ onto itself, then the linear system (3.1.1) transforms into a nonlinear system

$$\dot{y} = \left(\frac{\partial h}{\partial y}\right)^{-1} Ah(y) + \left(\frac{\partial h}{\partial y}\right)^{-1} Bu,$$

where $h = g^{-1}$.

In this section we shall consider the inverse problem, namely given a nonlinear system, when does there exist a diffeomorphism which transforms the system to a linear one? However, we shall allow a more general class of transformations than that considered above – we shall also let the controls enter into the diffeomorphism, thus obtaining *feedback equivalent* systems. Thus, consider again the linear system (3.1.1) and note first that, as is well-known, we can find an invertible matrix P, such that the equivalent system (3.1.3) has the 'companion form':

$$\dot{y} = \begin{bmatrix} 0 & 1 & 0 & 0 & \dots & 0 \\ 0 & 0 & 1 & 0 & \dots & 0 \\ \vdots & \vdots & \vdots & \vdots & & \vdots \\ 0 & 0 & 0 & 0 & \dots & 1 \\ \alpha_1 & \alpha_2 & \alpha_3 & \alpha_4 & \dots & \alpha_n \end{bmatrix} y + \begin{bmatrix} 0 \\ \vdots \\ 0 \\ 1 \end{bmatrix} u$$

for some $\alpha_1, \dots, \alpha_n$, provided (3.1.1) is controllable. If we now allow a state dependent control transformation

$$v = \alpha^{\mathsf{T}} y + u$$

where $\alpha = (\alpha_1, \dots, \alpha_n)$, then (3.1.1) has the feedback equivalent form

$$\dot{y} = \begin{bmatrix} 0 & 1 & 0 & & \dots & 0 \\ 0 & 0 & 1 & \dots & \dots & 0 \\ \vdots & \vdots & \vdots & & \dots & \vdots \\ 0 & 0 & 0 & & \dots & 1 \\ 0 & 0 & 0 & & \dots & 0 \end{bmatrix} y + \begin{bmatrix} 0 \\ \vdots \\ 0 \\ 1 \end{bmatrix} v \tag{3.1.4}$$

The diffeomorphism connecting (3.1.1) and (3.1.4) is

$$y = Px$$
$$v = \alpha^{\mathsf{T}} y + u = \alpha^{\mathsf{T}} Px + u.$$

In the general nonlinear case we shall allow state transformations to be control dependent and so we consider diffeomorphisms $T: V \to W$ of the form

$$y_1 = T_1(x, u), \dots, y_n = T_n(x, u), \quad v = T_{n+1}(x, u) \tag{3.1.5}$$

from some open set $V \subseteq \mathbb{R}^{n+1}$ to another open set $W \subseteq \mathbb{R}^{n+1}$ (with $0 \in V, 0 \in W$). Thus, following Su (1982), we say that the systems

$$\dot{x} = f(x, u) \tag{3.1.6}$$

and

$$\dot{y} = g(y, v) \tag{3.1.7}$$

are (locally) $\mathcal{T}$-*equivalent* at (x, u), (y, v) if there exist open sets V and W around (x, u) and (y, v), respectively, and a C^∞ diffeomorphism T of the form (3.1.5) such that $T(x, u) = (y, v)$ and

$$\dot{T}_1 = g_1(T_1, \dots, T_n, T_{n+1})$$
$$\vdots$$
$$\dot{T}_n = g_n(T_1, \dots, T_n, T_{n+1}). \tag{3.1.8}$$

Clearly $\mathscr{T}$-equivalence *is* an equivalence relation, and we are interested in the equivalence class $\mathscr{L}$ containing the linear system (3.1.4).

In order to characterize the class $\mathscr{L}$ we recall some basic properties of vector fields and differential forms (see Chapter 1). First, if $h: \mathbb{R}^n \to \mathbb{R}$ is a scalar field, then its exterior derivative is

$$\mathrm{d}h = \frac{\partial h}{\partial x_1}\,\mathrm{d}x_1 + \cdots + \frac{\partial h}{\partial x_n}\,\mathrm{d}x_n$$

and so can be represented by the row vector $(\partial h/\partial x_1, \ldots, \partial h/\partial x_n)$ consisting of the components of $\mathrm{d}h$ in the basis $\mathrm{d}x_1, \ldots, \mathrm{d}x_n$ of $T_x^*(\mathbb{R}^n)$. A vector field

$$f = f_1 \frac{\partial}{\partial x_1} + \cdots + f_n \frac{\partial}{\partial x_n}$$

and the differential form have the dual product $\langle \mathrm{d}h, f\rangle$ which is just the inner product of the corresponding row and column vectors, i.e.

$$\langle \mathrm{d}h, f\rangle = \sum_{i=1}^{n} \frac{\partial h}{\partial x_i} f_i.$$

If $\{X_1, \ldots, X_m\}$ is a set of linearly independent vector fields, then recall that this set is called completely integrable if there exists a maximal integral submanifold generated by the vector fields in the set. Since such a manifold is of dimension m it is clear that $\{X_1, \ldots, X_m\}$ is completely integrable if and only if there exist scalar functions $h_1, \ldots, h_{n-m}$ such that

$$\langle \mathrm{d}h_i, X_j\rangle = 0, \qquad 1 \leqslant i \leqslant n-m, \quad 1 \leqslant j \leqslant m.$$

Now returning to the linearization problem, note that for a general transformation T given by (3.1.5) the relations (3.1.8) imply that

$$\sum_{j=1}^{n} \frac{\partial T_i}{\partial x_j} f_j(x,u) + \frac{\partial T_i}{\partial u}\dot{u} = g_i(T(x,u)), \qquad 1 \leqslant i \leqslant n$$

and since $g_i(T(x,u))$ is independent of $\dot{u}$ we must have

$$\frac{\partial T_i}{\partial u} = 0, \qquad 1 \leqslant i \leqslant n.$$

Hence in order that the systems (3.1.6) and (3.1.7) be $\mathscr{T}$-equivalent, it is necessary that $T_1, \ldots, T_n$ be independent of u. We shall therefore restrict attention to the class of diffeomorphisms T of the form

$$y_1 = T_1(x), \ldots, y_n = T_n(x), \qquad v = T_{n+1}(x,u), \quad T(0) = 0. \qquad (3.1.9)$$

(Of course, T_{n+1} will still depend on u.)

We now consider the class $\mathscr{L}$ and show that nonlinear systems in $\mathscr{L}$ must have a special form. For, if

$$\dot{x} = f(x,u) \qquad (3.1.10)$$

is a nonlinear system with $f(0, 0) = 0$ which is $\mathscr{T}$-equivalent to a linear system

$$\dot{y} = Ay + bv \tag{3.1.11}$$

where $y = T(x, u)$ is of the form (3.1.9), then

$$\left(\frac{\partial T'}{\partial x}\right) f = Ay + bv$$

where $T' = (T_1, \ldots, T_n)$ and $\partial T'/\partial x$ is the Jacobian matrix of T'. Since T' is independent of u and T is a diffeomorphism, it follows that $\partial T'/\partial x$ is invertible. Hence,

$$f = \left(\frac{\partial T'}{\partial x}\right)^{-1} Ay(x) + \left(\frac{\partial T'}{\partial x}\right)^{-1} bv(x, u),$$

and we have shown that any system (3.1.10) in $\mathscr{L}$ must have the form

$$\dot{x} = f(x) + g(x)\phi(x, u) \tag{3.1.12}$$

where $f(0) = 0$, $\phi(0, 0) = 0$ and $\partial\phi/\partial u \neq 0$. Note that f, g are vector fields and ϕ is a scalar field.

For a controllable system (3.1.11), the system (3.1.4) is $\mathscr{T}$-equivalent to it and so we can find a transformation T which maps the system (3.1.12) into (3.1.4). Hence we can write

$$\sum_{j=1}^{n} \frac{\partial T_i}{\partial x_j} (f_j(x) + g_j(x) . \phi(x, u)) = \begin{cases} T_{i+1}(x), & 1 \leqslant i \leqslant n-1 \\ T_{i+1}(x, u), & i = n \end{cases}$$

As noted above we can write this in the form

$$\langle \mathrm{d}T_i, f + g . \phi \rangle = T_{i+1}, \qquad i = 1, \ldots, n. \tag{3.1.13}$$

Now, $T_1, \ldots, T_n$ are independent of u and $\partial T_{n+1}/\partial u \neq 0$ and so it follows from (3.1.13) that

$$\langle \mathrm{d}T_i, g \rangle = 0, \qquad \langle \mathrm{d}T_i, f \rangle = T_{i+1}, \qquad i = 1, \ldots, n-1 \tag{3.1.14}$$

and

$$\langle \mathrm{d}T_n, f + g . \phi \rangle = T_{n+1}, \qquad \langle \mathrm{d}T_n, g \rangle \neq 0. \tag{3.1.15}$$

Lemma 3.1.1
A necessary and sufficient condition for a system

$$\dot{x} = f(x) + g(x) . \phi(x, u) \tag{3.1.16}$$

with $f(0) = 0$, $\phi(0, 0) = 0$ to be in $\mathscr{L}$ is that

$$\langle \mathrm{d}T_1, (\mathrm{ad}\, f)^i g \rangle = 0, \qquad i = 0, 1 \ldots, n-2, \tag{3.1.17}$$

$$\mathrm{d}T_1, (\mathrm{ad}\, f)^{n-1} g \rangle \neq 0, \tag{3.1.18}$$

where $(\mathrm{ad}\, f)g = [f, g]$.

Proof
To prove necessity we show that (3.1.14) and (3.1.15) imply (3.1.17) and (3.1.18). First recall the Leibnitz identity‡

$$\langle dT, (\text{ad } f)g \rangle = \langle d\langle dT, g\rangle, f\rangle - \langle d\langle dT, f\rangle, g\rangle$$

(see Chapter 1). To prove (3.1.17) we shall prove even more; namely that

$$\langle dT_j, (\text{ad } f)^i g\rangle = 0, \qquad i = 0, 1, \ldots, n-1-j, \qquad j = 1, \ldots, n-1 \quad (3.1.19)$$

This is clearly true for $i=0$, $j=1$ by (3.1.14). Assume it is true for $j + i \leqslant k \leqslant n-2$. Then

$$\begin{aligned}
\langle dT_j, (\text{ad } f)^{i+1} g\rangle &= \langle d\langle dT_j, (\text{ad } f)^i g\rangle, f\rangle - \langle d\langle dT_j, f\rangle, (\text{ad } f)^i g\rangle \\
&= 0 - \langle dT_{j+1}, (\text{ad } f)^i g\rangle \\
&= -\langle d\langle dT_{j+1}, (\text{ad } f)^{i-1} g\rangle, f\rangle \\
&\quad + \langle d\langle dT_{j+1}, f\rangle, (\text{ad } f)^{i-1} g\rangle \\
&= \langle dT_{j+2}, (\text{ad } f)^{i-1} g\rangle \\
&\;\vdots \\
&= (-1)^{i+1}\langle dT_{j+i+1}, g\rangle \\
&= 0.
\end{aligned}$$

Similarly, $\langle dT_{j+1}, (\text{ad } f)^i g\rangle = 0$. Hence (3.1.19) is true for $j + i \leqslant k + 1$ and so it is true by induction, and (3.1.17) is true, *a fortiori*.

For (3.1.18) we have

$$\begin{aligned}
0 \neq \langle dT_n, g\rangle &= \langle d\langle dT_{n-1}, f\rangle, g\rangle \\
&= \langle d\langle dT_{n-1}, g\rangle, f\rangle - \langle dT_{n-1}, (\text{ad } f)g\rangle \\
&= -\langle dT_{n-1}, (\text{ad } f)g\rangle \\
&= -\langle d\langle dT_{n-2}, f\rangle, (\text{ad } f)g\rangle \\
&= -\langle d\langle dT_{n-2}, (\text{ad } f)g\rangle, f\rangle + \langle dT_{n-2}, (\text{ad } f)^2 g\rangle \\
&= \langle dT_{n-2}, (\text{ad } f)^2 g\rangle, \qquad \text{by (3.1.17)} \\
&\;\vdots \\
&= (-1)^{n+1}\langle dT_1, (\text{ad } f)^{n-1} g\rangle
\end{aligned}$$

and the result follows.

The sufficiency of (3.1.17) and (3.1.18) is proved similarly by defining $T_2, \ldots, T_{n+1}$ successively from T_1 by (3.1.14) and (3.1.15). □

Lemma 3.1.2
If the system (3.1.16) belongs to $\mathscr{L}$, then the vectors

$$g, (\text{ad } f)g, \ldots, (\text{ad } f)^{n-1} g \text{ span } \mathbb{R}^n$$

in some neighbourhood of the origin.

‡ This is a special case of the formula $L_f\langle \omega, g\rangle = \langle L_f(\omega), g\rangle + \langle \omega, [f, g]\rangle$

Proof
Suppose, on the contrary, that

$$(\text{ad } f)^i g = \sum_{k=0}^{i-1} \alpha_k (\text{ad } f)^k g$$

for some i with $0 < i \leqslant n-1$. Then

$$\begin{aligned}\langle dT_1, (\text{ad } f)^{n-1} g\rangle &= \sum_{k=0}^{i-1} \alpha_k \langle dT_1, (\text{ad } f)^{n-i-1+k} g\rangle \\ &= 0\end{aligned}$$

by (3.1.17). However, this contradicts (3.1.18) and so the lemma is true. □

We can now characterize the nonlinear systems which belong locally to $\mathscr{L}$.

Theorem 3.1.3
The following three conditions are necessary and sufficient for a nonlinear system $\dot{x} = \tilde{f}(x, u)$ to belong to $\mathscr{L}$ in a neighbourhood U of the origin of $\mathbb{R}^{n+1}$:

(a) $\tilde{f}(x, u) = f(x) + g(x) . \phi(x, u)$ with $f(0) = 0$, $\phi(0, 0) = 0$, $\partial\phi/\partial u \neq 0$ for $(x, u) \in U$
(b) $\text{span}\{g, (\text{ad } f)g, \ldots, (\text{ad } f)^{n-1} g\} = \mathbb{R}^n$, for $(x, u) \in U$.
(c) the set of vector fields $\{g, (\text{ad } f)g, \ldots, (\text{ad } f)^{n-2} g\}$ is involutive.

Proof
We have seen that a nonlinear system belongs to $\mathscr{L}$ if and only if there exists a scalar field T_1 such that (3.1.17) and (3.1.18) hold. Now Frobenius' theorem (Chapter 1) states that $n-1$ linearly independent vector fields locally define an $(n-1)$-dimensional integral submanifold if and only if they are involutive. Hence a differential form dT_1 can exist if and only if conditions (b) and (c) hold. □

3.1.2 Global theory (Hunt *et al.* (1983))

We have seen that a necessary and sufficient condition for a nonlinear system of the form (3.1.16) to be in $\mathscr{L}$ is that (3.1.17) and (3.1.18) hold. Hence we must prove the existence of a solution T_1 of the partial differential equations

$$\frac{\partial T_1}{\partial x_1}((\text{ad } f)^k g)_1 + \frac{\partial T_1}{\partial x_2}((\text{ad } f)^k g)_2 + \cdots + \frac{\partial T_1}{\partial x_n}((\text{ad } f)^k g)_n = 0 \qquad (3.1.20)$$

$$(k = 0, 1, \ldots, n-2)$$

and

$$\frac{\partial T_1}{\partial x_1}((\mathrm{ad}\, f)^{n-1}g)_1 + \frac{\partial T_1}{\partial x_2}((\mathrm{ad}\, f)^{n-1}g)_2 + \cdots + \frac{\partial T_1}{\partial x_n}((\mathrm{ad}\, f)^{n-1}g)_n \neq 0 \tag{3.1.21}$$

where $(.)_i$ denotes the ith component of a vector field. We shall parameterize the coordinates $x_1, \ldots, x_n$ as functions of the variables $s, t_1, \ldots, t_{n-1}$ and consider the (ordinary) differential equations

$$\frac{\partial x}{\partial s} = (\mathrm{ad}\, f)^{n-1}g \tag{3.1.22}$$

$$\frac{\partial x}{\partial t_i} = (\mathrm{ad}\, f)^{n-i-1}g, \qquad 1 \leqslant i \leqslant n-1. \tag{3.1.23}$$

We can solve these equations as follows. First solve $\partial x/\partial s = (\mathrm{ad}\, f)^{n-1}g$ with the initial condition $x(0) = 0$ to give $x(s)$. Then solve $\partial x/\partial t_1 = (\mathrm{ad}\, f)^{n-2}g$ with initial condition $x(s, 0) = x(s)$ to give $x(s, t_1)$. Solve $\partial x/\partial t_2 = (\mathrm{ad}\, f)^{n-3}g$ with initial condition $x(s, t_1, 0) = x(s, t_1)$ and continue until we have solved $\partial x/\partial t_{n-1} = g$ with initial condition $x(s, t_1, \ldots, t_{n-2}, 0) = x(s, t_1, \ldots, t_{n-2})$. The solution $x(s, t_1, \ldots, t_{n-1})$ of the last equation then generates an integral manifold $\mathscr{I}_s$ for the involutive set $\{g, (\mathrm{ad}\, f)g, \ldots, (\mathrm{ad}\, f)^{n-2}g)\}$ for each s as $t_1, \ldots, t_{n-1}$ vary in $\mathbb{R}^{n-1}$. Then equation (3.1.20) becomes

$$\begin{aligned} 0 &= \sum_{i=1}^{n} \frac{\partial T_1}{\partial x_i^k}((\mathrm{ad}\, f)^k g)_1 \\ &= \sum_{i=1}^{n} \frac{\partial T_1}{\partial x_i^k}\frac{x_i^k}{t_{n-1-k}} \\ &= \frac{\partial T_1}{\partial t_{n-1-k}}, \qquad k = 0, 1, \ldots, n-2, \end{aligned} \tag{3.1.24}$$

where $x^k = x(s, t_1, \ldots, t_{n-1-k})$. Hence solving the partial differential equations (3.1.20) is equivalent to solving the ordinary differential equations (3.1.22) and the equations (3.1.24). However, the equations (3.1.24) have the (nonunique) solution $T_1 = s$. Thus $\mathrm{d}T_1 = \mathrm{d}s$. Since T_1 is constant on $\mathscr{I}_s$ for fixed s, the equations (3.1.24) imply that equations (3.1.20) hold on $\mathscr{I}_s$ for any s. Consider the involutive set $\{g, (\mathrm{ad}\, f)g, \ldots, (\mathrm{ad}\, f)^{n-2}g\}$. Since, for each fixed s, $x(s, t_1, \ldots, t_{n-1})$ defines an integral manifold for this set as $t_1, \ldots, t_{n-1}$ vary in $\mathbb{R}^{n-1}$, if $\langle \mathrm{d}T_1, (\mathrm{ad}\, f)^{n-1}g\rangle = 0$, it follows that $(\mathrm{ad}\, f)^{n-1}g$ must be tangent to this integral manifold. This contradicts the fact that the controllability matrix has rank n and so (3.1.21) also holds.

Hence, showing the existence of a function T_1 satisfying (3.1.20) and (3.1.21) (as a function of x) is equivalent to showing that the map $F: \mathbb{R}^n \to \mathbb{R}^n$ given by

$$F(s, t_1, \ldots, t_{n-1}) = x(s, t_1, \ldots, t_{n-1}) \tag{3.1.25}$$

is invertible. The Jacobian matrix of F is

$$\begin{bmatrix} \dfrac{\partial x_1}{\partial_s} & \dfrac{\partial x_1}{\partial t_1} & \cdots & \dfrac{\partial x_1}{\partial t_{n-1}} \\ \vdots & \vdots & & \vdots \\ \dfrac{\partial x_n}{\partial s} & \dfrac{\partial x_n}{\partial t_1} & \cdots & \dfrac{\partial x_n}{\partial t_{n-1}} \end{bmatrix}$$

By (3.1.22) and (3.1.23), this is just the controllability matrix and we shall use this to derive global linearization results, which are merely global inverse function theorems for F, as we have just seen.

Let us note first that the Jacobian matrix of $(T_1, \ldots, T_n)$ with respect to $s, t_1, \ldots, t_{n-1}$ has zeros above the main diagonal. For, we already know that

$$\frac{\partial T_1}{\partial t_i} = 0, \qquad 1 \leqslant i \leqslant n-1.$$

Applying the Leibnitz formula to (3.1.17), as in Lemma 3.1.1, we have

$$\langle \mathrm{d}T_2, (\mathrm{ad}\, f)^k g \rangle = 0, \qquad k = 0, 1, \ldots, n-3$$
$$\langle \mathrm{d}T_2, (\mathrm{ad}\, f)^{n-2} g \rangle \neq 0,$$

and so

$$\frac{\partial T_2}{\partial t_i} = 0, \qquad 2 \leqslant i \leqslant n-1.$$

The stated result now follows by induction.

In order to derive linearizations we shall state three results of Kou *et al.* (1973) whose proofs are straightforward.

Theorem 3.1.4

If $f: \mathbb{R}^n \to \mathbb{R}^n$ is a differentiable map with Jacobian matrix $J(x)$ and if there exists a constant $\varepsilon > 0$ such that

$$|\Delta_1| \geqslant \varepsilon, \qquad \frac{|\Delta_2|}{|\Delta_1|} \geqslant \varepsilon, \ldots, \qquad \frac{|\Delta_n|}{|\Delta_{n-1}|} \geqslant \varepsilon \tag{3.1.26}$$

for all $x \in \mathbb{R}^n$, then f is one-to-one from $\mathbb{R}^n$ onto $\mathbb{R}^n$. □

Here Δ_i is the ith leading principal minor of $J(x)$ (i.e. the determinant of the matrix obtained by deleting the last $n-i$ rows and columns of $J(x)$). The proof is by induction.

Corollary 3.1.5
If there is a constant $n \times n$ matrix A such that the matrix $AJ(x)$ satisfies (3.1.26), then f is again one-to-one and onto $\mathbb{R}^n$. □

Theorem 3.1.6
Let $f: \Omega \subseteq \mathbb{R}^n \to \mathbb{R}^n$ be defined on an open convex subset Ω of $\mathbb{R}^n$. Then

(a) If f is differentiable and there exists a constant $n \times n$ nonsingular matrix A so that $AJ(x)$ is positive-definite for all $x \in \Omega$, then f is one-to-one from Ω onto $f(\Omega)$.
(b) If f is continuously differentiable, Ω is bounded and there exists a constant nonsingular matrix A such that $\det AJ(x) > 0$ for all $x \in \Omega$ and $AJ(x) + (AJ(x))^{\mathrm{T}}$ has nonnegative principal minors for all $x \in \Omega$, then f is one-to-one from Ω onto $f(\Omega)$. □

Returning to the linearization of the nonlinear system

$$\dot{x}(t) = f(x(t)) + u(t)g(x(t)), \tag{3.1.27}$$

we can now prove

Theorem 3.1.7
If the controllability matrix of (3.1.27) is nonsingular on $\mathbb{R}^n$, $\{g, (\operatorname{ad} f)g, \ldots, (\operatorname{ad} f)^{n-2}g\}$ is an involutive set on $\mathbb{R}^n$ and the Jacobian matrix J_{F} of the function F given by (3.1.25) satisfies the condition (3.1.26), then the system (3.1.27) is in $\mathscr{L}$ and there exists a C^∞ map $T: \mathbb{R}^{n+1} \to \mathbb{R}^{n+1}$ such that

(a) $T(0) = 0$
(b) $T_1, \ldots, T_n$ are independent of u
(c) $\partial T/\partial(x, u)$ is a nonsingular $(n+1) \times (n+1)$ matrix on $\mathbb{R}^n$
(d) T is one-to-one and onto $\mathbb{R}^{n+1}$.

Proof
By Theorem 3.1.4 F is one-to-one from $\mathbb{R}^n$ to $\mathbb{R}^n$, and so we can solve for $s, t_1, \ldots, t_{n-1}$ globally as functions of $x_1, \ldots, x_n$. Hence T is defined globally as shown above. We have seen that the map $(T_1, \ldots, T_n)$ is nonsingular if the controllability matrix has rank n. By construction, T_{n+1} is a function of $(x_1, \ldots, x_n, u)$ which can be inverted as a function of u for all $x \in \mathbb{R}^n$ and so $\partial T/\partial(x, u)$ is nonsingular.

Since $(T_1, \ldots, T_n)$ has a nonsingular Jacobian (with respect to $s, t_1, \ldots, t_{n-1}$) which has zeros above the main diagonal, as noted previously, we have $\partial T_1/\partial s = 1 \neq 0$, $\partial T_2/\partial t_1 \neq 0, \ldots, \partial T_n/\partial t_{n-1} \neq 0$ on $\mathbb{R}^n$. Hence $(T_1, \ldots, T_n)$ is one-to-one on $\mathbb{R}^n$. Finally, $T = (T_1, \ldots, T_{n+1})$ is one-to-one on $\mathbb{R}^{n+1}$ since $\partial T/\partial(x, u)$ is nonsingular. □

Using Corollary 3.1.5 it is clear that the conclusions of Theorem 3.1.7 hold if J_F is replaced by AJ_F for any nonsingular matrix A.

Example

Consider the system

$$\begin{bmatrix} \dot{x}_1 \\ \dot{x}_2 \end{bmatrix} = \begin{bmatrix} \frac{1}{2} x_1^2 + e^{x_2} + x_2 - 1 \\ x_1^2 \end{bmatrix} + u \begin{bmatrix} 0 \\ 1 \end{bmatrix} \tag{3.1.28}$$

The right-hand side is of the form $f(x(t)) + u(t)g(x(t))$ and so the system may be in $\mathscr{L}$. We have

$$(\mathrm{ad}\, f)g = [f, g] = \begin{bmatrix} -(e^{x_2} + 1) \\ 0 \end{bmatrix}$$

and so $(g, (\mathrm{ad}\, f)g)$ has rank 2 on $\mathbb{R}^2$.

Solving (3.1.22) and (3.1.23) by the method described above we first solve

$$\frac{\partial x}{\partial s} = (\mathrm{ad}\, f)g = \begin{bmatrix} -(e^{x_2} + 1) \\ 0 \end{bmatrix}$$

with initial condition $x(0) = 0$. This gives $x_1(s) = -2s$ and $x_2 = 0$. Then we solve $\partial x/\partial t_1 = g = (0 \quad 1)^{\mathrm{T}}$, with initial condition $x(s, 0) = (-2s, 0)$. This gives $x = (-2s, t)$. Hence the Jacobian of F is

$$\begin{bmatrix} \partial x_1/\partial s & \partial x_1/\partial t_1 \\ \partial x_2/\partial s & \partial x_2/\partial t_1 \end{bmatrix} = \begin{bmatrix} -2 & 0 \\ 0 & 1 \end{bmatrix}$$

which satisfies condition (3.1.26). Applying Theorem 3.1.7 we see that the system (3.1.28) belongs to $\mathscr{L}$ globally.

The following result is useful in proving that global linearizations do not exist, even though local ones may exist.

Lemma 3.1.8

If T is a transformation of $\mathbb{R}^{n+1}$, then the points where f and g are linearly dependent must map under T into points where the vectors $A\widetilde{T}$ and B are linearly dependent, where $A, \widetilde{T}$ and B are defined by the linear system

$$\begin{bmatrix} \dot{T}_1 \\ \dot{T}_2 \\ \vdots \\ \dot{T}_n \end{bmatrix} = \begin{bmatrix} T_2 \\ T_3 \\ \vdots \\ 0 \end{bmatrix} + T_{n+1} \begin{bmatrix} 0 \\ 0 \\ \vdots \\ 1 \end{bmatrix} = A\widetilde{T} + T_{n+1}B.$$

Proof

If $f(x) = cg(x)$ for some c, then by (3.1.14), we have

$$0 = c\langle \mathrm{d}T_i, g\rangle(x) = \langle \mathrm{d}T_i, f\rangle(x) = T_{i+1}(x), \qquad 1 \leqslant i \leqslant n-1$$

and so $T_2 = \cdots = T_n = 0$ at x; i.e. $A\widetilde{T}$ and B are linearly dependent. □

Example
The system

$$\begin{bmatrix}\dot{x}_1\\ \dot{x}_2\end{bmatrix} = \begin{bmatrix}x_2 - x_2^2\\ 0\end{bmatrix} + u\begin{bmatrix}0\\ 1\end{bmatrix}$$

is locally linearizable at the origin as can easily be seen. However, f and g are linearly dependent on the two distinct lines $x_2 = 0$, $x_2 = 1$, whereas the linearization

$$\dot{T}_1 = T_2, \qquad \dot{T}_2 = T_3$$

has the vectors $(T_2 \quad 0)^{\mathrm{T}}$ and $(0 \quad 1)^{\mathrm{T}}$ linearly independent when $T_2 = 0$. Since a one-to-one map cannot map two distinct lines to a single line, this system does not belong to $\mathscr{L}$ globally.

Of course, there is a maximal open neighbourhood of $0 \in \mathbb{R}^n$ on which a given system belongs to $\mathscr{L}$ – it is just the union of all open neighbourhoods of 0 on which the conditions of Theorem 3.1.7 hold. Note finally that Theorem 3.1.6 gives rise to a result similar to Theorem 3.1.7 when the Jacobian of F has the properties specified in Theorem 3.1.6. We leave the precise formulation to the reader.

3.1.3 Linear state equivalence

In the above theory we have allowed the control to enter into the transformation T, thereby obtaining feedback linear equivalence. If we consider only transformations on the state we obtain state equivalence. In this section we shall derive necessary and sufficient conditions for the systems

$$\dot{x} = a_0(x) + \sum_{i=1}^{k} u_i(t)a_i(x), \qquad x(0) = x^0 \tag{3.1.29}$$

and

$$\dot{y} = b_0(y) + \sum_{i=1}^{k} u_i(t)b_i(y), \qquad y(0) = y^0 \tag{3.1.30}$$

to be state equivalent, where $a_j(x)$, $b_j(x)$, $0 \leqslant j \leqslant k$ are analytic vector fields and $u(t) = (u_1(t), .., u_k(t))$ is a bounded measurable control. These results are due to Krener (1973). By state equivalence we mean that there exists a local diffeomorphism betwen the state spaces of the systems which maps a solution of one to a solution of the other for any given control.

Let $t \to \alpha_i(t)x$ be the integral curve of $a_i(x)$ through x and recall that

$$[a_i, a_j](x^0) = L_{a_i}(a_j)(x^0) = \frac{\mathrm{d}}{\mathrm{d}t}(\alpha_i(-t)_* a_j(\alpha_i(t)x^0))\Big|_{t=0}$$

where f_* denotes the differential of f. Since a_i and a_j are analytic we have the Taylor series

$$\alpha_i(-t)_* a_j(\alpha_i(t)x^0) = \sum_{l=0}^{\infty} (t^l/l!)(\text{ad } a_i)^l a_j(x^0).$$

Introduce the set of vector fields

$$D^0(A) = \{ a_i : i = 0, \dots, k \}$$
$$D^j(A) = D^{j-1}(A) \cup \{ [a_i, c] : i = 0, \dots, k, c \in D^{j-1}(A) \}, \ j \geq 1$$

and

$$D(A) = \bigcup_{j \geq 0} D^j(A).$$

Also let

$$D(A)_x = \{ c(x) : c \in D(A) \} \subseteq \mathbb{R}^m.$$

We call the dimension of span $D(A)_x$ the *rank* of $D(A)$ at x (or the *rank* of the system (3.1.29) at x). A theorem of Nagano (1966) shows that for a system of rank r at x^0 there is a submanifold M of dimension r through x^0 such that the solutions for any bounded measurable control belong to M for small $|t|$. We shall say that M *carries* the system (at x^0). Let $\mathscr{B}(a)$ denote the set of all possible brackets formed from the vector fields $a_0, \dots, a_k$ and similarly for $\mathscr{B}(b)$. Also let $B(a_0, \dots, a_k)$ denote a generic element of $\mathscr{B}(a)$. Then we have

Theorem 3.1.9

Let M and N be manifolds carrying (3.1.29) and (3.1.30) at x^0 and y^0, respectively. Then there exists a linear map $L : \text{span } D(A)_{x^0} \to \text{span } D(B)_{y^0}$ such that

(a) $L(a_i(x^0)) = b_i(y^0), \quad 0 \leq i \leq k$

(b) $L(B(a_0, \dots, a_k)) = B(b_0, \dots, b_k) \quad$ for all $B(a) \in \mathscr{B}(a)$

if and only if there exist neighbourhoods U and V of x^0 and y^0 in M and N and an analytic map $\lambda : U \to V$ such that λ carries (3.1.29) into (3.1.30).

Proof

Assume that L exists. Since this is a local result we can assume that $M = \mathbb{R}^m$ and $N = \mathbb{R}^n$, so that span $D(A)_{x^0} = \mathbb{R}^m$ and span $D(B)_{y^0} = \mathbb{R}^n$. Let $c_1, \dots, c_n$ be elements of $D^0(A)$ such that $c_1(x^0), \dots, c_n(x^0)$ is a maximal linearly independent subset of $D^0(A)_{x^0}$ and if $c_i(x) = a_j(x)$ then let $d_i(y) = b_j(y)$, $1 \leq i \leq h$. Now complete $c_1, \dots, c_h$ with $c_{h+1}, \dots, c_m \in D(A)$ so that span$\{c_1(x^0), \dots, c_m(x^0)\} = \mathbb{R}^m$. If $c_i(x) = B(a_0, \dots, a_k)(x)$, then let $d_i(y) = B(b_0, \dots, b_k)(y)$, for $h + 1 \leq i \leq m$. Let $\alpha_i(t)$ and $\beta_i(t)$ be the integral curves of $c_i(x)$ and $d_i(y)$, respectively, and for $s = (s_1, \dots, s_m)$

define the maps $g_1 : s \to x$ and $g_2 : s \to y$ by

$$g_1(s) = \alpha_m(s_m)\ldots\alpha_2(s_2)\alpha_1(s_1)x^0$$

$$g_2(s) = \beta_m(s_m)\ldots\beta_2(s_2)\beta_1(s_1)y^0.$$

Then $(\partial g_1/\partial s_i)(0) = c_i(x^0)$ and so $g_1^{-1} : x \to s$ exists for x in some neighbourhood U of x^0. We define the map $\lambda : x \to y$ on U by $\lambda = g_2 \circ g_1^{-1}$. Then we must prove that $\lambda(x(t)) = y(t)$ for any solutions of (3.1.29) and (3.1.30), respectively, which depend on the same control. However, $\lambda(x(0)) = \lambda(x^0) = y(0)$ and so it is sufficient to show that

$$(\mathrm{d}/\mathrm{d}t)\lambda(x(t)) = (\mathrm{d}/\mathrm{d}t)y(t) \qquad \text{or} \qquad \lambda_*(\dot{x}(t)) = \dot{y}(t).$$

The latter condition holds if $\lambda_*(a_i(x)) = b_i(\lambda(x))$, $i = 0, ,\ldots, k$ for all $x \in U$, which would follow from $\lambda_*(c_i(x)) = d_i(\lambda(x))$, $i = 1, \ldots, m$, for all $x \in U$.

To prove the last equality we introduce the notation

$$x = g_1(s), \quad x^i = g_1(s_1, \ldots, s_i, 0, \ldots, 0), \qquad i = 1, \ldots, m$$

$$y = \lambda(x) = g_2(s), \quad y^i = g_2(s_1, \ldots, s_i, 0, \ldots, 0), \qquad i = 1, \ldots, m.$$

Then $x^m = x$ and

$$\alpha_i(-s_i)(x^i) = \alpha_i(-s_i)\alpha_i(s_i)\ldots\alpha_1(s_1)x^0 = x^{i-1}.$$

Similarly, $y^m = y$ and $\beta_i(s_i)(y^{i-1}) = y^i$.

Now,

$$\begin{aligned} \partial g_1(s)/\partial s_i &= \alpha_m(s_m)_*\ldots\alpha_{i+1}(s_{i+1})_* c_i(\alpha_i(s_i)\ldots\alpha_1(s_1)x^0) \\ &= \alpha_m(s_m)\ldots\alpha_{i+1}(s_{i+1})_*\alpha_i(s_i)_* c_i(\alpha_{i-1})(s_{i-1})\ldots\alpha_1(s_1)x^0) \\ &= \alpha_m(s_m)\ldots\alpha_i(s_i)_* c_i(x^{i-1}) \end{aligned}$$

and similarly,

$$\partial g_2(s)/\partial s_i = \beta_m(s_m)_*\ldots\beta_i(s_i)_* d_i(y^{i-1}).$$

Hence,

$$\begin{aligned} &\beta_m(s_m)_*\ldots\beta_1(s_1)_* L\alpha_1(-s_1)_*\ldots\alpha_m(-s_m)_*\partial g_1(s)/\partial s_i \\ &= \beta_m(s_m)_*\ldots\beta_1(s_1)_* L\alpha_1(-s_1)_*\ldots\alpha_{i-1}(-s_{i-1})_* c_i(x^{i-1}) \\ &= \beta_m(s_m)_*\ldots\beta_1(s_1)_* L\Sigma \frac{(s_1)^{h_1}}{h_1!}\,\mathrm{ad}^{h_1}(c_1)\left(\ldots\Sigma\frac{(s_{i-1})^{h_{i-1}}}{(h_{i-1})!}\,\mathrm{ad}^{h_{i-1}}(c_{i-1})c_i\ldots\right)(x^0) \\ &= \beta_m(s_m)_*\ldots\beta_1(s_1)_*\Sigma \frac{(s_1)^{h_1}}{h_1!}\,\mathrm{ad}^{h_1}(d_1)\left(\ldots\Sigma\frac{(s_{i-1})^{h_{i-1}}}{(h_{i-1})!}\,\mathrm{ad}^{h_{i-1}}(d_{i-1})d_i\ldots\right)(y^0) \\ &= \beta_m(s_m)_*\ldots\beta_i(s_i)_*\mathrm{d}_i(y^{i-1}) \\ &= \frac{\partial g_2(s)}{\partial s_i} \\ &= \lambda_*(\partial g_1(s)/\partial s_i). \end{aligned}$$

However, the vectors $\partial g_1(s)/\partial s_i$ form a basis for $\mathbb{R}^m$, so

$$\lambda_* = \beta_m(s_m)_* \ldots \beta_1(s_1)_* L \alpha_1(-s_1)_* \ldots \alpha_m(-s_m)_*.$$

Just as above we can then show that

$$\lambda_*(c_i(x^m)) = d_i(y^m).$$

Now, $\lambda_*(c_i(x^0)) = \mathrm{d}_i(y^0)$ so $L = \lambda_*$ at x^0 and so the inverse function theorem implies that if L is a linear isomorphism then λ is a local diffeomorphism.

To prove the converse note that if λ exists with $\lambda(x(t)) = y(t)$ for the same control, then by uniqueness of the solutions of differential equations we have $\lambda_*(a_i(x)) = b_i(\lambda(x))$, $0 \leqslant i \leqslant k$. Hence‡

$$\lambda_*(B(a_0, \ldots, a_k)(x)) = B(b_0, \ldots, b_k)(\lambda x),$$

for any bracket B and so $L = \lambda_{*x^0}$ satisfies the required conditions. □

Now consider the linear system

$$\dot{y}(t) = F(t)y(t) + G(t)u(t) + h(t) \tag{3.1.31}$$

where $F(.), G(.)$ are matrix-valued functions. We wish to find conditions under which the system (3.1.29) can be transformed into this system. First we transform (3.1.31) so that it has the same formal structure as the system (3.1.29). To do this let $\Phi(t, 0)$ be the transition matrix of $F(t)$ (from $t = 0$). Then if we put $y = \Phi(t, 0)z$ we have

$$\begin{aligned} \dot{y} = \dot{\Phi} z + \Phi \dot{z} &= F\Phi z + \Phi \dot{z} \\ &= F\Phi z + Gu + h \end{aligned}$$

and so

$$\dot{z} = \Phi^{-1}(t, 0)G(t)u(t) + \Phi^{-1}(t, 0)h(t).$$

Hence

$$\dot{w} = \Phi^{-1}(t, 0)G(t)u(t) \tag{3.1.32}$$

where

$$w(t) = z(t) - \int_0^t \Phi^{-1}(\tau, 0)h(\tau)\,\mathrm{d}\tau.$$

If we now set $w_0 = t$ we can write (3.1.32) in the form

$$\dot{w} = b_0 + \sum_{i=1}^{k} u_i b_i(w_0), \tag{3.1.33}$$

where $b_0 = (1, 0, \ldots, 0)^{\mathrm{T}}$, $b_i = (0, *, \ldots, *)$, $1 \leqslant i \leqslant k$ and $w_0 = 0$, and where $*$ denotes a function of w_0. Clearly the map $y \to z \to w$ is a local diffeo-

‡ For $f_*[X, Y](p) = [f_*X, f_*Y](flp))$; for vector fields X, Y, see Section 1.1.5.

morphism at 0 and so we can consider the system (3.1.33) rather than (3.1.31).

Theorem 3.1.10
For the system (3.1.29) let $n = \text{rank } D(A)_{x^0}$ and let M be an n-dimensional manifold which carries (3.1.29). Then there exists a linear system (3.1.33) (and hence a system (3.1.31)) and a diffeomorphism $\lambda: U \to V$ for neighbourhoods $x_0 \in U \subseteq M$ and $w^0 = 0 \in V \subseteq \mathbb{R}^n$ carrying (3.1.29) into (3.1.33) if and only if $[a_i, \text{ad}^h(a_0)a_j] = 0$ in a neighbourhood of x^0 for all $h \geqslant 0$ and $1 \leqslant i, j \leqslant k$.

Proof
If the system (3.1.33) and the map λ exist, then

$$\lambda^*([a_i, \text{ad}^h(a_0)a_j](x)) = [b_i, \text{ad}^h(b_0)b_j](\lambda(x))$$

in some neighbourhoods of x^0. However, it is easy to see that $[b_i, \text{ad}^h(b_0)b_j] = 0$ everywhere for (3.1.33). Hence the necessity follows.

For sufficiency note that if $[a_i, \text{ad}^h(a_0)a_j](x) = 0$ we construct a system of the form (3.1.33) as follows. Let

$$b_0 = (1, 0, \ldots, 0)^{\mathrm{T}} \quad \text{and} \quad b_j(y_0) = \alpha_0(-y_0)_* a_j(\alpha_0(y_0)x^0), \qquad j = 1, \ldots, k$$

where α_0 is the flow of a_0. Then by the Taylor series

$$b_j(y_0) = \sum_{h=0}^{\infty} ((y_0)^h/h!)\text{ad}^h(a_0)a_j(x^0)$$

it follows that

$$\text{ad}^h(b_0)b_j(0) = \frac{\mathrm{d}^h}{\mathrm{d}y_0^h} b_j(0) = \text{ad}^h(a_0)a_i(x^0), \qquad j = 1, \ldots, k, \quad h \geqslant 0.$$

Since any bracket in $D(A)$ is either of the form $\text{ad}^h(a_0)a_j$ or $[a_i, [*, a_j]] = 0$, by assumption, it follows that the hypothesis of Theorem 3.1.9 holds with $L = $ identity and so λ exists. □

Example
Consider the system

$$\begin{aligned} \dot{x}_1 &= 1 + ux_3 \\ \dot{x}_2 &= x_1^2 x_2 + u \\ \dot{x}_3 &= x_3. \end{aligned}$$

Then

$$a_0 = \begin{bmatrix} 1 \\ x_1^2 x_2 \\ x_3 \end{bmatrix}, \quad a_1 = \begin{bmatrix} x_3 \\ 1 \\ 0 \end{bmatrix}$$

and so

$$[a_0, a_1] = \begin{bmatrix} x_3 \\ -2x_1x_2x_3 - x_1^2 \\ 0 \end{bmatrix}, \; [a_1, [a_0, a_1]] = \begin{bmatrix} 0 \\ -2x_2x_3^2 - 4x_1x_3 \\ 0 \end{bmatrix}.$$

Hence this system is not linearizable in general. However, on the submanifold $\{x: x_3 = 0\}$ we have the system

$$\begin{aligned} \dot{x}_1 &= 1 \\ \dot{x}_2 &= x_1^2x_2 + u \end{aligned}$$

and so

$$a_0 = \begin{bmatrix} 1 \\ x_1^2x_2 \end{bmatrix}, \; a_1 = \begin{bmatrix} 0 \\ 1 \end{bmatrix}, \; [a_0, a_1] = \begin{bmatrix} 0 \\ -x_1^2 \end{bmatrix}, \; [a_1, [a_0, a_1]] = \begin{bmatrix} 0 \\ 0 \end{bmatrix},$$

$$\mathrm{ad}^2(a_0)a_1 = \begin{bmatrix} 0 \\ -2x_1 \end{bmatrix}, \; [a_1, \mathrm{ad}^2(a_0)a_1] = \begin{bmatrix} 0 \\ 0 \end{bmatrix}, \; \mathrm{ad}^3(a_0)a_1 = \begin{bmatrix} 0 \\ -2 \end{bmatrix}.$$

The remaining brackets are zero and so the system is linearizable. To find the linear equivalent system we have

$$b_0 = \begin{bmatrix} 1 \\ 0 \end{bmatrix}, \qquad b_1(y_0) = \sum_{h=0}^{\infty} \frac{{y_0}^h}{h!} \mathrm{ad}^h(a_0)a_1(x^0)$$

and so if $x^0 = (0,0)$, for example, $y_0 = t$, $\dot{y}_1 = u(1 - t^3/3)$, $y_1(0) = 0$.

Note finally that the case of systems with outputs is considered by Hunt *et al.* (1986).

3.2 EXISTENCE OF LINEARIZING CONTROLLERS

3.2.1 Linearization by static state feedback

Consider a control system of the form

$$\begin{aligned} \dot{x} &= f(x) + g(x)u = f(x) + \sum_{i=1}^{m} g_i(x)u_i \\ y &= h(x) \end{aligned} \tag{3.2.1}$$

where $x \in \mathbb{R}^n, u \in \mathbb{R}^m$, $y \in \mathbb{R}^p$ and $f, g_1, \ldots, g_m$ are real analytic vector fields on $\mathbb{R}^n$ and h is a real analytic function on $\mathbb{R}^n$. In this section we shall follow Isidori (1985) and seek a control in the form

$$u = \alpha(x) + \beta(x)v$$

where $\alpha: U \subseteq \mathbb{R}^n \to \mathbb{R}^m$, $\beta: U \subseteq \mathbb{R}^n \to \mathbb{R}^{m \times m}$ are real analytic functions defined on an open set U such that the resulting system

$$\begin{aligned} \dot{x} &= \bar{f}(x) + \tilde{g}(x)v \\ y &= h(x), \end{aligned} \tag{3.2.2}$$

where

$$\bar{f}(x) = f(x) + g(x)\alpha(x), \qquad \tilde{g}(x) = g(x)\beta(x),$$

has a linear input–output map.

In order to produce linear input–output behaviour in the system (3.2.1) we must obtain an input–output map for this system. We shall prove in Chapter 4 that the input–output relation for the system (3.2.1) may be written in the form of a 'Volterra series'; namely,

$$\begin{aligned} y(t) = w(t, x) &+ \sum_{i=1}^{m} \int_0^t w_i(t, \tau_1, x) u_i(\tau_1)\, \mathrm{d}\tau_1 \\ &+ \sum_{i_1, i_2 = 1}^{m} \int_0^t \int_0^{\tau_1} w_{i_1 i_2}(t, \tau_1, \tau_2, x) u_{i_1}(\tau_1) u_{i_2}(\tau_2)\, \mathrm{d}\tau_1\, \mathrm{d}\tau_2 \\ &+ \cdots \end{aligned} \tag{3.2.3}$$

($x(0) = x$ is the initial condition of (3.2.1)), where

$$w(t, x) = \sum_{k=0}^{\infty} L_f^k h(x) \frac{t^k}{k!}$$

$$w_i(t, \tau_1, x) = \sum_{k_1, k_2 = 0}^{\infty} L_f^{k_2} L_{g_i} L_f^{k_1} h(x) \frac{(t - \tau_1)^{k_1}}{k_1!} \frac{\tau_1^{k_2}}{k_2!}$$

$$w_{i_1, i_2}(t, \tau_1, \tau_2, x) = \sum_{k_1, k_2, k_3 = 0}^{\infty} L_f^{k_3} L_{g_{i_2}} L_f^{k_2} L_{g_{i_1}} L_f^{k_1} h(x) \frac{(t - \tau_1)^{k_1} (\tau_1 - \tau_2)^{k_2} \tau_2^{k_3}}{k_1! k_2! k_3!}$$

etc. Here, as before, $L_X \gamma(x)$ is the Lie derivative of the function γ with respect to the vector field X, i.e.

$$L_X \gamma(x) = \langle \mathrm{d}\gamma, X \rangle = \sum_{i=1}^{n} \frac{\partial \gamma}{\partial x_i} X_i(x)$$

where $X(x) = \sum_{i=1}^{n} X_i(x)\, \partial / \partial x_i$. We can make the series (3.2.3) linear in u by making all terms other than the first two on the right-hand side zero. This will be the case if we impose the condition that

$$L_{g_i} L_f^k h(x) \quad \text{is independent of } x, \quad \text{for } k \geqslant 0, \quad 1 \leqslant i \leqslant m \tag{3.2.4}$$

If this condition holds we have

$$w_i(t, \tau_1, x) = \sum_{k_1 = 0}^{\infty} L_{g_i} L_f^{k_1} h(x) \frac{(t - \tau_1)^{k_1}}{k_1!}$$

which is a function of $t - \tau_1$ and is independent of x. We write this function as $w_i(t - \tau_1)$ and then we have

$$y(t) = w(t, x) + \sum_{i=0}^{m} \int_0^t w_i(t - \tau_1) u(\tau_1) \, \mathrm{d}\tau_1 \tag{3.2.5}$$

If, conversely, a Volterra series is of the form (3.2.5), then the condition (3.2.4) holds since the first-order kernels are independent of x and depend only on the difference $t - \tau_1$. Hence the input–ouput relation is affine in u if and only if the condition (3.2.4) is satisfied, and we can state the *input–output linearization problem* in the form: given a system (f, g, h) find an open dense set $U \subseteq \mathbb{R}^n$ and a pair of analytic functions α and β, with β invertible in U, such that $L_{(g\beta)_i} L^k_{(f+g\alpha)} h(x)$ is independent of $x (\in U)$ for $1 \leqslant i \leqslant m$ and all $k \geqslant 0$.

3.2.2 Solution of the input–output linearization problem

We shall derive necessary and sufficient conditions for the solution of this problem by using a certain Toeplitz matrix and establishing a rank condition for this matrix. First define the matrices

$$T_k(x) = [L_{g_1} L_f^k h(x) \cdots L_{g_m} L_f^k h(x)] = L_g L_f^k h$$

and similarly

$$\bar{T}_k(x) = L_{\bar{g}} L_{\bar{f}}^k h$$

where $\bar{f}$ and $\bar{g}$ are as defined above. (For example,

$$T_0(x) = \begin{bmatrix} \langle dh_1, g_1 \rangle & \langle dh_1, g_2 \rangle & \cdots & \langle dh_1, g_m \rangle \\ \langle dh_2, g_1 \rangle & \langle dh_2, g_2 \rangle & \cdots & \langle dh_2, g_m \rangle \\ \vdots & \vdots & \cdots & \vdots \\ \langle dh_p, g_1 \rangle & \langle dh_p, g_2 \rangle & \cdots & \langle dh_p, g_m \rangle \end{bmatrix}$$

in the case $k = 0$.) Then we have seen that we require $T_k(x)$ to be a constant matrix for all $k \geqslant 0$. We shall organize the matrices $T_k(x)$ into two forms: firstly we consider the formal power series

$$T(s, x) = \sum_{k=0}^{\infty} T_k(x) s^{-(k+1)}$$

and secondly we can store the information in the matrices T_k in the form of an infinite sequence of Toeplitz matrices

$$\mathcal{T}_k(x) = \begin{bmatrix} T_0(x) & T_1(x) & \cdots & T_k(x) \\ 0 & T_0(x) & \cdots & T_{k-1}(x) \\ \vdots & \vdots & & \vdots \\ 0 & 0 & \cdots & T_0(x) \end{bmatrix}.$$

Let $\rho(\mathcal{T}_k)$ denote the rank of $\mathcal{T}_k$ over the real number field $\mathbb{R}$. Since $\mathcal{T}_k$ is a matrix over the ring C^ω of real analytic functions defined on $\mathbb{R}^n$ we can also consider its rank, $\sigma(\mathcal{T}_k)$, over the quotient field $\mathbb{K}(C^\omega)$ of C^ω. Clearly,

$$\rho(\mathcal{T}_k) \geqslant \sigma(\mathcal{T}_k), \qquad k \geqslant 0.$$

It turns out that the inverse inequality is equivalent to the existence of a solution of the input–output linearization problem. In fact, we have (Isidori and Ruberti (1984):

Theorem 3.2.1
The following statements are equivalent:

(a) There exists a solution to the input–output linearization problem
(b) There exists a formal power series

$$W(s) = \sum_{k=0}^{\infty} W_k s^{-(k+1)},$$

with coefficients which are $p \times m$ matrices over $\mathbb{R}$ and an invertible formal power series

$$R(s,x) = R(x) + \sum_{k=0}^{\infty} R_k(x) s^{-(k+1)}$$

with coefficients which are $m \times m$ matrices of real analytic functions defined on an open dense subset U of $\mathbb{R}^n$ such that

$$T(s,x) = W(s)R(s,x).$$

(c) $\rho(\mathcal{T}_k) = \sigma(\mathcal{T}_k)$, $k \geqslant 0$.

Proof
(a) ⇒ (b): There exist functions α, β defined on U such that

$$L_{g\beta} L_{(f+g\alpha)}^k h = \bar{T}_k$$

is constant on U for all $k \geqslant 0$. We then prove by induction that

$$L_{\tilde{f}}^k h = L_f^k h + \bar{T}_{k-1}\hat{\alpha} + \bar{T}_{k-2} L_f \hat{\alpha} + \cdots + \bar{T}_0 L_f^{k-1} \hat{\alpha} \tag{3.2.6}$$

where

$$\hat{\alpha} = -\beta^{-1}\alpha.$$

In fact,

$$\begin{aligned} L_f^{k+1} h &= L_{(\tilde{f}+\tilde{g}\hat{\alpha})} L_{\tilde{f}}^k h + L_f(\bar{T}_{k-1}\hat{\alpha} + \cdots + \bar{T}_0 L_f^{k-1}\hat{\alpha}) \\ &= L_{\tilde{f}}^{k+1} + (L_{\tilde{g}} L_{\tilde{f}}^k h)\hat{\alpha} + \bar{T}_{k-1} L_f \hat{\alpha} + \cdots + \bar{T}_0 L_f^k \hat{\alpha}. \end{aligned}$$

Hence, by (3.2.6),

$$\begin{aligned} T_k(x) &= L_g L_f^k h + \bar{T}_{k-1} L_g \hat{\alpha}(x) + \bar{T}_{k-2} L_g L_f \hat{\alpha}(x) + \cdots + \bar{T}_0 L_g L_f^{k-1} \hat{\alpha}(x) \\ &= \bar{T}_k \beta^{-1}(x) + \bar{T}_{k-1} L_g \hat{\alpha}(x) + \bar{T}_{k-2} L_g L_f \hat{\alpha}(x) + \cdots + \bar{T}_0 L_g L_f^{k-1} \hat{\alpha}(x) \end{aligned} \tag{3.2.7}$$

Let $W_k = \bar{T}_k$, $R(x) = \beta^{-1}(x)$, $R_k(x) = L_g L_f^k \hat{\alpha}(x)$. Then

$$W(s) = \sum_{k=0}^{\infty} \bar{T}_k s^{-(k+1)}, \qquad R(s,x) = \beta^{-1}(x) + \sum_{k=0}^{\infty} L_g L_f^k \hat{\alpha}(x) s^{-(k+1)}.$$

By (3.2.7), the coefficient of $s^{-(k+1)}$ in $T(s, x)$ is precisely $T_k(x)$.

(b) ⇒ (c): If (b) holds then we have

$$\mathcal{T}_k(x) = \begin{bmatrix} W_0 & W_1 & \cdots & W_k \\ 0 & W_0 & \cdots & W_{k-1} \\ \vdots & \vdots & & \vdots \\ 0 & 0 & \cdots & W_0 \end{bmatrix} \begin{bmatrix} R(x) & R_0(x) & R_1(x) & \cdots & R_{k-1}(x) \\ 0 & R(x) & R_0(x) & \cdots & R_{k-2}(x) \\ \vdots & \vdots & \vdots & & \vdots \\ 0 & 0 & 0 & \cdots & R(x) \end{bmatrix}.$$

Since the power series $R(s, x)$ is invertible, it follows that $R(x)$ is nonsingular. Hence the right-hand matrix is nonsingular on U and the left-hand matrix is just a constant real matrix on U. Thus, the rank of $\mathcal{T}_k(x)$ is determined (over $\mathbb{R}$ or $\mathbb{K}(C^\omega)$) by the left-hand matrix and so (c) is true.

(c) ⇒ (a): In this part of the proof we shall follow Silverman's structure algorithm (Silverman (1969)) for the inversion of a linear system. Since (c) is assumed to hold, finding $\mathbb{K}(C^\infty)$-linearly independent rows of a matrix can be done by using only $\mathbb{R}$-linear elementary operations on the matrix. The algorithm can be described as follows:

Step (1) Let

$$V_1 = \begin{bmatrix} P_1 \\ K_1 \end{bmatrix}$$

be a real nonsingular $p \times p$ matrix, where P_1 performs row permutations, such that

$$V_1 T_0(x) = \begin{bmatrix} S_1(x) \\ 0 \end{bmatrix}$$

where the r_1 rows of $S_1(x)$ are linearly independent over $\mathbb{K}(C^\infty)$. If $T_0(x) = 0$ we shall take K_1^1 to be the $p \times p$ identity matrix and set $V_1 = K_1^1$. Let

$$\delta_1 = r_1, \qquad \gamma_1(x) = P_1 h(x), \qquad \bar{\gamma}_1(x) = K_1^1 h(x).$$

Since V_1 is a real (constant) matrix we have

$$L_g \gamma_1(x) = S_1(x), \qquad L_g \bar{\gamma}_1(x) = 0.$$

Step (i) Consider the matrix

$$\begin{bmatrix} L_g\gamma_1(x) \\ \vdots \\ L_g\gamma_{i-1}(x) \\ L_gL_f\bar{\gamma}_{i-1}(x) \end{bmatrix} = \begin{bmatrix} S_{i-1}(x) \\ L_gL_f\bar{\gamma}_{i-1}(x) \end{bmatrix}.$$

It is not difficult to see that searching for linearly independent rows of this matrix is equivalent to searching the Toeplitz matrix $\mathcal{T}_{i-1}(x)$ for linearly independent rows (see also Van Dooren *et al.* (1979)). Hence by (b) there exists a nonsingular real $p \times p$ matrix

$$V_k = \begin{bmatrix} I_{\delta_1} & \cdots & 0 & 0 \\ \vdots & & \vdots & \vdots \\ 0 & & I_{\delta i-1} & 0 \\ 0 & & 0 & P_i \\ K_1^i & \cdots & K_{i-1}^i & K_i^i \end{bmatrix}$$

where P_i performs row permutations, such that

$$V_i \begin{bmatrix} L_g\gamma_1(x) \\ \vdots \\ L_g\gamma_{i-1}(x) \\ L_gL_f\bar{\gamma}_{i-1}(x) \end{bmatrix} = \begin{bmatrix} S_i(x) \\ 0 \end{bmatrix}$$

and the rows of $S_i(x)$ are linearly independent over $\mathbb{K}(C^\omega)$. If $S_i(x)$ has r_i ($\geqslant r_{i-1}$) rows, then let

$$\begin{aligned} \delta_i &= r_i - r_{i-1} \\ \gamma_i(x) &= P_iL_f\bar{\gamma}_{i-1}(x), \\ \bar{\gamma}_i(x) &= K_1^i\gamma_1(x) + \cdots + K_{i-1}^i\gamma_{i-1}(x) + K_i^iL_f\bar{\gamma}_{i-1}(x). \end{aligned}$$

Then we clearly have

$$\begin{bmatrix} L_g\gamma_1(x) \\ \vdots \\ L_g\gamma_i(x) \end{bmatrix} = S_i(x), \tag{3.2.9}$$

$$L_g\bar{\gamma}_i(x) = 0.$$

Of course, if the rows of the matrix $L_gL_f\bar{\gamma}_{i-1}(x)$ in (3.2.8) depend linearly on those of $S_{i-1}(x)$, then P_i is taken to be the empty matrix, $K_i^i = I$, $\delta_i = 0$ and $S_i(x) = S_{i-1}(x)$.

Alternatively, if there exists an integer d such that the p rows of the matrix

$$\begin{bmatrix} L_g\gamma_1(x) \\ \vdots \\ L_g\gamma_{d-1}(x) \\ L_gL_f\bar{\gamma}_{d-1}(x) \end{bmatrix} \tag{3.2.10}$$

are linearly independent over $\mathbb{K}(C^\omega)$, then the algorithm is terminated at the dth step by defining $V_d = I$, $P_d = I$ and

$$\gamma_d = P_dL_f\bar{\gamma}_{d-1}(x), \qquad S_d(x) = \begin{bmatrix} S_{d-1}(x) \\ L_g\gamma_d(x) \end{bmatrix}.$$

($K_1^d, K_2^d, \ldots, K_d^d$ are empty matrices.) If the matrix (3.2.10) does not have rank p for any $d > 0$ then let d be the number of the last nondegenerate step of the algorithm. Then for any $j > d$ we can take $K_j^j = I$ and P_j to be empty.

To conclude the proof of the theorem, let

$$\Gamma(x) = \begin{bmatrix} \gamma_1(x) \\ \vdots \\ \gamma_m(x) \end{bmatrix}.$$

Then the equations

$$\begin{aligned} (L_g\Gamma(x))\alpha(x) &= -L_f\Gamma(x) \\ (L_g\Gamma(x))\beta(x) &= \text{constant} \end{aligned} \tag{3.2.11}$$

are soluble for analytic functions α and β on U since $S_d(x) = L_g\Gamma(x)$ has full rank over $\mathbb{K}(C^\omega)$. We can write equations (3.2.11) in the form

$$\begin{aligned} L_{(f+g\alpha)}\gamma_i &= L_{\tilde{f}}\gamma_i = 0, & 1 \leqslant i \leqslant m \\ L_{(g\beta)}\gamma_i &= L_{\tilde{g}}\gamma_i = \text{constant}, & 1 \leqslant i \leqslant m. \end{aligned}$$

By (3.2.9) we also have

$$L_{\tilde{g}}\bar{\gamma}_i = (L_g\bar{\gamma}_i)\beta = 0 \tag{3.2.12}$$

and

$$L_{\tilde{f}}\bar{\gamma}_i = L_f\bar{\gamma}_i + (L_g\bar{\gamma}_i)\alpha = L_f\bar{\gamma}_i, \tag{3.2.13}$$

for $i \geqslant 1$.

Consider the vector $P_1L_{\tilde{g}}L_{\tilde{f}}^kh(x) = L_{\tilde{g}}L_{\tilde{f}}^k\gamma_1(x)$. It is constant (or zero) for all $k \geqslant 0$ by (3.2.12), (3.2.13). Similarly,

$$\begin{aligned} P_2K_1^1L_{\tilde{g}}L_{\tilde{f}}^kh(x) = P_2L_{\tilde{g}}L_{\tilde{f}}^k\bar{\gamma}_1(x) &= L_{\tilde{g}}L_{\tilde{f}}^{k-1}P_2L_{\tilde{f}}\bar{\gamma}_1(x) \\ &= L_{\tilde{g}}L_{\tilde{f}}^{k-1}\gamma_2(x) = \text{constant}, \end{aligned}$$

by (3.2.12), (3.2.13). In general, it can be seen that if we define

$$H_1 = P_1$$
$$H_i = P_i K_{i-1}^{i-1} K_{i-2}^{i-2} \cdots K_1^1, \qquad 1 \leqslant i \leqslant m$$
$$H_{d+1} = K_d^d K_{d-1}^{d-1} \cdots K_1^1$$

then the matrix

$$\begin{bmatrix} H_1 \\ H_2 \\ \vdots \\ H_m \\ H_{m+1} \end{bmatrix}$$

is nonsingular and so $L_{\tilde{g}} L_f^k h(x) = \text{constant}$ for all $k \geqslant 0$. □

Using the above ideas one may derive an output feedback which will diagonalize a system for which the input–output linearization problem is solvable. Moreover, under certain conditions one can match a prescribed linear input–output behaviour by using state feedback. For details we refer the reader to Isidori (1985).

3.3 INFINITE-DIMENSIONAL TENSOR REPRESENTATIONS OF NONLINEAR SYSTEMS

3.3.1 Nonlinear differential equations

In this section we shall consider the nonlinear differential equation

$$\dot{x} = f(x) \tag{3.3.1}$$

where $f: \mathbb{R}^n \to \mathbb{R}^n$ is assumed to be globally analytic. We shall show that this equation is equivalent to a linear equation on a Hilbert space of tensors. The method which we shall use has its origins in the work of Carleman (1932) and was developed for use in systems theory by Brockett (1976a).

First we shall introduce some notation. Let $\mathbb{N}^n$ denote the set of n-tuples of natural numbers and denote a typical element of $\mathbb{N}^n$ by

$$i = (i_1, \ldots, i_n), \; i_j \in \mathbb{N}.$$

Then we shall consider the Taylor monomial

$$\phi_i = x^i = x_1^{i_1} \cdots x_n^{i_n}, \qquad x = (x_1, \ldots, x_n) \in \mathbb{R}^n.$$

for any $i \in \mathbb{N}^n$. Finally we shall denote by $1(k)$ the n-tuple with a 1 in the kth place and zeros elsewhere.

Differentiating each ϕ_i along the system (3.3.1) we have

$$\dot{\phi}_i = \frac{\partial \phi_i}{\partial x} \frac{\mathrm{d}x}{\mathrm{d}t} = \sum_{k=1}^{n} i_k x^{(i-1(k))} \dot{x}_k$$

$$= \sum_{k=1}^{n} i_k x^{(i-1(k))} f_k(x).$$

Since f is assumed to be analytic, we can write f_k in a Taylor series of the form

$$f_k(x) = \sum_{i \in \mathbb{N}^n} \alpha_i^k x^i, \qquad 1 \leqslant k \leqslant n \tag{3.3.2}$$

and so

$$\dot{\phi}_i = \sum_{k=1}^{n} i_k x^{(i-1(k))} \sum_{j \in \mathbb{N}^n} \alpha_j^k x^j$$

$$= \sum_{j \in \mathbb{N}^n} \sum_{k=1}^{n} i_k \alpha_j^k x^{i+j-1(k)}$$

$$= \sum_{l \in \mathbb{N}^n} \sum_{k=1}^{n} i_k \alpha_{l-i+1(k)}^k x^l$$

where we set $\alpha_i^k = 0$ if any element of i is negative. Hence,

$$\dot{\phi}_i = \sum_{l \in \mathbb{N}^n} a_i^l \phi_l \tag{3.3.3}$$

where

$$a_i^l = \sum_{k=1}^{n} i_k \alpha_{l-i+1(k)}^k.$$

Let A denote the tensor operator defined by

$$(A\Phi)_i = \sum_{l \in \mathbb{N}^n} a_i^l \phi_l$$

where Φ is the tensor with ith component ϕ_i, then the equation (3.3.3) may be written

$$\dot{\Phi} = A\Phi. \tag{3.3.4}$$

Note that in the application of this method, most authors choose to order the tensor Φ into an infinite-dimensional vector rather than a tensor and then the operator A becomes an ordinary (infinite-dimensional) matrix.

However, it turns out to be better for most purposes to leave Φ in the form of a tensor. Before proceeding with the general theory of systems of the form (3.3.4) we shall present two examples to illustrate the above ideas.

Examples

1. Consider the simple one-dimensional equation

$$\dot{x} = x^2, \qquad x(0) = x_0. \tag{3.3.5}$$

and write

$$\phi_i = x^i.$$

Then

$$\dot{\phi}_i = ix^{i-1}\dot{x} = ix^{i+1}.$$

Hence if we define $\Phi = (\phi_0 \quad \phi_1 \quad \phi_2 \quad \ldots)^{\mathrm{T}}$, then we obtain the linear system

$$\dot{\Phi} = A\Phi, \Phi(0) = \Phi_0 \tag{3.3.6}$$

where

$$A = \begin{bmatrix} 0 & 0 & 0 & 0 & 0 & 0 & \cdots \\ 0 & 0 & 1 & 0 & 0 & 0 & \cdots \\ 0 & 0 & 0 & 2 & 0 & 0 & \cdots \\ 0 & 0 & 0 & 0 & 3 & 0 & \cdots \\ \vdots & \vdots & \vdots & \vdots & \vdots & \vdots & \end{bmatrix}$$

We shall show later that e^{At} is well-defined in an appropriate space, and so, ignoring convergence questions for now, we simply compute

$$e^{At} \triangleq I + At + (At)^2/2! + \cdots$$

$$= \begin{bmatrix} 1 & 0 & 0 & 0 & 0 & 0 & 0 & \cdots \\ 0 & 1 & t & t^2 & t^3 & t^4 & t^5 & \cdots \\ 0 & 0 & 1 & 2t & \frac{2.3}{2!}t^2 & \frac{2.3.4}{3!}t^3 & \frac{2.3.4.5}{4!}t^4 & \cdots \\ 0 & 0 & 0 & 1 & 3t & \frac{3.4}{2!}t^2 & \frac{3.4.5}{3!}t^3 & \cdots \\ \vdots & \vdots & \vdots & \vdots & \vdots & \vdots & \vdots & \end{bmatrix}$$

The solution $\Phi(t)$ of (3.3.6) is given by $\Phi(t) = e^{At}\Phi_0$, where $\Phi_0 = (1, x_0, x_0^2, \ldots)^{\mathrm{T}}$. Since $\Phi(t) = (1, x(t), x^2(t), \ldots)$ where $x(t)$ is the solution of (3.3.5) we have

$$x(t) = (0 \quad 1 \quad t \quad t^2 \quad t^3 \quad \ldots)(1 \quad x_0 \quad x_0^2 \quad \ldots)^{\mathrm{T}}$$
$$= x_0 \sum_{n=0}^{\infty} (x_0 t)^n$$
$$= \frac{x_0}{1 - x_0 t}$$

provided $|x_0 t| < 1$. This is, of course, the solution of (3.3.5) which can be obtained by elementary means. Hence $e^{At}\Phi_0$, if evaluated as a column of formal power series in t, gives the Taylor series of the solution of the original equation. Of course, it also contains redundant information since all the elements of the column are just powers of this Taylor series. If we truncate A by retaining only the first m rows and columns we obtain the finite-dimensional system

$$\dot{\Phi}_m = A_m \Phi_m$$

with solution $e^{A_m t}\Phi_{0m}$, which gives a truncated Taylor series solution to the original equation which will be valid close to x_0 for small t. The larger we take m, the larger will be the region of close approximation to the correct solution.

2. Consider the Van der Pol oscillator

$$\dot{x}_1 = x_2 - x_1^3 + x_1$$
$$\dot{x}_2 = -x_1.$$

Define

$$\phi_{ij} = x_1^i x_2^j.$$

Then

$$\dot{\phi}_{ij} = i x_1^{i-1} x_2^j \dot{x}_1 + j x_1^i x_2^{j-1} \dot{x}_2$$
$$= i x_1^{i-1} x_2^{j-1} - i x_1^{i+2} x_2^j + i x_1^i x_2^j - j x_1^{i+1} x_2^{j-1}$$
$$= i \sum_{k=0}^{\infty} \sum_{l=0}^{\infty} \delta_k^{i-1} \delta_l^{j+1} x_1^k x_2^l - i \sum_{k=0}^{\infty} \sum_{l=0}^{\infty} \delta_k^{i+2} \delta^j x_1^k x_2^l$$
$$+ i \sum_{k=0}^{\infty} \sum_{l=0}^{\infty} \delta_k^i \delta_l^j x_1^k x_2^l - j \sum_{k=0}^{\infty} \sum_{l=0}^{\infty} \delta_k^{i+1} \delta^{j-1} x_1^k x_2^l$$
$$= \sum_{k=0}^{\infty} \sum_{l=0}^{\infty} a_{ij}^{kl} \phi_{kl},$$

where

$$a_{ij}^{kl} = \delta_k^{i-1} \delta_l^{j+1} - \delta_k^{i+2} \delta_l^j + \delta_k^i \delta_l^j - \delta_k^{i+1} \delta_l^{j-1}. \tag{3.3.7}$$

If we order the functions ϕ_{ij} as a vector Φ, for example,

$$\Phi = (\phi_{00}, \phi_{01}, \phi_{10}, \phi_{02}, \phi_{11}, \phi_{20}, \phi_{03}, \phi_{12}, \phi_{21}, \phi_{30}, \ldots)^{\mathrm{T}}$$

then writing down the matrix A in the equation $\dot{\Phi} = A\Phi$ is considerably more difficult than if we leave ϕ_{ij} in tensor form, when we obtain the relatively simple expression (3.3.7). The reader is invited to try writing down the matrix A in the vector formulation. General polynomial systems are considered in Banks and Ashtiani (1985).

Returning now to the general linearized equation (3.3.4) we shall discuss next the existence theory of such an equation and the space on which it is naturally defined. Let l^1 denote the standard Banach space of absolutely summable sequences and let l^1_e denote the Banach space of sequences $(\alpha_0, \alpha_1, \ldots)$ such that the sequence $(\alpha_0, \alpha_1/1!, \alpha_2/2!, \ldots)$ belongs to l^1. Define a norm on l^1_e by

$$\| \alpha \|_e = \sum_{n=0}^{\infty} \frac{|\alpha_n|}{n!}, \qquad (\alpha) \in l^1_e.$$

It is clear that l^1_e is, in fact, a Banach space and the map

$$E : l^1_e \to l^1$$

defined by

$$E(\alpha) = (\alpha_n/n!), \qquad \alpha = (\alpha_n) \in l^1_e$$

is an isometric isomorphism. We can extend this definition to the space of infinite-dimensional tensors of rank n in the following way. Let

$$\mathscr{L}_n = \bigotimes_n l^1_e$$

denote the algebraic tensor product of n copies of l^1_e and let $\| \cdot \|$ be any cross norm on $\mathscr{L}_n$ (see Chapter 1). Thus, if $\Phi \in \mathscr{L}_n$ is a simple tensor of the form

$$\Phi = (\alpha_{1i_1}\alpha_{2i_2} \ldots \alpha_{ni_n}) = \alpha_1 \otimes \ldots \otimes \alpha_n$$

where

$$\alpha_k = (\alpha_{k0}, \alpha_{k1}, \alpha_{k2}, \ldots) \in l^1_e,$$

then we have

$$\| \Phi \| = \prod_{k=1}^{n} \| \alpha_k \|_e. \tag{3.3.8}$$

This norm has particularly nice properties on simple tensors of the form $(x_1^{i_1} \ldots x_n^{i_n})$ as shown by the following result.

Lemma 3.3.1
For any n-vector $x = (x_1, \ldots, x_n)$, the tensor $(x_1^{i_1} \ldots x_n^{i_n})$ belongs to $\otimes_n l^1_e$ and we have

$$\| (x_1^{i_1} \ldots x_n^{i_n}) \| = \exp\left(\sum_{k=1}^{n} | x_k | \right).$$

Proof

$$\begin{aligned}\| (x_1^{i_1} \dots x_n^{i_n}) \| &= \prod_{k=1}^{n} \| (x_k^{i_k}) \|_e \\ &= \prod_{k=1}^{n} \left\{ \sum_{l=0}^{\infty} \frac{|x_k|^l}{l!} \right\} \\ &= \prod_{k=1}^{n} \mathrm{e}^{|x_k|} \\ &= \exp\left(\sum_{k=1}^{n} |x_k| \right).\end{aligned}$$

□

We shall now denote by $\mathscr{L}_n^{\mathrm{T}}$ the subset of $\mathscr{L}_n$ consisting of tensors of the form $\Phi = (x^i)$. Of course, $\mathscr{L}_n^{\mathrm{T}}$ is *not* a linear subspace of $\mathscr{L}_n$.

Theorem 3.3.2

On the space $\mathscr{L}_n^{\mathrm{T}}$ we have

$$\| A\Phi \| \leqslant \sum_{k=1}^{n} | f_k(x) | \, \| \Phi \|. \tag{3.3.9}$$

Proof

We have

$$(A\Phi)_i = \sum_{k=1}^{n} i_k x^{(i-1(k))} f_k(x).$$

Consider the term

$$(A_1\Phi)_i \triangleq i_1 x_1^{i_1 - 1} x_2^{i_2} \cdots x_n^{i_n} f_1(x).$$

Then, if $\Phi \in \mathscr{L}_n^{\mathrm{T}}$, we have

$$\begin{aligned}\| A_1\Phi \| &= \| i_1 x_1^{i_1-1} \|_{\mathrm{e}} \prod_{k=2}^{n} \| (x_k^{i_k}) \|_{\mathrm{e}} \, | f_1(x) | \\ &= \sum_{l=1}^{\infty} i_1 \frac{|x_1|^{i_1-1}}{i_1!} \exp\left(\sum_{k=2}^{n} |x_k| \right) | f_1(x) | \\ &= \exp\left(\sum_{k=1}^{n} |x_k| \right) | f_1(x) | \\ &= \| \Phi \| \, . \, | f_1(x) |\end{aligned}$$

by Lemma 3.3.1. The result now follows directly. □

Note that A can be extended to the space of simple tensors by defining

$$A\Phi = \sum_{k=1}^{n} i_k \alpha_{1i_1} \ldots \alpha_{ki_k-1} \ldots \alpha_{ni_n} f_k(x),$$

where $\Phi = (\alpha_{1i_1} \ldots \alpha_{ni_n}) = \alpha_1 \otimes \cdots \otimes \alpha_n$. The inequality (3.3.9) holds on this space and so we can define the iterates $A^k\Phi = A(A^{k-1}\Phi)$ inductively on the space of simple tensors.

Corollary 3.3.3

$$\| A^l \Phi \| \leqslant \left(\sum_{k=1}^{n} | f_k(x) | \right)^l \| \Phi \|. \qquad \square$$

Corollary 3.3.4
$e^{At}\Phi$ exists for all t and for all $\Phi \in \mathscr{L}_n^{\mathrm{T}}$ and we have

$$\| e^{At}\Phi \| \leqslant \exp\left\{ \left(\sum_{k=1}^{n} | f_k(x) | \right) t \right\} \| \Phi \|. \qquad \square$$

In other words, the system

$$\dot{\Phi} = A\Phi$$

is soluble in $\mathscr{L}_n^{\mathrm{T}}$ and if $\Phi_0 = (x_0^i)$ for some fixed $x_0 \in \mathbb{R}^n$, then the solution $\Phi(t)$ satisfies

$$\| \Phi(t) \| \leqslant \exp\left\{ \left(\sum_{k=1}^{n} | f_k(x_0) | \right) t \right\} \| \Phi_0 \|. \tag{3.3.10}$$

REMARKS (a) It should be noted that Corollary 3.3.4 only holds on the nonlinear subspace $\mathscr{L}_n^{\mathrm{T}}$ of tensors of the form $\Phi = (x^i)$ and it does not follow that

$$\| e^{At}\Phi \| \leqslant \exp\left\{ \left(\sum_{k=1}^{n} | f_k(x) | \right) t \right\} \| \Phi \|$$

for Φ in the closed linear span of tensors of this form. Hence, although A is a linear tensor operator, e^{At} cannot be extended to a linear semigroup on such a linear subspace of $\mathscr{L}_n$. However, e^{At} is a nonlinear semigroup, since strong continuity at $t = 0$ follows from Corollary 3.3.4.

(b) It also follows from Corollary 3.3.4 that e^{At} may be obtained from the usual series $\sum_{k=0}^{\infty} (At)^k/k!$. The ith component of $e^{At}\Phi$, where $\Phi = (x_0^i)$, is then just the ith power of the Taylor series of the solution $x(T)$ of the equation (3.3.1) with initial condition x_0.

(c) Note that one can expand the solution of (3.3.1) in terms of polynomials other than the Taylor monomials. In fact, one can also use any set of orthogonal polynomials; see Takata (1979).

3.3.2 Nonlinear analytic control systems

Suppose we now consider the general nonlinear analytic system

$$\dot{x} = f(x, u), \qquad x \in \mathbb{R}^n, \quad u \in \mathbb{R}^m. \tag{3.3.11}$$

If we try to apply the above method with $\Phi = (x^i)$ then the resulting system is not linear in u. However, we can obtain a bilinear system by using the extended system, as introduced in Chapter 2. Thus we consider the system

$$\begin{aligned} \dot{x} &= f(x, u) \\ \dot{u} &= v \end{aligned} \tag{3.3.12}$$

so that u is treated as a state and the control v now appears linearly. Since the state is now (x, u) we consider the tensors

$$\Phi = (\phi_{(i,j)}) = (x^i u^j), \tag{3.3.13}$$

where $i = (i_1, .., i_n)$, $j = (j_1, .., j_m)$ belong to $\mathbb{N}^n$ and $\mathbb{N}^m$, respectively. Then, proceeding as before, we have

$$\dot{\phi}_{(i,j)} = \sum_{i=1}^{n} i_k x^{(i-1(k))} u^j \dot{x}_k + \sum_{k=1}^{m} j_k x^i u^{(j-1(k))} \dot{u}_k$$

(where the first occurrence of $1(k)$ is of dimension n and the second is of dimension m). Since f is (globally) analytic we may write

$$f_k(x, u) = \sum_{(i,j) \in I} \alpha^k_{(i,j)} x^i u^j, \qquad 1 \leqslant k \leqslant n$$

for some constants $\alpha^k_{(i,j)}$, where $I = \mathbb{N}^{n+m}$. Hence we have

$$\begin{aligned} \dot{\phi}_{(i,j)} &= \sum_{k=1}^{n} i_k x^{(i-1(k))} u^j \sum_{(i',j') \in I} \alpha^k_{(i',j')} x^{i'} u^{j'} + \sum_{k=1}^{m} j_k x^i u^{(j-1(k))} v_k \\ &= \sum_{(i',j') \in I} \sum_{k=1}^{n} i_k \alpha^k_{(i',j')} x^{i+i'-1(k)} u^{j+j'} + \sum_{k=1}^{m} j_k x^i u^{(j-1(k))} v_k \\ &= \sum_{(i'',j'') \in I} \sum_{k=1}^{n} i_k \alpha^k_{(i''-i+1(k), j''-j')} x^{i''} u^{j''} + \sum_{k=1}^{m} j_k x^i u^{(j-1(k))} v_k \end{aligned}$$

where we set $\alpha^k_{(i,j)} = 0$ for each k if $(i, j) < (0, 0)$, and it follows that

$$\dot{\phi}_{(i,j)} = \sum_{(i',j') \in I} a^{i'j'}_{ij} \phi_{(i',j')} + \sum_{(i',j') \in I} b^{i'j'}_{ij,k} \phi_{(i',j')} v_k,$$

where

$$a^{i'j'}_{ij} = \sum_{k=1}^{n} i_k \alpha^k{}_{(i'-i+1(k), j''-j')}, \qquad b^{i'j'}_{ij,k} = \sum_{k=1}^{m} j_k \delta^{i'j'}_{i,j-1(k)}.$$

Here, δ^{ij}_{kl} is the tensor defined by

$$\delta^{ij}_{kl} = \delta^{i_1}_{k_1} \delta^{i_2}_{k_2} \cdots \delta^{i_n}_{k_n} \delta^{j_1}_{l_1} \cdots \delta^{j_m}_{l_m} \tag{3.3.14}$$

Hence, defining the tensor operators A and B_μ by

$$(A\Phi)_{(i,j)} = \sum_{(k,l)\in I} a_{ij}^{k}\phi_{(k,l)} \tag{3.3.15}$$

$$(B_\mu\Phi)_{(ij)} = \sum_{(k,l)\in I} b_{ij,\mu}^{k}\phi_{(k,l)}, \qquad 1 \leqslant \mu \leqslant m, \tag{3.3.16}$$

where $(\Phi)_{(i,j)} = \phi_{(i,j)}$, we have

$$\dot{\Phi} = A\Phi + \sum_{\mu=1}^{m} v_\mu B_\mu \Phi = A\Phi + vB\Phi, \tag{3.3.17}$$

where $vB \triangleq \sum_{i=1}^{m} v_i B_i$. In the same way as for Theorem 3.3.2, we can prove

Theorem 3.3.5
Under the cross norm $\|\,.\,\|$on $\mathscr{L}_{n+m}^{\mathrm{T}}$ introduced above, we have

$$\|A\Phi\| \leqslant \sum_{k=1}^{n} |f_k(x,u)|\,\|\Phi\| \tag{3.3.18}$$

and

$$\|B\Phi\| \triangleq \sum_{i=1}^{m} \|B_i\Phi\| \leqslant m\,\|\Phi\|. \qquad \square \tag{3.3.19}$$

Corollary 3.3.6
$e^{At}\Phi_0$ exists for all t and for all $\Phi_0 \in \mathscr{L}_{n+m}^{\mathrm{T}}$, and is the solution of the system

$$\dot{\Phi} = A\Phi, \qquad \Phi(0) = \Phi_0$$

related to the nonlinear differential equation

$$\begin{bmatrix}\dot{x}\\ \dot{u}\end{bmatrix} = \begin{bmatrix} f(x,u)\\ 0\end{bmatrix}, \qquad x(0) = x_0, \quad u(0) = u_0 \tag{3.3.20}$$

in the above way with $\Phi_0 = (x_0^i u_0^j)$. Moreover, we have

$$\|e^{At}\Phi_0\| \leqslant \exp\left\{\left(\sum_{k=1}^{n} |f_k(x,u)|\right)t\right\}\|\Phi\|. \qquad \square$$

It follows that the nonlinear system (3.3.12) has a bilinear (infinite-dimensional) representation (3.3.17) and in the next chapter we shall see how to write such a representation in input–ouput form (as a Volterra series).

3.3.3 Global theory

We shall now consider the generalization of the above results to the case of a system defined on a manifold. For simplicity we shall assume that the input spaces are independent of the state – it is easy to extend the ideas below to cover the general case. Thus, let M and U be real analytic manifolds of dimensions n and m, respectively, and let

$$X(.) : U \to D(M) \tag{3.3.21}$$

be an analytic map from U to the set $D(M)$ of analytic vector fields on M. Then if $p \in M$ we may express (3.3.21) in the local coordinates x near p in the form of a differential equation

$$\dot{x} = f_p(x, u).$$

As in the previous section we shall reformulate the problem in the following way (cf. the extended system defined in Chapter 2). Consider the product manifold $M \times U$ with the tangent bundle $T(M \times U)$ given by

$$T_{(p,u)}(M \times U) \cong T_pM \oplus T_uU$$

for each $(p, u) \in M \times U$. If $D(M \times U)$ denotes the set of analytic vector fields on $M \times U$, then we consider, instead of (3.3.21), the analytic map

$$Y(.) : T(U) \to D(M \times U)$$

such that, for each $Z \in T(U)$,

$$Y(Z)_{(p,u)} = (X(u)_p, v),$$

where $v \in T_uU$. Hence, in a neighbourhood of (p, u), we can write Y in the form

$$\begin{aligned} \dot{x} &= f_p(x, u) \\ \dot{u} &= v. \end{aligned} \tag{3.3.22}$$

The control space is now the tangent bundle of U rather than U. Using the theory of Section 3.3.2, we can replace the local system (3.3.22) by the tensor-valued system

$$\dot{\Phi}_p = A_p\Phi_p + vB_p\Phi_p \tag{3.3.23}$$

where $vB_p = \sum_{i=1}^{m} v_iB_{i,p}$. Similar equations hold at each point $p \in M$, and to relate the systems arising from two intersecting coordinate neighbourhoods we must consider the effect of a coordinate transformation on the tensor space $\mathscr{L}^{\mathrm{T}}_{n+m}$. Thus, let $(y, w) = g(x, u)$ be a bi-analytic coordinate transformation from (x, u)-coordinates to (y, w)-coordinates. Then we can write

$$g(x, u) = \sum_{i=0}^{\infty} \sum_{j=0}^{\infty} g_{ij}x^iu^j$$

and we have

$$\begin{aligned} y^\alpha w^\beta &= \left(\sum_{i=0}^{\infty} \sum_{j=0}^{\infty} g_{ij}x^iu^j \right)^{(\alpha,\beta)} \\ &= \sum_{i=0}^{\infty} \sum_{j=0}^{\infty} g_{ij}^{\alpha\beta}x^iu^j \end{aligned} \tag{3.3.24}$$

for some numbers $g_{ij}^{\alpha\beta}$. Let G be the tensor operator with the representation $(g_{ij}^{\alpha\beta})$. Then we have

$$\Psi = G\Phi,$$

where

$$\Psi = (y^\alpha w^\beta), \qquad \Phi = (x^i u^j),$$

and

$$\Phi = G^{-1}\Psi,$$

where G^{-1} is the inverse tensor operator, which is defined in the same way from g^{-1} as G is from g.

If

$$(y, w) = g(x, u), \qquad (z, q) = h(y, v),$$

then

$$(z, q) = h \circ g(x, u)$$

and

$$\Xi = HG\Phi,$$

where

$$\Xi = (z^\alpha q^\beta), \qquad \Phi = (x^i u^j)$$

and H, G are defined as above. Hence the set of tensor operators of the type defined by (3.3.24) is a group and operates as a transformation group on $\mathscr{L}^{\mathsf{T}}_{n+m}$. Thus, if we assign a space of tensors $\mathscr{L}^{\mathsf{T}}_{n+m,p}$ of type $(x^i u^j)$ at each point p of M (with local coordinates (x, u)), then we can make $\bigcup_{p\in M}\mathscr{L}^{\mathsf{T}}_{n+m,p}$ with the projection

$$\pi : \bigcup_{p\in M} \mathscr{L}^{\mathsf{T}}_{n+m,p} \to M$$

into a fibre bundle over M. We shall denote the bundle $\bigcup_{p\in M}\mathscr{L}^{\mathsf{T}}_{n+m,p}$ by Γ_{n+m}.

We can now extend the definition of a bilinear system as follows. Let $X(.): U \to D(M)$ be a system as defined above. We shall say that $m+1$ sections $\mathscr{A}, \mathscr{B}_1, \ldots, \mathscr{B}_m$ of the fibre bundle Γ_{n+m} form a *bilinear system* on M if they are of the form $\mathscr{A}_p = A_p\Phi_p$, $\mathscr{B}_{ip} = B_{ip}\Phi_p$, for tensor operators A_p, B_{ip}, and if the local representation of $X(.)$ given by

$$\dot{x} = f_p(x, u)$$
$$\dot{u} = v$$

at p is related to the bilinear system

$$\dot{\Phi}_p = A_p\Phi_p + v_1 B_{1p}\Phi_p + \cdots + v_m B_{mp}\Phi_p \tag{3.3.25}$$

as above.

Note that the action of the group of transformations of type (3.3.24) implies that local representations of the form (3.3.25) are related by

$$\dot{\Psi}_q = GA_pG^{-1}\Psi_q + v_1GB_{1p}G^{-1}\Psi_q + \cdots + v_mGB_{mp}G^{-1}\Psi_q$$

where $\Psi_q = G\Phi_p$ and G is a transformation of type (3.3.24) between the coordinates (y, w) at q and (x, u) at p where

$$y(q) = x(p) = 0, \qquad w(q) = u(p) = 0.$$

For a given section $\mathscr{A}$ of Γ_{n+m}, which belongs to a bilinear system, we can define an exponential map for this 'tensor field' by defining locally

$$(e^{\mathscr{A}t})_p = e^{A_pt}. \tag{3.3.26}$$

This is well defined since A_p is just a linear tensor operator and so e^{A_pt} can be defined by the usual series locally and we have, under a change of coordinates G,

$$G(e^{A_pt})G^{-1} = e^{GA_pG^{-1}t}.$$

3.4 APPLICATIONS OF THE TENSOR REPRESENTATION TO NONLINEAR STABILITY THEORY

3.4.1 Existence of limit cycles (Banks, 1986b)

Consider the nonlinear differential equation

$$\dot{x} = f(x) \tag{3.4.1}$$

and its linear infinite-dimensional tensor representation

$$\dot{\Phi} = A\Phi \tag{3.4.2}$$

associated with it as in the last section, where $\Phi_i = x_1^{i_1} \cdots x_n^{i_n}$. A *limit cycle* of (3.4.1) is just a periodic orbit; i.e. if $\phi_t(x)$ is the flow of the vector field f, then x is on a limit cycle if

$$\phi_{t+p}(x) = \phi_t(x),$$

for some finite $p > 0$ and all t.

For a linear finite-dimensional system $\dot{x} = Ax$, a limit cycle is just a circular orbit in phase space (in appropriate coordinates) which corresponds to a pure imaginary eigenvalue of the matrix A. We can now generalize this to nonlinear systems as follows.

Theorem 3.4.1

The equation (3.4.1) has a limit cycle if and only if the corresponding tensor operator A in (3.4.2) has a pure imaginary eigenvalue ($\neq 0$) with an eigenvector in the space $\mathscr{L}_n^{\mathrm{T}}$.

Proof

If (3.4.1) has a limit cycle γ and $x_0 \in \gamma$, then let Φ^0 be the tensor in $\mathscr{L}_n^{\mathrm{T}}$ with components

$$\Phi^0_{i_1 \ldots i_n} = x_{01}^{i_1} \cdots x_{0n}^{i_n}.$$

Let T be the period of γ. Then

$$\exp(AT)\Phi^0 = \Phi^0,$$

since $\Phi = \exp(AT)\Phi^0$ is the solution of (3.4.2). Thus, $\exp(AT)$ has the eigenvalue 1 with eigenvector $\Phi^0 \in \mathscr{L}_n^{\mathrm{T}}$. Using the same arguments as in the spectral mapping theorem (Yosida (1974), Banks (1983)), it follows that the spectrum of $\exp(AT)$ equals the exponential of the spectrum of AT. Hence

$$e^{\lambda T} = 1$$

where $\lambda \in \sigma(A)$, and so λ is imaginary.

The converse is simply a reversal of this argument. □

REMARKS (a) Let $\sigma_N(A)$ denote the set of eigenvalues of A with eigenvectors in N and define

$$S = \{|\lambda| : \lambda = i\mu \in \sigma_N(A),\ \mu \neq 0\},$$
$$T_{\mathrm{i}} = \inf S$$
$$T_{\mathrm{s}} = \sup S.$$

If $0 < T_{\mathrm{i}}, T_{\mathrm{s}} < \infty$ then $2\pi/T_{\mathrm{s}}$ is the minimum period of oscillation and $2\pi/T_{\mathrm{i}}$ is the maximum period of oscillation.

(b) In Theorem 3.4.1, the condition that $\Phi^0 \in \mathscr{L}_n^{\mathrm{T}}$ is crucial. Indeed, if $\exp(AT)\Phi = \Phi$ for some $\Phi \in l^2 \otimes \cdots \otimes l^2$, then a limit cycle does not necessarily exist since Φ may not be of the form $x_1^{i_1} \cdots x_n^{i_n}$ for some x. This is, of course, where the nonlinearity of the original problem manifests itself in the apparently linear eigenvalue problem.

(c) The main difficulty with this approach to the limit cycle behaviour of nonlinear systems is the computation of the spectrum of a tensor operator. However, we can use the theory in a negative way, for Theorem 3.4.1 has the following obvious corollary.

Corollary 3.4.2

If the spectrum of A does not intersect the imaginary axis, then the system (3.4.1) cannot have a limit cycle. □

In order to examine the spectrum of A in more detail we shall derive an infinite-dimensional tensorial version of the Gersgorin circle theorem for the approximate localization of the eigenvalues of a matrix. To do this we shall truncate the tensor A by restricting each index to the range 0 to k. Thus we write A_k for the tensor operator with components $A^{j_1 \dots j_n}_{i_1 \dots i_n}$ where $0 < i_l, \dots, j_l \leqslant k$. Since A is bounded on $\mathscr{L}^{\mathrm{T}}_n$ it is easy to see that $\| A_k \| \to \| A \|$ as $k \to \infty$ and $\sigma_N(A_k) \to \sigma_N(A)$ (in the Hausdorff metric – see p. 2).

Note that λ is an eigenvalue of A if and only if

$$A\Phi = \lambda\Phi,$$

which implies that

$$\sum_{j_1, \dots, j_n} A^{j_1 \dots j_n}_{i_1 \dots i_n} \Phi_{j_1 \dots j_n} = \lambda \delta^{j_1}_{i_1} \dots \delta^{j_n}_{i_n} \Phi_{j_1 \dots j_n}.$$

It is natural to call $A^{j_1 \dots j_n}_{i_1 \dots i_n}$ the *diagonal* elements of A. Although the generalization of Gersgorin's theorem which we require can be proved directly, it is simpler to write a tensor operator in the form of a matrix and apply the classical result.

Thus, for example, given a rank-2 tensor $A^{j_1 j_2}_{i_1 i_2}$ of dimension $k+1$, we can organize it into a matrix in the following way:

$$\begin{bmatrix}
A^{00}_{00} & A^{10}_{00} & A^{20}_{00} & \dots & A^{k0}_{00} & A^{01}_{00} & A^{11}_{00} & \dots & A^{k1}_{00} & \dots & A^{0k}_{00} & \dots & A^{kk}_{00} \\
A^{00}_{10} & A^{10}_{10} & A^{20}_{10} & \dots & A^{k0}_{10} & A^{01}_{10} & A^{11}_{10} & \dots & A^{k1}_{10} & \dots & A^{0k}_{10} & \dots & A^{kk}_{10} \\
\vdots & \vdots & \vdots & & \vdots & \vdots & \vdots & & \vdots & & \vdots & & \vdots \\
A^{00}_{k0} & A^{10}_{k0} & A^{20}_{k0} & \dots & A^{k0}_{k0} & A^{01}_{k0} & A^{11}_{k0} & \dots & A^{k1}_{k0} & \dots & A^{0k}_{k0} & \dots & A^{kk}_{k0} \\
\vdots & \vdots & \vdots & & \vdots & \vdots & & & \vdots & & \vdots & & \vdots \\
A^{00}_{0k} & A^{10}_{0k} & A^{20}_{0k} & \dots & A^{k0}_{0k} & A^{01}_{0k} & A^{11}_{0k} & \dots & A^{k1}_{0k} & \dots & A^{0k}_{0k} & \dots & A^{kk}_{0k} \\
\vdots & \vdots & & & \vdots & \vdots & \vdots & & \vdots & & \vdots & & \vdots \\
A^{00}_{kk} & A^{10}_{kk} & A^{20}_{kk} & \dots & A^{k0}_{kk} & A^{01}_{kk} & A^{11}_{kk} & \dots & A^{k1}_{kk} & \dots & A^{0k}_{kk} & \dots & A^{kk}_{kk}
\end{bmatrix}$$

which corresponds to the ordering of $\Phi_{j_1 \dots j_n}$ in the rank-1 array

$$(\Phi_{00\dots0} \quad \Phi_{10\dots0} \quad \dots \quad \Phi_{k0\dots0} \, \Phi_{01\dots0} \quad \Phi_{110\dots0} \dots \Phi_{k10\dots0} \quad \dots \quad \Phi_{kk\dots k})$$

If A_k denotes the matrix associated with the tensor A_k as above, then the eigenvalues of A are clearly the same as those of A_k. Hence the classical Gersgorin theorem gives

Theorem 3.4.3

Let $A^{j_1 j_2}_{i_1 i_2}$ be a rank-2 tensor of dimension $k+1$. Define the numbers $\mathscr{A}^{pq}_{pq}$ by

$$\mathscr{A}^{pq}_{pq} = \sum_{m,n=0}^{k} | A^{mn}_{pq} | - | A^{pq}_{pq} |, \qquad 0 \leqslant p, q \leqslant k.$$

Then the eigenvalues of A_k are contained in the union of the interiors of the circles with centres A_{pq}^{pq} and radii $\mathscr{A}_{pq}^{pq}$.

REMARK An obvious analogue holds for 'columns'.

Corollary 3.4.4.
If $A_{i_1 \dots i_n}^{j_1 \dots j_n}$ is a rank-n tensor of dimension $k+1$ and we define

$$\mathscr{A}_{p_1 \dots p_n}^{p_1 \dots p_n} = \sum_{m_1, \dots, m_n = 0}^{k} |A_{p_1 \dots p_n}^{m_1 \dots m_n}| - |A_{p_1 \dots p_n}^{p_1 \dots p_n}|, \qquad 0 \leqslant p_1, \dots, p_n \leqslant k,$$

then the eigenvalues of A_k are contained in the union of the interiors of the $(k+1)^n$ circles with centres $A_{p_1 \dots p_n}^{p_1 \dots p_n}$ and radii $\mathscr{A}_{p_1 \dots p_n}^{p_1 \dots p_n}$. □

From the representation (3.4.2) of the nonlinear system (3.4.1) we therefore obtain the following sufficient condition for the non-existence of limit cycles.

Corollary 3.4.5
If the inequality

$$2\left|\sum_{l=1}^{n} p_l \alpha_{0, \dots, 1, \dots, 0}^{l}\right| > \sum_{m_1, \dots, m_n = 0}^{\infty} \left|\sum_{l=1}^{n} p_l \alpha'_{m_1 - p_1, \dots, m - p + 1, \dots, m_n - p_n}\right| \tag{3.4.3}$$

holds for all $p_1, \dots, p_n \in \{0, 1, 2, \dots\}$ (assuming the sum on the right exists) then the system (3.4.1) has no limit cycles. □

REMARK The sum on the right of (3.4.3) is finite if f is a polynomial function, since then only a finite number of the Taylor polynomials have non-zero coefficients.

To be more precise, consider the polynomic system

$$\dot{x}_1 = \sum_{i=0}^{l_1} \sum_{j=0}^{m_1} \alpha_{ij} x_1^i x_2^j$$

$$\dot{x}_2 = \sum_{i=0}^{l_2} \sum_{j=0}^{m_2} \beta_{ij} x_1^i x_2^j$$

Then defining, as before, $\phi_{\lambda\mu} = x^\lambda x^\mu$ we have

$$\begin{aligned}
\dot{\phi}_{\lambda\mu} &= \lambda x_1^{\lambda-1} \dot{x}_1 x_2^\mu + \mu x_1^\lambda x_2^{\mu-1} \dot{x}_2 \\
&= \lambda \sum_{i=0}^{l_1} \sum_{j=0}^{m_1} \alpha_{ij} x_1^{i+\lambda-1} x_2^{j+\mu} + \mu \sum_{i=0}^{l_2} \sum_{j=0}^{m_2} \beta_{ij} x_1^{i+\lambda} x_2^{\mu+j-1} \\
&= \lambda \sum_{i=0}^{l_1} \sum_{j=0}^{m_1} \alpha_{ij} \phi_{i+\lambda-1, j+\mu} + \mu \sum_{i=0}^{l_2} \sum_{j=0}^{m_2} \beta_{ij} \phi_{i+\lambda, \mu+j-1}
\end{aligned}$$

$$= \left(\lambda \sum_{i=0}^{l_1} \sum_{j=0}^{m_1} \alpha_{ij} \delta^p_{i+\lambda-1} \delta^q_{j+\mu} + \mu \sum_{i=0}^{l_2} \sum_{j=0}^{m_2} \beta_{ij} \delta^p_{i+\lambda} \delta^q_{\mu+j-1} \right) \phi_{pq}$$

$$= \sum_{p,q=0}^{\infty} A^{pq}_{\lambda\mu} \phi_{pq}$$

where

$$A^{pq}_{\lambda\mu} = \lambda \sum_{i=0}^{l_1} \sum_{j=0}^{m_1} \alpha_{ij} \delta^p_{i+\lambda-1} \delta^q_{j+\mu} + \mu \sum_{i=0}^{l_2} \sum_{j=0}^{m_2} \beta_{ij} \delta^p_{i+\lambda} \delta^p_{\mu+j-1}.$$

Now,

$$|A^{pq}_{pq}| = \left| p \sum_{i=0}^{l_1} \sum_{j=0}^{m_1} \alpha_{ij} \delta^p_{i+p-1} \delta^q_{j+q} + q \sum_{i=0}^{l_2} \sum_{j=0}^{m_2} \beta_{ij} \delta^p_{i+p} \delta^q_{q+j-1} \right| = |p\alpha_{10} + q\beta_{01}|.$$

and

$$\sum_{\lambda,\mu=0}^{\infty} |A^{\lambda\mu}_{pq}| = \sum_{\lambda,\mu=0}^{\infty} \left| p \sum_{i=0}^{l_1} \sum_{j=0}^{m_1} \alpha_{ij} \delta^\lambda_{i+p-1} \delta^\mu_{j+q} + q \sum_{i=0}^{l_2} \sum_{j=0}^{m_2} \beta_{ij} \delta^\lambda_{i+p} \delta^\mu_{q+j-1} \right|$$

$$= \sum_{\lambda,\mu=0}^{\infty} \left| p\alpha_{\lambda-p+1,\mu-q} + q\beta_{\lambda-p,\mu-q+1} \right|$$

where we take $\alpha_{kl} = 0$ or $\beta_{kl} = 0$ if k and l are outside the respective ranges 0 to l_1, 0 to m_1 or 0 to l_2, 0 to m_2. If we put

$$l = \max\{l_1, l_2\}, \qquad m = \max\{m_1, m_2\},$$

then we have

$$\sum_{\lambda,\mu=0}^{\infty} |A^{\lambda\mu}_{pq}| = \sum_{\lambda=p-1}^{p+l} \sum_{\mu=q-1}^{q+m} |p\alpha_{\lambda-p+1,\mu-q} + q\beta_{\lambda-p,\mu-q+1}|$$

$$= \sum_{\lambda=-1}^{l} \sum_{\mu=-1}^{m} |p\alpha_{\lambda+1,\mu} + q\beta_{\lambda,\mu+1}|.$$

Using (3.4.3) we now see that the system will have no limit cycles if

$$2|p\alpha_{10} + q\beta_{01}| > \sum_{\lambda=-1}^{l} \sum_{\mu=-1}^{m} |p\alpha_{\lambda+1,\mu} + q\beta_{\lambda,\mu+1}|$$

for all $p, q \geqslant 0$. In particular, if α_{10} and β_{01} have the same sign it is sufficient that

$$2|\alpha_{10}| > \sum_{\lambda=-1}^{l_1} \sum_{\mu=0}^{m_1} |\alpha_{\lambda+1,\mu}|$$

$$2|\beta_{01}| > \sum_{\lambda=0}^{l_2} \sum_{\mu=-1}^{m_2} |\beta_{\lambda,\mu+1}|$$

$$\operatorname{sgn} \alpha_{10} = \operatorname{sgn} \beta_{01}.$$

3.4.2 Extension of Lyapunov's equation to nonlinear systems

In this section we shall generalize the familiar Lyapunov equation

$$A^{\mathrm{T}}P + PA = -I \tag{3.4.4}$$

for a positive definite symmetric matrix P, which characterizes the stability of the linear equation

$$\dot{x} = Ax$$

through the Lyapunov function $V = \langle Px, x \rangle$. Consider, therefore, the nonlinear differential equation

$$\dot{x} = f(x) \tag{3.4.5}$$

and replace it by the tensor equation

$$\dot{\Phi} = A\Phi \tag{3.4.6}$$

developed above. It will be necessary to use a Hilbert space rather than the Banach space $\mathscr{L}_n = \bigotimes_n l_{\mathrm{e}}^1$ introduced to discuss the existence theory of (3.4.6). Therefore instead of $\mathscr{L}_n$ we shall use the space $\mathscr{L}_n^2 \triangleq \bigotimes_n l_{\mathrm{e}}^2$, where l_{e}^2 is the Hilbert space of sequences $\{\alpha_i\}$ which are square summable relative to the sequence $\{1/i!\}$, i.e. l_{e}^2 has the inner product

$$\langle \alpha, \beta \rangle = \sum_{i=0}^{\infty} \frac{\langle \alpha_i, \beta_i \rangle}{(i!)^2}, \qquad \alpha = \{\alpha_i\}, \quad \beta = \{\beta_i\} \in l_{\mathrm{e}}^2. \tag{3.4.7}$$

$\mathscr{L}_n^2$ is then given the cross norm as before. It is easy to see that if Φ is a tensor of the form $(x_1^{i_1} \cdots x_n^{i_n}) = (x^i)$ then

$$\| \Phi \| \leqslant \exp(\tfrac{1}{2} \| x \|^2).$$

Moreover the results of Section 3.4.1 can be modified to prove that the solution $\Phi(t)$ of (3.4.6) satisfies the inequality

$$\| \Phi(t) \| \leqslant \exp\left\{ \left(\sum_{k=1}^{n} | f_k(x_0) | \right) t \right\} \| \Phi_0 \|. \tag{3.4.8}$$

We turn now to the Lyapunov equation

$$A^{\mathrm{T}}P + PA = -I.$$

It turns out that if A is a tensor operator which is derived from equation (3.4.5) by Carleman linearization, then this equation does not have a solution, in general. In fact, it does not even have a 'weak' solution P in the sense that

$$\langle (A^{\mathrm{T}}P + PA)\Phi, \Phi \rangle = -\langle \Phi, \Phi \rangle \tag{3.4.9}$$

for all $\Phi \in \mathscr{L}_n^{2,\mathrm{T}} \triangleq \{\Phi \in \mathscr{L}_n^2 : \Phi = (x_1^{i_1} \ldots x_n^{i_n}), \text{ some } x \in \mathbb{R}^n\}$.

Before discussing this equation in detail we note first that we may assume that 0 is an equilibrium point of equation (3.4.5) by changing coordinates, if necessary. (In the following discussion we shall assume that 0 is a globally asymptotically stable equilibrium point. If other isolated equilibria exist then the discussion only applies to the domain of attraction of the point in question.) Note also that in equation (3.4.6), $\Phi_{00\ldots0} = 1$ and so the zero tensor $\Phi = (0)$ is not an equilibrium point of (3.4.6). This can be remedied by defining the tensor **1** with

$$\mathbf{1}_0 = 1, \quad \mathbf{1}_i = 0 \quad \text{for } i \neq 0.$$

Then we shall write

$$\bar{\mathscr{L}}_n^{2,T} = \mathscr{L}_n^{2,T} - \mathbf{1}$$

i.e. for any $\Phi \in \mathscr{L}_n^{2,T}$ there corresponds $\Psi \in \bar{\mathscr{L}}_n^{2,T}$ such that

$$\Psi = \Phi - \mathbf{1}.$$

Now, although the equation (3.4.9) does not have a solution we shall say that the Lyapunov equation (3.4.4) has an *ε-approximate, weak solution* if there exists a positive definite symmetric tensor P_ε, for each $\varepsilon > 0$, such that

$$\langle (A^{\mathsf{T}} P_\varepsilon + P_\varepsilon A)\Psi, \Psi \rangle = -\langle \Psi, \Psi \rangle + \varepsilon, \tag{3.4.10}$$

for all $\Psi \in \bar{\mathscr{L}}_n^{2,T}$.

Theorem 3.4.6

If there exists a strictly positive definite, symmetric tensor P_ε with lower bound independent of ε, for each (bounded) $\varepsilon > 0$, such that the Lyapunov equation (3.4.4) has an ε-approximate weak solution (i.e. (3.4.10) holds), then the system (3.4.5) is asymptotically stable at the origin.

Conversely, if the system (3.4.5) is (globally) asymptotically stable at the origin, then if any solution $x(t; x_0)$ of the system satisfies

$$\| x(t; x_0) \|^2 \leqslant \log(1 + C/t^\alpha) \tag{3.4.11}$$

for some constants C, α and for sufficiently large t, then there exists an ε-approximate weak solution of (3.4.4).

Proof

Suppose that an ε-approximate solution P_ε exists and that the system (3.4.5) is not asymptotically stable. Then there exists $\delta > 0$ such that $\| \Psi(t_i) \|^2 > \delta$ for some sequence $t_i \to \infty$, where $\Psi(t) = (x^i(t)) - 1 \in \mathscr{L}_n^{2,T}$ is the tensor associated with the solution $x(t)$. Let $\varepsilon < \delta$ and consider the function

$$V = \langle \Phi, P_\varepsilon \Psi \rangle, \ (\Psi + \mathbf{1}) \in \mathscr{L}_n^{2,T}.$$

Then

$$\dot{V} = \langle A\Psi(t), P_\varepsilon\Psi(t)\rangle + \langle \Psi(t), P_\varepsilon A\Psi(t)\rangle$$
$$= -\langle \Psi, \Psi\rangle + \varepsilon.$$

Since P_ε is strictly positive definite we can write

$$\langle \Psi, P_\varepsilon\Psi\rangle \geqslant \alpha \| \Psi \|^2,$$

for some α, independent of ε. Hence,

$$\alpha \int \frac{d}{dt} \| \Psi(t) \|^2 \leqslant -\| \Psi(t) \|^2 + \varepsilon$$
$$< 0$$

if $\| \Psi(t) \|^2 > \varepsilon$. This is a contradiction.

To prove the converse we shall follow the classical proof for the linear case. Thus, consider the tensor operator-valued differential equation

$$\dot{X} = A^{\mathrm{T}}X + XA, \qquad X(0) = I. \tag{3.4.12}$$

This equation has the unique solution

$$X(t) = e^{A^{\mathrm{T}}t}e^{At}.$$

Let Ψ_0 be such that $\Psi_0 + \mathbf{1} \in \mathscr{L}_n^{2,T}$. Then

$$\Psi_0^T X \Psi_0 = \Psi_0^T e^{A^{\mathrm{T}}t}e^{At}\Psi_0.$$

However, as we have seen, $e^{At}\Psi_0$ is the solution of (3.4.6) with initial value Ψ_0 and so $e^{At}\Psi_0$ is of the form $\{x^i(t)\} - \mathbf{1}$, where $x(t)$ is the solution of (3.4.5) with initial condition x_0 for which

$$\{x_0^i\} - \mathbf{1} = \Psi_0.$$

Hence,

$$\Psi_0^T X \Psi_0 = \langle \Psi^T(t), \Psi(t)\rangle$$

where $\Psi(t) = e^{At}\Psi_0$ and so

$$\Psi_0^T X \Psi_0 = \| \Psi(t) \|^2 = \| \{x^i(t)\} \|^2 - 1$$
$$\leqslant \exp(\tfrac{1}{2} \| x(t) \|^2) - 1.$$

By (3.4.12),

$$\langle X(t)\Psi_0, \Psi_0\rangle - \langle \Psi_0, \Psi_0\rangle = \int_0^t \langle (A^T X(t) + X(t)A)\Psi_0, \Psi_0\rangle \, dt.$$

Using (3.4.11) we see that

$$\langle X(t)\Psi_0, \Psi_0\rangle \leqslant \exp(\tfrac{1}{2} \| x(t; x_0) \|^2) - 1 \to 0$$

as $t \to \infty$, so if $\varepsilon > 0$ choose $t(\varepsilon)$ so that $\langle X(t(\varepsilon))\Psi_0, \Psi_0\rangle \leqslant \varepsilon$ and define

$$P_\varepsilon = \int_0^{t(\varepsilon)} X(t)\,\mathrm{d}t.$$

Then.

$$\varepsilon - \langle \Psi_0, \Psi_0\rangle = \langle (A^T P_\varepsilon + P_\varepsilon A)\Psi_0, \Psi_0\rangle.$$

so that (3.4.4) has an ε-approximate weak solution. P_ε is clearly symmetric and it is strictly positive definite since

$$\langle P_\varepsilon \Psi_0, \Psi_0\rangle = \int_0^{t(\varepsilon)} \{E(x(t; x_0)) - 1\}\,\mathrm{d}t$$

$$\geqslant c \,\| x(t; x_0) \|^2$$

for some constant c, if ε is bounded above, where

$$E(x) = \prod_{k=1}^{n} \left\{ \sum_{l=0}^{\infty} \frac{x_k^{2l}}{(l!)^2} \right\}^{1/2}. \qquad \square$$

REMARKS (a) We can replace I in (3.4.4) by an arbitrary positive definite symmetric tensor, just as in the classical case.

(b) By using different norms on $\mathscr{L}_n^{2,T}$ it is possible to relax the condition (3.4.11). In fact, if ρ is an analytic function such that

$$\int_0^\infty \rho(\| x(t; x_0) \|)\,\mathrm{d}t < \infty$$

for each solution $x(t; x_0)$ of the equation, then, instead of the inner product (3.4.7) we can choose the inner product

$$\langle \alpha, \beta\rangle = \sum_{i=0}^{\infty} \rho_i^2 \alpha_i \beta_i, \tag{3.4.13}$$

where

$$\rho(x) = \sum_{i=0}^{\infty} \rho_i x^i$$

is the Taylor series expansion of ρ.

(c) If an equation has several stable equilibria, then the result can be applied with $\Psi_0 = \{x_0^i\}$ where x_0 is in the region of attraction of a particular stable equilibrium.

We have seen above that, given an analytic system (3.4.5) which has $x = 0$ as an asymptotically stable equilibrium point, we can construct a Lyapunov function of the form

$$V = \int_0^\infty \langle \Psi, \mathrm{e}^{A^T t}\mathrm{e}^{At}\Psi\rangle\,\mathrm{d}t, \tag{3.4.14}$$

which we have seen exists provided the condition (3.4.11) holds. Since e^{At} is the solution of the system with initial condition $\Psi = \{x_0^i\} - \mathbf{1}$, evaluating (3.4.14) requires us to solve the equation (3.4.5). Hence finding a Lyapunov function for a general system of the form (3.4.5) can only be achieved by solving the original equation. Of course the Lyapunov function (3.4.14) can be expressed in the form

$$V(x_0) = \int_0^\infty \{E(x(t; x_0)) - 1\}\, \mathrm{d}t$$

where $x(t; x_0)$ is the solution of (3.4.5) with initial condition $x(0) = x_0$. This is not surprising since, for any system of the form (3.4.5) which is asymptotically stable, we can write down the Lyapunov function

$$V(x_0) = \int_0^\infty \rho(\|x(t; x_0)\|)\, \mathrm{d}t$$

where ρ is any function which makes the integral exist.

The advantage of the expression (3.4.14) is that, since the formal evaluation of $e^{A^{\mathrm{T}}t}\Psi$ as a series just gives the Taylor series of the solution $x(t; x_0)$ as a function of t and x_0, if we truncate Ψ and A in the obvious way to a finite-dimensional tensor and tensor operator, respectively, then we can write the finite-dimensional approximation

$$\dot{\Psi}_m = A_m \Psi_m$$

to the exact equation $\dot{\Psi} = A\Psi$, for each $m \geqslant 1$. The solutions $e^{A_m t}\Psi$ to these equations clearly converge uniformly to the solution $e^{At}\Psi$ of the exact equation on compact sets since the former are just approximations of Taylor series.

If we solve the finite-dimensional Lyapunov equations

$$A_m^{\mathrm{T}} P_m + P_m A_m = -I_m,$$

then we obtain a sequence of Lyapunov functions

$$V_m(x) = \langle P\Psi_m, \Psi_m \rangle_m \tag{3.4.15}$$

which are valid in expanding neighbourhoods of 0 as m increases. Here, $\langle .,. \rangle_m$ is not the usual inner product in finite-dimensional space, but the obvious truncation of the one introduced above on $\mathscr{L}_n^{2,T}$. Each function V_m in (3.4.15) is a polynomial in x, the zeros of which move away from the origin to ∞ as $m \to \infty$.

4 BILINEAR SYSTEMS AND VOLTERRA SERIES

4.1 VOLTERRA SERIES

4.1.1 Bilinear systems

In this section we shall generalize the linear variation of constants formula

$$x(t) = \mathrm{e}^{At}x_0 + \int_0^t \mathrm{e}^{A(t-s)}Bu(s)\,\mathrm{d}s, \tag{4.1.1}$$

which is obtained from the linear system

$$\dot{x} = Ax + Bu, \qquad x(0) = x_0 \in \mathbb{R}^n,$$

to the corresponding formula for the general bilinear system

$$\dot{x} = Ax + u_1B_1x + \cdots + u_mB_mx, \qquad x(0) = x_0 \in \mathbb{R}^n. \tag{4.1.2}$$

To obtain an input–output relation for this system we use the well-known Picard iteration technique. First we define the sequence $\xi_k(t)$ by

$$\begin{aligned}
\xi_0(t) &= \mathrm{e}^{At}x_0 \\
\xi_1(t) &= \int_0^t \mathrm{e}^{A(t-s)} \sum_{i=1}^m u_i(s)B_i\xi_0(s)\,\mathrm{d}s \\
&\ \ \vdots \\
\xi_k(t) &= \int_0^t \mathrm{e}^{A(t-s)} \sum_{i=1}^m u_i(s)B_i\xi_{k-1}(s)\,\mathrm{d}s, \qquad k \geqslant 1.
\end{aligned} \tag{4.1.3}$$

Then we define

$$x(t) = \sum_{k=0}^{\infty} \xi_k(t) \tag{4.1.4}$$

assuming the series exists.

Theorem 4.1.1

The series in (4.1.4) exists on compact intervals for any control‡ $u \in L^{\infty}_{\text{loc}}(\mathbb{R}^m; \mathbb{R})$ and satisfies equation (4.1.2).

Proof

Assume first that the series converges uniformly on compact intervals of $\mathbb{R}$. Then

$$\begin{aligned} x(t) &= \xi_0(t) + \sum_{k=1}^{\infty} \xi_k(t) \\ &= e^{At}x_0 + \sum_{k=1}^{\infty} \int_0^t e^{A(t-s)} \sum_{i=1}^{m} u_i(s)B_i\xi_{k-1}(s)\,ds \\ &= e^{At}x_0 + \int_0^t e^{A(t-s)} \sum_{i=1}^{m} u_i(s)B_i \sum_{k=1}^{\infty} \xi_{k-1}(s)\,ds \\ &= e^{At}x_0 + \int_0^t e^{A(t-s)} \sum_{i=1}^{m} u_i(s)B_i x(s)\,ds \end{aligned}$$

and so $x(t)$ satisfies (4.1.2).

To show that the series (4.1.4) converges on compact intervals first recall that for any matrix A we can write

$$\| e^{At} \| \leqslant Ke^{\omega t}$$

for some real constants K, ω. Hence by (4.1.3) we have

$$\| \xi_k(t)\| \leqslant \int_0^t Ke^{\omega(t-s)} \sum_{i=1}^{m} | u_i |_T \| B_i \| \,.\, \|\xi_{k-1}(s)\|\,ds,$$

for $t \in [0, T]$, where

$$| u_i |_T = \sup_{0 \leqslant t \leqslant T} | u_i(t)| \,.$$

Thus, if $\beta = \max_{1 \leqslant i \leqslant m} \{\| B_i \|\}$, and $\nu = \sum_{i=1}^{m} | u_i |_T$, then we have

$$\begin{aligned} \| \xi_k(t)\| &\leqslant \int_0^t Ke^{\omega(t-s)} \nu\beta \, \| \xi_{k-1}(s)\|\,ds \\ &\leqslant \int_0^t Ke^{\omega(t-s)} \nu\beta \int_0^s Ke^{\omega(s-\tau)} \nu\beta \, \| \xi_{k-2}(\tau)\|\,d\tau\,ds, \end{aligned}$$

‡ i.e. $u^T \in L^{\infty}(\mathbb{R}^m; \mathbb{R})$ for any $t > 0$, where $u^T(t) = u(t)$, $0 \leqslant t \leqslant T$, $u^T(t) = 0$, $t > T$.

by iteration, and so

$$\begin{aligned}\|\xi_k(t)\| &\leqslant K^2(\nu\beta)^2 \int_0^t (t-\tau)\mathrm{e}^{\omega(t-\tau)}\|\xi_{k-2}(\tau)\|\,\mathrm{d}\tau \\ &\leqslant K^k(\nu\beta)^k \int_0^t \frac{(t-\tau)^{k-1}}{(k-1)!}\mathrm{e}^{\omega(t-\tau)}\,\mathrm{e}^{\omega\tau}\|x_0\|\,\mathrm{d}\tau \\ &= (K\nu\beta)^k\frac{t^k}{k!}\mathrm{e}^{\omega t}\|x_0\|.\end{aligned}$$

It follows from (4.1.4) that

$$\|x(t)\| \leqslant \mathrm{e}^{(\omega+K\nu\beta)t}\|x_0\|,$$

and the result is proved. □

REMARK Uniqueness of the solution of equation (4.1.2) follows by standard arguments – see, for example, Coddington and Levinson (1955) or Lefschetz (1977). We can write the solution (4.1.4) in a closed form by iterating the expressions in (4.1.3). In fact we have

$$\xi_k(t) = \sum_{i_1=1}^{m}\cdots\sum_{i_k=1}^{m}\int_0^t\int_0^{\tau_1}\cdots\int_0^{\tau_{k-1}} \mathrm{e}^{A(t-\tau_1)}B_{i_1}\,\mathrm{e}^{A(\tau_1-\tau_2)}\,B_{i_2}\,\mathrm{e}^{A(t_2-\tau_3)}B_{i_3}\cdots$$
$$B_{i_k}\,\mathrm{e}^{A\tau_k}x_0.u_{i_1}(\tau_1)u_{i_2}(\tau_2)\cdots u_{i_k}(\tau_k)\,\mathrm{d}\tau_k\cdots\mathrm{d}\tau_1$$

and so we can write $x(t)$ in the form

$$x(t) = w_0(t) + \sum_{k=1}^{\infty}\sum_{i_1=1}^{m}\cdots\sum_{i_k=1}^{m}\int_0^t\int_0^{\tau_1}\cdots\int_0^{\tau_{k-1}} w_{k;\,i_1,\dots,i_k}(t,\tau_1,\dots,\tau_k)$$
$$\cdot u_{i_1}(\tau_1)\cdots u_{i_k}(\tau_k)\,\mathrm{d}\tau_k\cdots\mathrm{d}\tau_1, \qquad (4.1.5)$$

where

$$w_0(t) = \mathrm{e}^{At}x_0$$

and

$$w_{k;\,i_1,\dots,i_k} = \mathrm{e}^{A(t-\tau_1)}B_{i_1}\,\mathrm{e}^{A(\tau_1-\tau_2)}B_{i_2}\dots B_{i_k}\,\mathrm{e}^{A\tau_k}x_0.$$

In a more compressed notation we can regard $w_{k;\,i_1,\dots,i_k}$ and $u_{i_1}\dots u_{i_k}$ as tensors and write

$$\sum_{i_1=1}^{m}\cdots\sum_{i_k=1}^{m} w_{k;\,i_1,\dots,i_k}(t,\tau_1,\dots,\tau_k)u_{i_1}(\tau_1)\cdots u_{i_k}(\tau_k)$$
$$= w_k(t,\tau_1,\dots,\tau_k)u(\tau_1)\otimes\cdots\otimes u(\tau_k)$$

and then (4.1.5) becomes

$$x(t) = w_0(t) + \sum_{k=1}^{\infty}\int_0^t\int_0^{\tau_1}\cdots\int_0^{\tau_{k-1}} w_k(t,\tau_1,\dots,\tau_k)u(\tau_1)\otimes$$
$$\cdots\otimes u(\tau_k)\,\mathrm{d}\tau_k\cdots\mathrm{d}\tau_1. \qquad (4.1.6)$$

The (tensor-valued) function $w_k(t, \tau_1, \ldots, \tau_k)$ is called the kth*-order kernel* of the *Volterra series*‡ (4.1.6). Knowing the kernels of a nonlinear operator allows one to evaluate the input–output map from (4.1.6). Further discussion of the Volterra series for a bilinear system can be found in Brockett (1976b) and Di Pillo *et al.* (1974).

4.1.2 Linear analytic systems

In this section we shall derive a Volterra series expansion for the linear analytic system

$$\dot{x} = f(x) + ug(x), \qquad x(0) = x_0, \qquad y = h(x) \tag{4.1.7}$$

where the functions f, g and h are (real) analytic and u and y are scalar-valued. We shall follow Lesiak and Krener (1978) who apply, essentially, the nonlinear variation of constants formula due to Alekseev (1961) and Brauer (1966). A similar approach is discussed by Fliess (1980).

First we show that (4.1.7) has a solution provided the unperturbed equation $\dot{x} = f(x)$, $x(0) = x_0$ is soluble. To do this we need the following version of Gronwall's inequality:

Lemma 4.1.2

Let $\mu, \nu \in L^1[0, T]$ be such that $\mu(s) \geqslant 0$, $\nu(s) \geqslant 0$ and suppose that f is continuous on $[0, T]$ and satisfies

$$0 \leqslant f(t) \leqslant \int_0^t [\mu(s)f(s) + \nu(s)] \, ds$$

for all $t \in [0, T]$. Then

$$0 \leqslant f(t) \leqslant e^{\int_0^t \mu(r) \, dr} \int_0^T \nu(r) \, dr.$$

Proof

The function $g(s) = f(s)e^{-\int_0^s \mu(r) \, dr}$ is nonnegative and continuous and so has a maximum γ when $s = \alpha$, say. Then

$$0 \leqslant f(t_0) = \gamma e^{\int_0^\alpha \mu(r) \, dr} \leqslant \int_0^\alpha [\mu(s)\gamma e^{\int_0^s \mu(r) \, dr} + \nu(s)] \, ds$$

$$= \gamma(e^{\int_0^\alpha \mu(r) \, dr} - 1) + \int_0^\alpha \nu(s) \, ds.$$

‡ See also Rugh (1981).

Hence

$$f(t)\mathrm{e}^{-\int_0^t \mu(r)\,\mathrm{d}r} \leqslant \gamma \leqslant \int_0^{\alpha} \nu(s)\,\mathrm{d}s \leqslant \int_0^{T} \nu(r)\,\mathrm{d}r. \qquad \square$$

Theorem 4.1.3

If f and g are C^1 (i.e. continuously differentiable) vector fields on $\mathbb{R}^n$, and if the equation

$$\dot{x} = f(x), \qquad x(0) = x_0 \tag{4.1.8}$$

has a solution on $[0, T]$, then there exists $\delta > 0$ such that the equation

$$\dot{x} = f(x) + ug(x), \qquad x(0) = x_0 \tag{4.1.9}$$

has a unique solution on $[0, T]$ for all u for which $\| u \|_1 \triangleq \int_0^T | u(s)|\,\mathrm{d}s < \delta$.

Proof

Let $\bar{x}(t)$ be the solution of (4.1.8) and let B be the set defined by

$$B = \{ x\colon \inf_{0 \leqslant t \leqslant T} \| x - \bar{x}(t)\| \leqslant 1 \}.$$

Since $\bar{x}(t)$ is bounded on $[0, T]$, B is clearly compact. Hence there exists $M > 0$ such that

$$\| f(x_1) - f(x_2)\| \leqslant M \| x_1 - x_2 \|, \| g(x_1) - g(x_2)\| \leqslant M \| x_1 - x_2 \|,$$
$$x_1, x_2 \in B$$

for, by hypothesis, f and g are uniformly continuous on compact sets. Let

$$\gamma = \max_{0 \leqslant t \leqslant T} \| g(\bar{x}(t))\|$$

and

$$\delta = \min\{ T, (2\gamma\, \mathrm{e}^{2TM})^{-1} \}.$$

Then let $x(t)$ be the solution of (4.1.9) corresponding to u which satisfies $\| u \|_1 \leqslant \delta$. (Such a solution exists for some time interval $[0, \varepsilon]$ by the elementary theory of differential equations.) Let $[0, \tau)$ be the maximal time interval on which $x(t)$ exists and belongs to B. Then, for $s \in [0, \tau)$,

$$\| \bar{x}(s) - x(s)\| = \left\| \int_0^s \{ f(x(t)) - f(x(t)) - u(t)g(x(t))\}\,\mathrm{d}t \right\|$$

$$\leqslant \int_0^s \{ \| f(x(t)) - f(x(t))\| + | u(t)|\, \| g(x(t)) - g(x(t))\| \}\,\mathrm{d}t$$

$$+ \int_0^s | u(t)|\, \| g(x(t))\|\,\mathrm{d}t$$

$$\leqslant \int_0^s \{ M(1 + | u(t)|) \| \bar{x}(t) - x(t)\| + | u(t)|\, \| g(x(t))\| \}\,\mathrm{d}t.$$

By Lemma 4.1.2 it follows that

$$\|\bar{x}(s) - x(s)\| \leqslant \gamma \exp\left\{\int_0^T M(1 + |u(t)|)\, dt\right\} \int_0^T |u(t)|\, dt$$
$$\leqslant \gamma\, e^{2TM_\delta} \leqslant \tfrac{1}{2}.$$

By continuity of $x(s)$ and the definition of B it follows that $\tau \geqslant T$. Uniqueness is proved similarly. □

Before continuing with the derivation of the Volterra series for (4.1.9) we must first recall some elementary facts from the theory of ordinary differential equations (see Coddington and Levinson (1955) and Brauer (1966)). Given a differential equation

$$\dot{x} = f(x, t) \tag{4.1.10}$$

where f is continuously differentiable on a region $D \times [t_0, \infty)$ in $\mathbb{R}^{n+1}$, it is well-known that the solution $x(t; t_0, x_0)$ of (4.1.10) is differentiable in (t, t_0, x_0) on $[t_0, \infty) \times [t_0, \infty) \times D$. Clearly the matrix function Φ given by

$$\Phi(t; t_0, x_0) = \frac{\partial}{\partial x_0}\,[x(t; t_0, x_0)] \tag{4.1.11}$$

is the fundamental solution of the 'variational system'

$$\dot{Z} = f_x(x(t; t_0, x_0), t) Z. \tag{4.1.12}$$

Note that the vector

$$y(t; t_0, x_0) = \frac{\partial}{\partial t_0}\, x(t; t_0, x_0) \tag{4.1.13}$$

is a solution of (4.1.12) and since $x(t; t_0, x_0) = x_0$ we have

$$\frac{\partial}{\partial t_0}\, x(t_0; t_0, x_0) = 0$$
$$= \frac{\partial}{\partial t}\, x(t; t_0, x_0)\bigg|_{t=t_0} + \frac{\partial}{\partial t_0}\, x(t; t_0, x_0)\bigg|_{t=t_0}$$
$$= f(x_0, t_0) + y(t_0; t_0, x_0).$$

Hence

$$y(t_0; t_0, x_0) = -f(x_0, t_0)$$

and it follows that

$$y(t; t_0, x_0) = -\Phi(t; t_0, x_0) f(x_0, t_0). \tag{4.1.14}$$

Next we shall describe the formal development of the Volterra series for (4.1.9); the convergence questions will be considered below. Let $\gamma(t; s, x)$ denote the solution of the equation (4.1.8) for which $\gamma(s; s, x) = x$, and for $u \in L^1([0, T]; \mathbb{R})$ let $\gamma_u(t; s, x)$ be the solution of (4.1.9) for which $\gamma_u(t; s, x) = x$. Define the function $\rho(s)$, for each fixed t, by

$$\rho(s) = \gamma(t; s, \gamma_u(s; 0, x_0)).$$

Then

$$\rho(0) = \gamma(t; 0, x_0), \rho(t) = \gamma_u(t; 0, x_0)$$

and so

$$h(\gamma_u(t; 0, x_0)) = h(\gamma(t; 0, x_0)) + \int_0^t \frac{\mathrm{d}}{\mathrm{d}s} h(\rho(s))\, \mathrm{d}s. \tag{4.1.15}$$

Now,

$$\rho(s) = \gamma(t - s; 0, \gamma_u(s; 0, x_0))$$

and so

$$\begin{aligned}
\frac{\mathrm{d}}{\mathrm{d}s} h(\rho(s)) &= \frac{\partial}{\partial s_1} h(\gamma(t - s_1; 0, \gamma_u(s_2; 0, x_0)))\bigg|_{s_1 = s_2 = s} \\
&\quad + \frac{\partial}{\partial s_2} h(\gamma(t - s_1; 0, \gamma_u(s_2; 0, x_0)))\bigg|_{s_1 = s_2 = s} \\
&= \left(\frac{-\partial h(x)}{\partial x} f(x)\right)\bigg|_{x = \gamma(t - s; 0, \gamma_u(s; 0, x_0))} \\
&\quad + \frac{\partial h(\gamma(t - s; 0, x))}{\partial x} (f(x) + ug(x))\bigg|_{x = \gamma_u(s; 0, x_0)}.
\end{aligned}$$

However,

$$\begin{aligned}
\frac{\partial h(\gamma(t - s; 0, x))}{\partial x} f(x) &= \frac{\partial h(\gamma(t - s; 0, x))}{\partial \gamma} \frac{\partial \gamma(t - s; 0, x)}{\partial x} f(x) \\
&= \frac{\partial h(\gamma(t - s; 0, x))}{\partial \gamma} \dot{\gamma}(t - s; 0, x) \\
&= \frac{\partial h(\gamma(t - s; 0, x))}{\partial \gamma} f(\gamma(t - s; 0, x))
\end{aligned}$$

by (4.1.13) and (4.1.14). Hence

$$\left(\frac{\partial h(x)}{\partial x} f(x)\right)\bigg|_{x = \gamma(t - s; 0, \gamma_u(s; 0, x_0))} = \left(\frac{\partial h(\gamma(t - s; 0, x))}{\partial x} f(x)\right)\bigg|_{x = \gamma_u(s; 0, x_0)}$$

and it follows that

$$\frac{\mathrm{d}}{\mathrm{d}s} h(\rho(s)) = u(s)\left[\frac{\partial h(\gamma(t; s, x))}{\partial x} g(x)\right]_{x=\gamma_u(s; 0, x_0)}.$$

Defining

$$w_0(t) = h(\gamma(t; 0, x_0))$$

$$\bar{w}_1(t; s, x) = \frac{\partial h(\gamma(t; s, x))}{\partial x} g(x)$$

we have, from (4.1.15),

$$h(\gamma_u(t; 0, x_0)) = w_0(t) + \int_0^t u(s)\bar{w}_1(t; s, \gamma_u(s; 0, x_0))\, \mathrm{d}s. \quad (4.1.16)$$

(This is called the *nonlinear variation constants formula.*) We can apply (4.1.16) with $h(.)$ replaced by $\bar{w}_1(t; s,.)$ to obtain

$$\bar{w}_1(t; s, \gamma_u(s; 0, x_0)) = w_1(t; s) + \int_0^s u(r)\bar{w}_2(t; s, r, \gamma_u(r; 0, x_0))\, \mathrm{d}r, \quad (4.1.17)$$

where

$$w_1(t; s) = \bar{w}_1(t; s, \gamma(s; 0, x_0))$$

$$\bar{w}_2(t; s, r, x) = \frac{\partial w_1(t; s, \gamma(s; r, x))}{\partial x} g(x).$$

Substituting (4.1.17) into (4.1.16) gives

$$h(\gamma_u(t; 0, x_0)) = w_0(t) + \int_0^t u(s)w_1(t; s)\, \mathrm{d}s$$

$$+ \int_0^t \int_0^s \bar{w}_2(t; s, r, \gamma_u(r; 0, x_0))u(s)u(r)\, \mathrm{d}r\, \mathrm{d}s.$$

Repeating this procedure gives

$$y(t) = w_0(t) + \sum_{i=1}^{k} \int_0^t \int_0^{s_1} \cdots \int_0^{s_{i-1}} w_i(t; s_1, \ldots, s_i)u(s_1) \cdots u(s_i)\, \mathrm{d}s_i \cdots \mathrm{d}s_1$$

$$+ \int_0^t \int_0^{s_1} \cdots \int_0^{s_k} \bar{w}_{k+1}(t; s_1, \ldots, s_{k+1}, \gamma_u(s_{k+1}; 0, x_0))u(s_1) \cdots u(s_{k+1})$$

$$\cdot\, \mathrm{d}s_{k+1} \cdots \mathrm{d}s_1 \quad (4.1.18)$$

where

$$\bar{w}_i(t; s_1, \ldots, s_i, x) = \frac{\partial \bar{w}_{i-1}(t; s_1, \ldots, s_{i-1}, \gamma(s_{i-1}; s_i, x))}{\partial x} g(x)$$

$$w_i(t; s_1, \ldots, s_i) = \bar{w}_i(t; s_1, \ldots, s_i, \gamma(s_i; 0, x_0)).$$

Note that, in order to obtain this expression we only require that $h \in C^{k+1}$, $f \in C^{k+1}$ and $g \in C^k$. Furthermore, note that w_0 is defined and continuous on $[0, T]$ and w_i is defined (and continuous) on $\{(t, s_1, \ldots, s_i): 0 \leqslant s_i \leqslant \cdots \leqslant s_1 \leqslant t \leqslant T\}$.

To prove that, for analytic functions f, g, h, we can continue the process indefinitely and that the series in (4.1.18) converges, we need an estimate for the functions w_i. For this we require some elementary theory of analytic functions of n complex variables (see Grauert and Fritzsche (1976)).

Let $P_r(a) = P_{r_1,\ldots,r_n}(a_1, \ldots, a_n)$ be the subset of $\mathbb{C}^n$ defined by

$$P_r(a) = P_{r_1,\ldots,r_n}(a_1, \ldots, a_n) = \{(z_1, \ldots, z_n) \in \mathbb{C}^n : |z_i - a_i| < r_i\}.$$

$P_r(a)$ is called the (open) *polydisc* with centre $a = (a_1, \ldots, a_n)$ and radius $r = (r_1, \ldots, r_n)$. The set

$$\Gamma_r(a) = \{(z_1, \ldots, z_n) \in \mathbb{C}^n : |a_i - z_i| = r_i\}$$

is called the *distinguished boundary* of P. A function $f(z) = f(z_1, \ldots, z_r)$ is analytic on an open set $U \subseteq \mathbb{C}^n$ if, for each $a \in U$, there exists a polydisc $P_r(a) \subseteq U$ such that f has a Taylor series expansion

$$f(z) = \sum_{i_1,\ldots,i_n \geqslant 0} b_{i_1,\ldots,i_n} (z_1 - a_1)^{i_1} \cdots (z_n - a_n)^{i_n}, \qquad z \in P_r(a).$$

For such a function, a generalization of the familiar one-dimensional Cauchy formula proves the estimate

$$\left| \frac{\partial^{i_1}}{\partial z_1^{i_1}} \cdots \frac{\partial^{i_n}}{\partial z_n^{i_n}} \right| f(a) \leqslant \frac{i_1! \ldots i_n!}{r_1^{i_1} \ldots r_n^{i_n}} \max_{z \in \Gamma_r(a)} |f(z)|. \tag{4.1.19}$$

Lemma 4.1.4

If $G_1, \ldots, G_k$, h are analytic functions of n variables on a closed polydisc $P = P_r(a)$ with $r = (R, \ldots, R)$ and

$$\max\left\{ \sup_{z \in P} |G_1(z)|, \ldots, \sup_{z \in P} |G_k(z)|, \sup_{z \in P} |h(z)| \right\} \leqslant M,$$

then we have the estimate

$$\left| \left(G_k \frac{\partial}{\partial z_{i_k}}\right) \cdots \left(G_1 \frac{\partial}{\partial z_{i_1}}\right) h(a) \right| \leqslant k! \left(\frac{2^{n+1} M}{R}\right)^k M.$$

Proof

If we expand the expression

$$\left(G_k \frac{\partial}{\partial z_{i_k}}\right) \cdots \left(G_1 \frac{\partial}{\partial z_{i_1}}\right) h(a)$$

by the product formula for differentiation and apply (4.1.19) to each term we obtain a bound which consists of a sum of terms each containing the factor $M(M/R)^k$ and a product of factorials. If we remove the factor $M(M/R)^k$ the remaining sum of these factorials is exactly the same as that obtained by evaluating the expression

$$A = \left(\frac{\partial}{\partial z_{i_k}}\right)\left(\left[\prod_{j=1}^{n} \frac{1}{1-z_j}\right] \frac{\partial}{\partial z_{i_{k-1}}}\right) \cdots \left(\left[\prod_{j=1}^{n} \frac{1}{1-z_j}\right] \frac{\partial}{\partial z_{i_1}}\right)\left[\prod_{j=1}^{n} \frac{1}{1-z_j}\right](0).$$

If we define

$$B = \left(\frac{\partial}{\partial z_{i_k}} \cdots \frac{\partial}{\partial z_{i_1}}\right)\left(\left[\prod_{j=1}^{n} \frac{1}{1-z_j}\right]\right)^k (0)$$

then we can show that

$$A \leqslant B.$$

Indeed, evaluating each expression by the product formula for differentiation we see that each term is positive and every term in the expansion for A appears in that for B. Hence applying (4.1.19) to A in the polydisc with centre 0 and radius $r = (\frac{1}{2}, \frac{1}{2}, \ldots, \frac{1}{2})$, we obtain

$$A \leqslant B \leqslant k_1! \cdots k_q! 2^{nl} 2^l \leqslant k! 2^{(n+1)l},$$

where q is the number of distinct elements of $\{i_1, \ldots, i_k\}$ and $k_1 + \cdots + k_q = k$. The result now follows. □

Now consider the term

$$\bar{w}_1(t; s, x) = \frac{\partial h(\gamma(t-s; 0, x))}{\partial x} g(x)$$

$$= \left.\frac{\partial h(y)}{\partial y}\right|_{y=\gamma(t-s;0,x)} \frac{\partial \gamma(t-s; 0, x)}{\partial x} g(x).$$

For fixed $t - s$ let $x \to \bar{\gamma}(x)$ be the inverse function of the local diffeomorphism $x \to \gamma(t-s; 0, x))$ and define

$$G(t-s, y) = \left.\frac{\partial \gamma(t-s; 0, x)}{\partial x}\right|_{x=\bar{\gamma}(y)} g(\bar{\gamma}(y)).$$

Writing $x(t) = \gamma(t; 0, x_0)$, we have

$$\bar{w}_1(t; s, x) = \left(\sum_{i=1}^{n} G_i(t-s, y) \frac{\partial}{\partial y_i}\right)(h)(\gamma(t-s; 0, x))$$

and so

$$w_1(t;s) = \left(\sum_{i=1}^{n} G_i(t-s,y)\frac{\partial}{\partial y_i}\right)(h)(x(t)).$$

In general we have

$$w_k(t;s_1,\ldots,s_k) = \left(\sum_{i_k=1}^{n} G_{i_k}(t-s_k,y)\frac{\partial}{\partial y_{i_k}}\right)\cdots$$

$$\left(\sum_{i_1=1}^{n} G_{i_1}(t-s_1,y)\frac{\partial}{\partial y_{i_1}}\right)(h)(x(t)) \qquad (4.1.20)$$

We can now prove

Theorem 4.1.5
If f and g are analytic vector fields, h is an analytic function and the equation (4.1.8) has a solution on $[0, T]$, then the input–output map of the system (4.1.9) has a Volterra series expansion.

Proof
Since f is analytic, $\gamma(t-s;0,x)$ is an analytic function of $t-s$ and x and so the vector field $G(t-s,y)$ is analytic in a neighbourhood of $x(t)$. Hence, for all $s \in [0,t]$, G is analytic in some closed ball with centre $x(t)$ and radius $R(x(t))$ and h is analytic in a ball of radius R_h about $x(t)$. Let $R_t = \min\{R(x(t)), R_h\}$ and let P_t be the closed polydisc (in $\mathbb{C}^n$) with centre $x(t)$ and radii $r_i = R_t/n$. By analytic continuation we can extend G and h to P_t and we have

$$|G(z)| \leqslant M_t, \quad |h(z)| \leqslant M_t \qquad \text{for } z \in P_t$$

($s \in [0, T]$). By (4.1.20) and Lemma 4.1.4 we have

$$|w_k(t;s_1,\ldots,s_k)| \leqslant M_t\left(\frac{n^2 M_t 2^{n+1}}{R_t}\right)^k k!$$

and so

$$\left|\int_0^t w_k(t;s_1,\ldots,s_k)u(s_1)\cdots u(s_k)\,\mathrm{d}s_k\cdots\mathrm{d}s_1\right| \leqslant \left(\frac{\|u\|_1 n^2 M_t 2^{n+1}}{R_t}\right)^k M_t.$$

Hence the Volterra series converges for $\|u\|_1$ sufficiently small and the result follows since M_t/R_t is continuous in t in the interval $[0, T]$. □

Uniqueness of the Volterra series of an input–output map can also be proved quite easily; see Lesiak and Krener (1978). Note finally that the representation (3.2.3) of the Volterra series used in Chapter 3 can be obtained from that derived above by forming the Taylor series expansions of h and γ – details are left to the reader.

4.1.3 General nonlinear systems

In Section 3.3 we obtained the local extended system

$$\dot{x} = f(x, u)$$
$$\dot{u} = v$$

and obtained the bilinear representation

$$\dot{\Phi} = A\Phi + \sum_{\mu=1}^{m} v_\mu B_\mu \Phi = A\Phi + vB\Phi. \tag{4.1.21}$$

Hence, as above, we can formally define a Volterra series for (4.1.21) in the following way:

$$\Phi(t) = w_0(t) + \sum_{\nu=1}^{\infty} \int_0^t \cdots \int_0^t W_\nu(t, \sigma_1, \ldots, \sigma_\nu) v(\sigma_1) \otimes \cdots \otimes v(\sigma_\nu)\, \mathrm{d}\sigma_1 \cdots \mathrm{d}\sigma_\nu \tag{4.1.22}$$

where

$$w_\nu(t, \sigma_1, \ldots, \sigma_\nu) = \mathrm{e}^{A(t-\sigma_1)} B \mathrm{e}^{A(\sigma_1 - \sigma_2)} B \cdots B \mathrm{e}^{A\sigma_\nu} \Phi_0$$

for $t \geqslant \sigma_1 \geqslant \sigma_2 \geqslant \cdots \geqslant \sigma_\nu$ and $w = 0$ otherwise. Here, we define

$$\mathrm{e}^{A(t-\sigma_1)} B \mathrm{e}^{A(\sigma_1-\sigma_2)} B \cdots B \mathrm{e}^{A\sigma_\nu} \Phi_0 (v(\sigma_1) \otimes \cdots \otimes v(\sigma_\nu))$$

as

$$\mathrm{e}^{A(t-\sigma_1)} v(\sigma_1) B \mathrm{e}^{A(\sigma_1-\sigma_2)} v(\sigma_2) B \cdots v(\sigma_\nu) B \mathrm{e}^{A\sigma_\nu} \Phi_0.$$

Then we have

Theorem 4.1.6
The Volterra series in (4.1.22) converges (in $\mathscr{L}_{n+m}$) and is the unique solution of equation (4.1.21) provided $v(t) \in L^\infty(0, \infty)$.

Proof
The only nontrivial part is to prove convergence of the Volterra series. Let

$$\| v \|_\infty = \max_{i=1,\ldots,m} \operatorname*{ess\,sup}_{s \in [0,\infty)} |v_i(s)|.$$

Then,

$$\| \Phi(t) \| \leqslant \| w_0(t) \| + \sum_{\nu=1}^{\infty} \| v \|_\infty^\nu \int_0^t \cdots \int_0^{\sigma_{\nu-2}} \int_0^{\sigma_{\nu-1}} \| w_\nu(t, \sigma_1, \ldots, \sigma_\nu) \| \, \mathrm{d}\sigma_1 \cdots \mathrm{d}\sigma_\nu.$$

However, by Corollary 3.3.6 and (3.3.19) we have

$$\| w_\nu(t, \sigma_1, \ldots, \sigma_\nu) \| \leqslant m^\nu \exp\left\{ \left(\sum_{k=1}^{n} | f_k(x_0, u_0) | \right) t \right\} \| \Phi_0 \|,$$

where $\Phi_0 = (x_0^i u_0^j)$. Hence,

$$\|\Phi(t)\| \leqslant \|w_0(t)\| + \sum_{\nu=1}^{\infty} \|v\|_\infty^\nu m^\nu \int_0^t \cdots \int_0^{\sigma_{\nu-1}} \exp\left\{\sum_{k=1}^{n} |f_k(x_0, u_0)|)t\right\} \|\Phi_0\| \, d\sigma_1 \cdots d\sigma_\nu$$

$$= \|w_0(t)\| + \sum_{\nu=1}^{\infty} \|v\|_\infty^\nu m^\nu \exp\left\{\left(\sum_{k=1}^{n} |f_k(x_0, u_0)|\right)t\right\} \|\Phi_0\| \, t^\nu/\nu!$$

$$\leqslant \exp\left\{\left(\sum_{k=1}^{n} |f_k(x_0, u_0)|\right)t + \|v\|_\infty mt\right\} \|\Phi_0\|. \qquad \square$$

Moving now to a global system $X(.): U \to D(M)$ we consider the extended system $Y(.): T(U) \to D(M \times U)$ which gives rise to the local systems defined by (3.3.22); i.e.

$$\begin{aligned} \dot{x} &= f_p(x, u) \\ \dot{u} &= v. \end{aligned} \qquad (4.1.23)$$

If $\{\mathscr{A}, \mathscr{B}_1, \ldots, \mathscr{B}_m\}$ is a global bilinear system consisting of $m+1$ sections of the fibre bundle Γ_{n+m} for which

$$\mathscr{A}_p = A_p \Phi_p, \quad \mathscr{B}_{ip} = B_{ip} \Phi_p, \qquad 1 \leqslant i \leqslant m$$

where A_p, B_{ip} are obtained from (4.1.23) by Carleman linearization, then we can prove the following result, which follows directly from Theorem 4.1.6.

Theorem 4.1.7

Given a nonlinear system $X(.): U \to D(M)$ on an analytic manifold M, we can associate with this system a Volterra series

$$\Phi(t) = w_0(t) + \sum_{\nu=1}^{\infty} \int_0^t \cdots \int_0^t w_\nu(t, \sigma_1, \ldots, \sigma_\nu) v(\sigma_1) \otimes \cdots \otimes v(\sigma_\nu) \, d\sigma_1 \cdots d\sigma_\nu$$

where the kernels are given locally by

$$w_\nu(t, \sigma_1, \ldots, \sigma_\nu; p) = e^{A_p(t-\sigma_1)} B_p e^{A_p(\sigma_1 - \sigma_2)} B_p \ldots B_p e^{A_p \sigma_\nu} \Phi_{0p}$$

with $B_p = (B_{1p}, \ldots, B_{mp})$. Moreover the kernels transform according to

$$\begin{aligned} w_\nu(t, \sigma_1, \ldots, \sigma_\nu; q) &= e^{A_q(t-\sigma_1)} B_q e^{A_q(\sigma_1-\sigma_2)} B_q \cdots B_q e^{A_q \sigma_\nu} \Psi_{0q} \\ &= G e^{A_p(t-\sigma_1)} G^{-1} G B_p G^{-1} G e^{A_p(\sigma_1-\sigma_2)} \ldots G e^{A_p \sigma_\nu} G^{-1} G \Phi_{0p} \\ &= G w_\nu(t, \sigma_1, \ldots, \sigma_\nu; p) \end{aligned}$$

where G is a matrix representation of the coordinate transformation from p to q. $\square$

4.2 REALIZATION THEORY OF BILINEAR SYSTEMS AND VOLTERRA SERIES

4.2.1 Realization of bilinear systems

In Chapter 2 we have considered the realization of general nonlinear systems. Here we shall obtain more precise results for bilinear systems following the work of D'Alessandro *et al.* (1974). This will appear as a direct generalization of the linear case. We shall therefore consider the bilinear system

$$\begin{aligned}\dot{x}(t) &= Ax(t) + Nx(t)u(t) + Bu(t), \quad x(0) = x_0 \in \mathbb{R}^n \\ y(t) &= Cx(t)\end{aligned} \tag{4.2.1}$$

where, for simplicity of exposition, we assume a scalar control u and a scalar output y. Note that we explicitly include the linear term in u, which is slightly more general than (4.1.2).

Using the methods of Section 4.1 we can write the response $y(t)$ as a function of u in the form

$$y(t) = y_x(t) + y_u(t) + y_{xu}(t),$$

where

$$y_x(t) = Ce^{At}x_0$$

$$y_u(t) = \sum_{i=1}^{\infty} \frac{1}{i!} \int_0^t \cdots \int_0^t w_i(t_1, \ldots, t_i) \left[\prod_{k=1}^{i} u(t - t_k) \right] \mathrm{d}t_1 \cdots \mathrm{d}t_i.$$

and

$$y_{xu}(t) = \sum_{i=1}^{\infty} \frac{1}{i!} \int_0^t \cdots \int_0^t z_i(t_1, \ldots, t_i) Ne^{A(t-t_1)}x_0 \left[\prod_{k=1}^{i} u(t - t_k) \right] \mathrm{d}t_1 \cdots \mathrm{d}t_i.$$

In this form of the input–output map for the system (4.2.1) we have used a slightly different form from that in (4.1.5). In fact, we have put

$$w_i(t_1, \ldots, t_i) = \sum_{\mathscr{C}_i} v_i(t_1, \ldots, t_i)$$

where $\mathscr{C}_i$ is the set of permutations of $\{t_1, \ldots, t_i\}$ and

$$v_1(t_1) = Ce^{At_1}B,$$

$$v_i(t_1, \ldots, t_i) = Ce^{At_i}Ne^{A(t_{i-1} - t_i)} \ldots Ne^{A(t_1 - t_2)}B \left[\prod_{k=0}^{i-2} S(t_{k+1} - t_{k+2}) \right]$$

where S is the unit step function. Similar expressions for z_i can be written with B replaced by the identity. The advantage of the above expression is that the kernels w_i, z_i are symmetric in $t_1, \ldots, t_i$.

We define realizability in terms of the zero state response y_u. Thus, if $W = \{w_i(t_1, \ldots, t_i),\ 1 \leqslant i \leqslant \infty\}$ is a sequence of symmetric kernels of a Volterra series, then we say that W is *realizable* (by a constant bilinear system with finite-dimensional state space) if there exist matrices A, N, B, C of appropriate dimensions such that

$$Ce^{At_1}B = w_1(t_1) \qquad \text{for all } t_1 \tag{4.2.2}$$

$$\sum_{\mathscr{C}_i} Ce^{At_i}Ne^{A(t_{i-1}-t_i)} \ldots Ne^{A(t_1-t_2)}B\left[\prod_{k=0}^{i-2} S(t_{k+1}-t_{k+2})\right]$$
$$= w_i(t_1, \ldots, t_i) \qquad \text{for all } t_k,\ \ k = 1, \ldots, i \ \text{ and } \ \text{all } i > 1. \tag{4.2.3}$$

From (4.2.3) it follows that if w_i is realizable (by a bilinear system) then it must be of a very special form consisting of a product of various factors. We can formalize this in

Theorem 4.2.1

The sequence $\{w_i(t_1, \ldots, t_i)\}_{1 \leqslant i \leqslant \infty}$ of symmetric kernels has a bilinear realization if and only if the following two conditions hold:

(a) $w_1(t_1)$ has a proper rational Laplace transform;
(b) we can write

$$w_i(t_1, \ldots, t_i) = H(t_i)F(t_{i-1}-t_i) \cdots F(t_2-t_3)G(t_1-t_2) \tag{4.2.4}$$

on the set

$$S_i = \{(t_1, \ldots, t_i) : t_1 > t_2 > \cdots > t_i\}, \quad i > 1,$$

for some matrix functions $F(t), G(t), H(t)$, of orders $m \times m$, $m \times 1$, $1 \times m$, respectively, which have proper rational Laplace transforms.

Proof

If the sequence $\{w_i\}$ is realizable then (4.2.2) and (4.2.3) hold for matrices A, N, B, C. (4.2.2) proves condition (a). On the set S_i the only nonzero term in w_i is the term

$$Ce^{At_i}Ne^{A(t_{i-1}-t_i)} \ldots Ne^{A(t_i-t_2)}B$$

and so w_i can be written in the form of (4.2.4) by factorizing N in an arbitrary way as $N = N_1N_2$ with N_1 and N_2 having orders $n \times m$ and $m \times n$, respectively.

To prove the converse, assume that (a) and (b) hold. Then the matrix

$$L(t) = \begin{bmatrix} w_1(t) & H(t) \\ G(t) & F(t) \end{bmatrix}$$

has a proper rational Laplace transform and so may be written in the form

$$L(t) = Se^{At}R$$

where A, R, S have respective orders $n \times n$, $n \times (m+1)$, $(m+1) \times n$. Partitioning

$$S = \begin{bmatrix} S_1 \\ S_2 \end{bmatrix} \quad \text{and} \quad R = [R_1 \quad R_2]$$

in the obvious way, we have

$$\begin{aligned} w_1(t) &= S_1 e^{At} R_1, & H(t) &= S_1 e^{At} R_2 \\ G(t) &= S_2 e^{At} R_1, & F(t) &= S_2 e^{At} R_2. \end{aligned}$$

Substituting these into (4.2.4) and defining $B = R_1$, $C = S_1$, $N = R_2 S_2$, the result follows. □

A sequence of kernels satisfying (4.2.4) is said to be *factorizable* and $\{F(t), G(t), H(t)\}$ is called a *factorization* of the sequence. Now introduce the matrices

$$P_k[F, G](t_1, \ldots, t_k) = [G(t_1) \quad F(t_2)G(t_1) \quad \ldots \quad F(t_k)F(t_{k-1})\cdots G(t_1)]$$

$$Q_k[F, H](t_1, \ldots, t_k) = \begin{bmatrix} H(t_1) \\ H(t_1)F(t_2) \\ \vdots \\ H(t_1)\cdots F(t_{k-1})F(t_k) \end{bmatrix}$$

and define

$$\mathscr{P}_k[F, G] = \int_{\Delta_k} P_k P_k^T \, dt_1 \ldots dt_k,$$

$$\mathscr{Q}_k[F, H] = \int_{\Delta_k} Q_k Q_k^T \, dt_1 \cdots dt_k,$$

where $\Delta_k = \{(t_1, \ldots, t_k) : 0 \leqslant t_i \leqslant 1, \text{ for all } i\}$. $\mathscr{P}_k$ and $\mathscr{Q}_k$ are generalizations of the classical Gramian matrices for linear systems. It is easy to check that $\mathscr{R}[\mathscr{P}_k] \subseteq \mathscr{R}[\mathscr{P}_{k+1}]$ for all k, and

$$\mathscr{R}[\mathscr{P}_{k-1}] = \mathscr{R}[\mathscr{P}_k] \quad \text{for some } k$$

implies

$$\mathscr{R}[\mathscr{P}_k] = \mathscr{R}[\mathscr{P}_{k+1}],$$

where $\mathscr{R}[M]$ is the *range* of the matrix M. Hence there exists an integer k' such that

$$\mathscr{R}[\mathscr{P}_k] \subset \mathscr{R}[\mathscr{P}_{k+1}] \text{ for all } k < k'$$

and

$$\mathscr{R}[\mathscr{P}_k] = \mathscr{R}[\mathscr{P}_{k+1}].$$

Of course, $k' \leqslant m$ since $\mathscr{R}[\mathscr{P}_k] \subseteq \mathscr{R}^m$ for all k.

Similarly, for $\mathcal{Q}_k$ there exists an integer $k'' \leqslant m$ such that

$$\mathcal{R}[\mathcal{Q}_k] \subset \mathcal{R}[\mathcal{Q}_{k+1}] \qquad \text{for all } k < k''$$

and

$$\mathcal{R}[\mathcal{Q}_k] = \mathcal{R}[\mathcal{Q}_{k+1}] \qquad \text{for all } k \geqslant k''.$$

We are interested in the rank of the matrices P_k and Q_k. In particular, we have

Lemma 4.2.2

If $P_m[F,G](t_1,\ldots,t_m)$ has row rank $m' < m$ over Δ_m, then there exists a (constant) nonsingular matrix T such that

$$TF(t)T^{-1} = \begin{bmatrix} F_{11}(t) & F_{12}(t) \\ 0 & F_{22}(t) \end{bmatrix}, \qquad TG(t) = \begin{bmatrix} G_1(t) \\ 0 \end{bmatrix} \tag{4.2.5}$$

where the matrices $F_{11}(t)$ and $G_1(t)$ are of orders $m' \times m'$ and $m' \times 1$, respectively and $P_{m'}[F_{11}, G_1](t_1,\ldots,t_{m'})$ is of full row rank over $\Delta_{m'}$.

Similarly, if $Q_m[F,H](t_1,\ldots,t_m)$ has column rank $m'' < m$ over Δ_m, then there exists T such that

$$TF(t)T^{-1} = \begin{bmatrix} F_{11}(t) & 0 \\ F_{21}(t) & F_{22}(t) \end{bmatrix}, \; H(t)T^{-1} = [H_1(t) \quad 0], \tag{4.2.6}$$

and $Q_{m''}[F_{11}, H_1](t_1,\ldots,t_{m''})$ has full column rank over $\Delta_{m''}$.

Proof

We shall prove only the first part. The second part follows similarly. Thus if m' rows of $P_m[F,G]$ are linearly independent over Δ_m, then we can find a nonsingular matrix T such that

$$TP_m[F,G](t_1,\ldots,t_m) = \begin{bmatrix} P'(t_1,\ldots,t_m) \\ 0 \end{bmatrix}$$

where P' has full row rank (m') over Δ_m. By definition of P_m this proves the second equation in (4.2.5).

Next write $TF(T)T^{-1}$ in the partitioned form

$$\begin{bmatrix} F_{11}(t) & F_{12}(t) \\ F_{21}(t) & F_{22}(t) \end{bmatrix}.$$

Then we have

$$TP_m = [TG(t_1)\; TF(t_2)T^{-1}TG(t_1) \cdots TF(t_k)T^{-1}TF(t_{k-1}) \cdots TG(t_1)] = \begin{bmatrix} P' \\ 0 \end{bmatrix}.$$

However,

$$TG(t) = \begin{bmatrix} G_1(t) \\ 0 \end{bmatrix} \quad \text{and} \quad TF(t_2)T^{-1}TG(t_1) = \begin{bmatrix} F_{11}(t_2)G_1(t_1) \\ F_{21}(t_2)G_1(t_1) \end{bmatrix}$$

and so $F_{21}(t_2)G_1(t_1) = 0$. Proceeding in this way we obtain

$$P'(t_1, \ldots, t_m) = P_m[F_{11}, G_1](t_1, \ldots, t_m) \tag{4.2.7}$$

and

$$F_{21}(t)P_{m-1}[F_{11}, G_1](t_1, \ldots, t_{m-1}) = 0. \tag{4.2.8}$$

By (4.2.7), $\mathcal{P}_m[F_{11}, G_1]$ has rank m' and so, by the above remarks on the range of $\mathcal{P}_k$, $\mathcal{P}_{m'}[F_{11}, G_1]$ also has rank m'. Since this implies that the m' rows of $P_{m'}[F_{11}, G_1](t_1, \ldots, t_{m'})$ are linearly independent and since $m' < m$, the rows of $P_{m-1}[F_{11}, G_1]$ must be linearly independent and (4.2.8) implies that $F_{21}(t) = 0$. □

The integer m in the factorization (4.2.4) is called its *dimension* and the factorization is *minimal* if the dimension of any other factorization is no smaller than m.

Lemma 4.2.2 leads to the following characterization of minimality:

Theorem 4.2.3
Let $\{F(t), G(t), H(t))$ be an m-dimensional factorization of a factorizable sequence of kernels. Then it is minimal if and only if the rows of $P_m[F, G]$ and the columns of $Q_m[F, H]$ are linearly independent over Δ_m. Moreover, any two minimal factorizations $\{F_1(t), G_1(t), H_1(t)\}$ and $\{F_2(t), G_2(t), H_2(t)\}$ are related by

$$F_1(t) = TF_2(t)T^{-1}$$
$$G_1(t) = TG_2(t)$$
$$H_1(t) = H_2(t)T^{-1}$$

for some nonsingular (constant) matrix T.

Proof
Let $\{F(t), G(t), H(t)\}$ be a minimal factorization and suppose that $P_m[F, G]$ has $m' < m$ linearly independent rows. By Lemma 4.2.2 we can write

$$TF(t)F^{-1} = \begin{bmatrix} F_{11}(t) & F_{12}(t) \\ 0 & F_{22}(t) \end{bmatrix}, \qquad TG(t) = \begin{bmatrix} G_1(t) \\ 0 \end{bmatrix}$$

for some nonsingular T. Writing $H(t)T^{-1} = (H_1(t)\ H_2(t))$ it is clear that $\{F_{11}(t), G_1(t), H_1(t)\}$ is also a factorization of the same sequence and has dimension m'. This is a contradiction. A similar argument holds for $Q_m[F, H]$.

Conversely, if the matrices $P_m[F, G]$ and $Q_m[F, H]$ satisfy the specified rank conditions for a nonminimal factorization $\{F(t), G(t), H(t)\}$, then let $\{\bar{F}(t), \bar{G}(t), \bar{H}(t)\}$ be a factorization of the same sequence with dimension $m' < m$. Since the sequence is given by (4.2.4) it follows that

$$Q_m[F,G](t_1,\ldots,t_m)P_m[F,G](t_{m+1},\ldots,t_{2m})$$
$$= Q_m[\bar{F},\bar{H}](t_1,\ldots,t_m)P_m[\bar{F},\bar{G}](t_{m+1},\ldots,t_{2m})$$

and so

$$\mathcal{Q}_m\mathcal{P}_m = \int_{\Delta_m} Q_m^{\mathrm{T}}[F,H]Q_m[\bar{F},\bar{H}]\ \mathrm{d}t_1\cdots\mathrm{d}t_m$$

$$\cdot\int_{\Delta_m} P_m[\bar{F},\bar{G}]\,P^{\mathrm{T}}[F,G]\ \mathrm{d}t_{m+1}\cdots\mathrm{d}t_{2m}.$$

Considering the ranks of the matrices on each side of this equality it follows that $m \leqslant m'$, which is again a contradiction.

For the last part note that if T is nonsingular, then $\{F(t), G(t), H(t)\}$ is minimal if and only if $\{TF(t)T^{-1}, TG(t), H(t)T^{-1}\}$ is minimal. Conversely, if $\{F_1(t), G_1(t), H_1(t)\}$ and $\{F_2(t), G_2(t), H_2(t)\}$ are minimal, then we have

$$Q_m[F_1,H_1](t_1,\ldots,t_m)\,.\,P_m[F_1,G_1](t_{m+1},\ldots,t_{2m})$$
$$= Q_m[F_2,H_2](t_1,\ldots,t_m)\,.\,P_m[F_2,G_2](t_{m+1},\ldots,t_{2m}).$$

Since the factorizations are minimal, the first part of the theorem shows that the columns of $Q_m[F_1,H_1]$ and $Q_m[F_2,H_2]$ and the rows of $P_m[F_1,G_1]$ and $P_m[F_2,G_2]$ are linearly independent over Δ_m. It is not difficult to show that we must therefore have

$$\begin{aligned} Q_m[F_1,G_1] &= Q_m[F_2,H_2]T^{-1} \\ P_m[F_1,G_1] &= TP_m[F_2,G_2] \end{aligned} \tag{4.2.9}$$

for some constant nonsingular matrix T. (See also Youla (1966).) By definition of P and Q we have

$$H_1(t) = H_2(t)T^{-1}, \qquad G_1(t) = TG_2(t).$$

Finally, again because the kernels of $\{F_1, G_1, H_1\}$ and $\{F_2, G_2, H_2\}$ are the same we have

$$Q_m[F_1,H_1](t_1,\ldots,t_m)\,.\,F_1(t_{m+1})\,.\,P_m[F_1,G_1](t_{m+2},\ldots,t_{2m+1})$$
$$= Q_m[F_2,H_2](t_1,\ldots,t_m)\,.\,F_2(t_{m+1})\,.\,P_m[F_2,G_2](t_{m+2},\ldots,t_{2m+1})$$

and by (4.2.9) and the linear independence of the rows and columns of P_m, Q_m respectively we obtain

$$F_1(t) = TF_2(t)T^{-1}. \qquad \square$$

Note that this theorem shows that the dimension m_0 of a minimal factorization (of a factorizable sequence of kernels) can be computed from the relation

$$m_0 = \text{rank}\{\mathscr{Q}_m[F,G]\mathscr{P}_m[F,G]\},$$

for any factorization $\{F, G, H\}$ of the sequence. Moreover, using the results of Lemma 4.2.2, successively, it follows that any factorization of a sequence of kernels can be written in the form

$$F(t) = T\begin{bmatrix} F_0(t) & 0 & F_{13}(t) & 0 \\ F_{21}(t) & F_{22}(t) & F_{23}(t) & F_{24}(t) \\ 0 & 0 & F_{33}(t) & 0 \\ 0 & 0 & F_{43}(t) & F_{44}(t) \end{bmatrix} T^{-1} \tag{4.2.10}$$

$$G(t) = T\begin{bmatrix} G_0(t) \\ G_2(t) \\ 0 \\ 0 \end{bmatrix}, \qquad H(t) = (H_0(t) \;\; 0 \;\; H_3(t) \;\; 0)T^{-1}$$

for some nonsingular constant matrix T, where $\{F_0(t), G_0(t), H_0(t)\}$ is a minimal factorization of the sequence.

Returning now to the theory of minimal bilinear realizations, we shall call a bilinear realization $\{A, N, B, C\}$ *minimal* if the dimension of the state space is no greater than that of any other realization. In order to obtain a realization theory for bilinear systems we shall use the theory of linear realizations. In particular, if $L(t)$ is an input–output relation (or *weighting pattern*) for a linear system we shall denote by $\delta\{L(t)\}$ its *order*, i.e. the dimension of the state space of a minimal realization (as a linear system). Then we have

Theorem 4.2.4

If $\{w_i(t_1, \ldots, t_i)\}_{1 \leqslant i < \infty}$ is a sequence of bilinearly realizable kernels and $\{F_0(t), G_0(t), H_0(t)\}$ is a minimal factorization of the sequence $\{w_i(t_1, \ldots, t_i)\}_{2 \leqslant i < \infty}$, then a minimal bilinear realization $\{A, N, B, C\}$ has the properties:

(a) the dimension of the state space is $n_0 = \delta\{L_0(t)\}$ where

$$L_0(t) = \begin{bmatrix} w_1(t) & H_0(t) \\ G_0(t) & F_0(t) \end{bmatrix}. \tag{4.2.11}$$

(b) the rank r_0 of N is given by

$$r_0 = \dim\{F_0(t), G_0(t), H_0(t)\} \qquad (= m_0)$$

and is minimal.

(c) any two minimal bilinear realizations $\{A_i, N_i, B_i, C_i\}$, $i = 1, 2$ are related by

$$A_1 = TA_2T^{-1}, \qquad N_1 = TN_2T^{-1}, \qquad B_1 = TB_2, \qquad C_1 = C_2T^{-1},$$

for some nonsingular matrix T.

Proof

By Theorem 4.2, bilinear realizations are in a one-to-one correspondence with linear systems which have a weighting pattern of the form (4.2.11) and so (a) follows easily by substituting the expressions (4.2.10) into an arbitrary weighting pattern

$$L(t) = \begin{bmatrix} w_1(t) & H(t) \\ G(t) & F(t) \end{bmatrix}.$$

To prove (b) note that, for any minimal bilinear realization $\{A, N, B, C\}$, where N has dimension $n \times n$ and rank r, if we write N in the form

$$N = N'\, N''$$

where N', N'' have orders $n \times r$, $r \times n$, respectively, and full rank, then

$$F(t) = N''e^{At}N', \qquad G(t) = N''e^{At}B, \qquad H(t) = Ce^{At}N' \quad (4.2.12)$$

is an r-dimensional factorization of the sequence of kernels. Hence $r \geqslant m_0$. Suppose that $r > m_0$. Then the factorization (4.2.12) is nonminimal and hence reducible. If the matrix $P_r[N''e^{At}N', N''e^{At}B]$ has only $r_1(< r)$ rows linearly independent (a similar argument applying if Q_r has lower column rank) then, as in Lemma 4.2.2, we can find an $r \times r$ matrix T such that

$$TN'' = \begin{bmatrix} \bar{N}_1'' \\ \bar{N}_2'' \end{bmatrix}, \qquad N'T^{-1} = (\bar{N}_1' \quad \bar{N}_2')$$

and

$$F_{11}(t) = \bar{N}_1''\, e^{At}\bar{N}_1', \qquad G_1(t) = \bar{N}_1''\, e^{At}B, \qquad H(t) = Ce^{At}\bar{N}_1'$$

is an r_1-dimensional factorization. Then $\{A, \bar{N}_1'\bar{N}_1'', B, C\}$ is clearly still a minimal bilinear realization and so

$$\left\{A, (B \quad \bar{N}_1'), \begin{bmatrix} C \\ \bar{N}_1'' \end{bmatrix}\right\}$$

is a minimal linear realization of

$$L_1(t) = \begin{bmatrix} w_1(t) & H_1(t) \\ G_1(t) & F_{11}(t) \end{bmatrix}.$$

Hence the rows of the matrix $e^{At}(B\;\bar{N}_1')$ are linearly independent. However, we also have

$$\bar{N}_2''\, e^{At}B = 0, \qquad \bar{N}_2''\, e^{At}\bar{N}_1' = 0$$

by Lemma 4.2.2 and so $\bar{N}_2'' = 0$. It follows that N'' does not have full rank and this is a contradiction.

To prove (b) let $\{A_i, N_i, B_i, C_i\}$, $i = 1, 2$, be two minimal bilinear realizations. Since rank $N_1 =$ rank $N_2 = m_0$, these realizations give rise to two minimal factorizations $\{F_i(t), G_i(t), H_i(t)\}$, $i = 1, 2$, which must be equivalent, as seen above. Hence, there exists a nonsingular matrix M (of order $m_0 \times m_0$) such that

$$\begin{bmatrix} w_1(t) & H_1(t) \\ G_1(t) & F_1(t) \end{bmatrix} = \begin{bmatrix} 1 & 0 \\ 0 & M \end{bmatrix} \begin{bmatrix} w_1(t) & H_2(t) \\ G_2(t) & F_2(t) \end{bmatrix} \begin{bmatrix} 1 & 0 \\ 0 & M^{-1} \end{bmatrix}.$$

Thus, by (4.2.12), we have

$$\begin{bmatrix} C_1 \\ N_1'' \end{bmatrix} e^{A_1 t} [B_1 \quad N_1'] = \begin{bmatrix} 1 & 0 \\ 0 & M \end{bmatrix} \begin{bmatrix} C_2 \\ N_2'' \end{bmatrix} e^{A_2 t}(B_2 \quad N_2') \begin{bmatrix} 1 & 0 \\ 0 & M^{-1} \end{bmatrix}.$$

Since these are two minimal linear realizations we must have

$$\begin{bmatrix} C_1 \\ N_1'' \end{bmatrix} = \begin{bmatrix} 1 & 0 \\ 0 & M \end{bmatrix} \begin{bmatrix} C_2 \\ N_2'' \end{bmatrix} T^{-1}, \quad (B_1 \quad N_1') = T(B_2 \quad N_2') \begin{bmatrix} 1 & 0 \\ 0 & M^{-1} \end{bmatrix},$$

$$A_1 = TA_2T^{-1}$$

and the result follows. □

The theory of minimal realizations naturally leads to the consideration of canonical forms for bilinear systems. For this we need to consider the reachable set from the origin (which is not generally a linear subspace). In fact, we have

Theorem 4.2.5

If we define, inductively, the sequence

$$V_i = B$$

$$V_i = (AV_{i-1} \quad NV_{i-1}), \qquad i = 2, 3, \ldots, \tag{4.2.13}$$

then the reachable set from the origin of the bilinear system (4.2.1) spans a linear subspace $\mathscr{X}$ of $\mathbb{R}^n$ given by

$$\mathscr{X} = \mathscr{R}[V_1 \quad V_2 \quad \ldots \quad V_n]. \tag{4.2.14}$$

Moreover, $\mathscr{X}$ is the intersection of all subspaces of $\mathbb{R}^n$ which are invariant under A and N and contain $\mathscr{R}[B]$.

Proof

First note that if $\mathscr{L} \subseteq \mathbb{R}^n$ is a linear subspace and $x(t)$ is a differentiable curve in $\mathscr{L}$, then $\dot{x}(t) \in \mathscr{L}$ for all t. If $B = 0$ then $\mathscr{X} = \{0\}$ and the result is true. Thus, assume $B \neq 0$ and then the reachable set (from the origin) is clearly $\neq \{0\}$, and so its linear span $\mathscr{X}$ has a nontrivial basis, say

$\{x_1, \ldots, x_r\}$. Thus, by the above remark

$$(A + Nux_j + Bu) \in \mathscr{X} \qquad \text{for all } u \in \mathbb{R} \quad \text{and} \quad 1 \leqslant j \leqslant r.$$

Similarly we must have

$$Bu \in \mathscr{X} \qquad \text{for all } u \in \mathbb{R}$$

and so

$$(A + Nu)\mathscr{X} \subseteq \mathscr{X} \qquad \text{for all } u \in \mathbb{R}.$$

Hence $\mathscr{X}$ contains the range of B and is invariant under A and N. If $\mathscr{L}$ is any subspace of $\mathbb{R}^n$ which is invariant under A and N and contains $\mathscr{R}[B]$, then any trajectory starting from zero can be shown to belong to $\mathscr{L}$ and so $\mathscr{X} \subseteq \mathscr{L}$. Hence $\mathscr{X}$ is the intersection of all such subspaces.

To prove (4.2.14) note that

$$\mathscr{R}[V_1 \quad \ldots \quad V_{i-1}] \subseteq \mathscr{R}[V_1 \quad \ldots \quad V_i] \subseteq \mathbb{R}^n$$

and

$$\mathscr{R}[V_1 \quad \ldots \quad V_{k-1}] = \mathscr{R}[V_1 \quad \ldots \quad V_k]$$

for some k implies that

$$\mathscr{R}[V_1 \quad \ldots \quad V_k] = \mathscr{R}[V_1 \quad \ldots \quad V_{k+1}].$$

Hence, we must have

$$\mathscr{R}[V_1 \quad \ldots \quad V_n] = \mathscr{R}[V_1 \quad \ldots \quad V_{n+1}]$$

and so $\mathscr{R}[V_1 \quad \ldots \quad V_n]$ is invariant under A and N and contains $\mathscr{R}[B]$. Thus $\mathscr{R}[V_1 \quad \ldots \quad V_n] \supseteq \mathscr{X}$. To prove the converse, we have

$$\mathscr{X} \supseteq \mathscr{R}[B] = \mathscr{R}[V_1],$$
$$\mathscr{X} \supseteq A\mathscr{X} \supseteq A\mathscr{R}[V_1], \qquad \mathscr{X} \supseteq N\mathscr{X} \supseteq N\mathscr{R}[V_1] \Rightarrow \mathscr{X} \supseteq \mathscr{R}[V_1 \quad V_2]$$

and so by induction we have

$$\mathscr{X} \supseteq \mathscr{R}[V_1 \quad \ldots \quad V_n]. \qquad \square$$

For the dual concept of observability we shall say that a state x is *unobservable* if the component of the response which depends on the initial state is identically zero. By examining the output responses y_x and y_{xu} it is clear that the set of unobservable states is a linear subspace of $\mathbb{R}^n$. Dually to (4.2.13) we define

$$W_1 = C,$$
$$W_i = \begin{bmatrix} W_{i-1}A \\ W_{i-1}N \end{bmatrix}, \qquad i = 2, 3, \ldots. \tag{4.2.15}$$

Then the dual of Theorem 4.2.5 can be proved in a similar way:

Theorem 4.2.6

Let $\mathscr{Y}$ be the set of unobservable states of (4.2.1). Then

$$\mathscr{Y} = \mathscr{N}\begin{bmatrix} W_1 \\ W_2 \\ \vdots \\ W_n \end{bmatrix} \tag{4.2.16}$$

where $\mathscr{N}(W)$ is the null space of W, and $\mathscr{Y}$ is the largest subspace invariant under A and N and contained in $\mathscr{N}[C]$. □

Using a similar approach to that developed for linear systems, Theorems 4.2.5 and 4.2.6 lead directly to

Corollary 4.2.7

With the above notation let the state space $\mathbb{R}^n$ of the system (4.2.1) be decomposed in the form

$$\mathbb{R}^n = \mathscr{A} \oplus \mathscr{B} \oplus \mathscr{C} \oplus \mathscr{D}$$

where

$$\mathscr{A} = \mathscr{X} \cap \mathscr{Y}, \qquad \mathscr{B} = \mathscr{X} \ominus \mathscr{A}, \qquad \mathscr{C} = \mathscr{Y} \ominus \mathscr{A}.$$

Then, if $x = (x_a^{\mathrm{T}}\ x_b^{\mathrm{T}}\ x_c^{\mathrm{T}}\ x_d^{\mathrm{T}})^{\mathrm{T}}$ is the corresponding partition of the state vector, then we can write the system (4.2.1) in the form

$$\begin{bmatrix} \dot{x}_a(t) \\ \dot{x}_b(t) \\ \dot{x}_c(t) \\ \dot{x}_d(t) \end{bmatrix} = \begin{bmatrix} A_{aa} & A_{ab} & A_{ac} & A_{ad} \\ 0 & A_{bb} & 0 & A_{bd} \\ 0 & 0 & A_{cc} & A_{cd} \\ 0 & 0 & 0 & A_{dd} \end{bmatrix}\begin{bmatrix} x_a(t) \\ x_b(t) \\ x_c(t) \\ x_d(t) \end{bmatrix}$$

$$+ \begin{bmatrix} N_{aa} & N_{ab} & N_{ac} & N_{ad} \\ 0 & N_{bb} & 0 & N_{bd} \\ 0 & 0 & N_{cc} & N_{cd} \\ 0 & 0 & 0 & N_{dd} \end{bmatrix}\begin{bmatrix} x_a(t) \\ x_b(t) \\ x_c(t) \\ x_d(t) \end{bmatrix} u(t) + \begin{bmatrix} B_a \\ B_b \\ 0 \\ 0 \end{bmatrix} u(t) \tag{4.2.17}$$

$$y(t) = (0 \quad C_b \quad 0 \quad C_d)\begin{bmatrix} x_a(t) \\ x_b(t) \\ x_c(t) \\ x_d(t) \end{bmatrix}.$$

Moreover, such canonical forms are equivalent under a transformation of the form

$$T = \begin{bmatrix} T_{aa} & T_{ab} & T_{ac} & T_{ad} \\ 0 & T_{bb} & 0 & T_{bd} \\ 0 & 0 & T_{cc} & T_{cd} \\ 0 & 0 & 0 & T_{dd} \end{bmatrix}. \qquad \square$$

(For further details on the theory of linear canonical forms, see D'Alessandro *et al.* (1973).)

It is easy to demonstrate that the kernels $w_i(t_1, ..., t_i)$ of the zero-state response depend only on the matrices A_{bb}, N_{bb}, B_b, C_b, and so we expect that reachable and observable realizations of Volterra series can be represented using only these terms. To prove that this is the case we first derive alternative forms for the spaces $\mathscr{X}$, $\mathscr{Y}$ defined by (4.2.14) and (4.2.16), respectively.

Lemma 4.2.8

$\mathscr{X}$ and $\mathscr{Y}$ can be expressed in the forms

$$\mathscr{X} = \mathscr{R}\{\mathscr{P}_n[e^{At}N, e^{At}B]\},$$
$$\mathscr{Y} = \mathscr{N}\{\mathscr{Q}_n[Ne^{At}, Ce^{At}]\}.$$

Proof

By definition,

$$\mathscr{P}_k[e^{At}N, e^{At}B] = \mathscr{P}_1[e^{At}N, e^{At}B] + \int_0^1 e^{At}N\mathscr{P}_{k-1}[e^{At}N, e^{At}B]N^* e^{A^*t}\, dt \tag{4.2.18}$$

and

$$\mathscr{R}\{\mathscr{P}_1[e^{At}N, e^{At}B]\} = \mathscr{R}[B \quad AB \quad \dots \quad A^{n-1}B].$$

Hence, iterating (4.2.18) and using the Cayley–Hamilton theorem, we have

$$\begin{aligned} \mathscr{R}\{\mathscr{P}_2[e^{At}N, e^{At}B]\} &= \mathscr{R}[B \quad AB \quad \dots \quad A^{n-1}B] \\ &\quad + \mathscr{R}[N(B \quad AB \quad \dots \quad A^{n-1}B) \quad \dots \quad A^{n-1}N(B \quad AB \quad \dots \quad A^{n-1}B)] \\ &\quad \vdots \\ \mathscr{R}\{\mathscr{P}_n[e^{At}N, e^{At}B]\} &= \mathscr{R}[V_1 \quad V_2 \quad \dots \quad V_n] = \mathscr{X}. \end{aligned} \qquad \square$$

This lemma leads to the main theorem on minimal realizations.

Theorem 4.2.9

A realization $\{A, N, B, C\}$ of a bilinearly realizable sequence of kernels is minimal if and only if the state space of the realization is observable and is spanned by the states which are reachable from the origin.

Proof
If a realization is not observable or is not spanned by the reachable states from the origin, then the canonical form (4.2.17) implies that the realization is not minimal.

Conversely if we can show that any two realizations which are observable and spanned by the reachable states from the origin have the same dimension, then any such realization (of dimension n) must be minimal. For, if there exists a minimal realization of dimension $n' < n$, then by the first part this realization is observable and spanned by the reachable states from the origin and so $n' = n$ by the previous statement. Thus, let $\{A_1, N_1, B_1, C_1\}$ and $\{A_2, N_2, B_2, C_2\}$ be realizations with state dimensions n_1 and n_2 which are observable and spanned by the reachable states from the origin. If $\bar{n} = \max\{n_1, n_2\}$ then, as before, we can show that

$$Q_{\bar{n}}[N_1 e^{A_1 t}, C_1 e^{A_1 t}](t_1, \dots, t_{\bar{n}}) P_{\bar{n}}[e^{A_1 t} N_1, e^{A_1 t} B_1](t_{\bar{n}+1}, \dots, t_{2\bar{n}})$$
$$= Q_{\bar{n}}[N_2 e^{A_2 t}, C_2 e^{A_2 t}](t_1, \dots, t_{\bar{n}}) P_{\bar{n}}[e^{A_2 t} N_2, e^{A_2 t} B_2](t_{\bar{n}+1}, \dots, t_{2\bar{n}}).$$

By Lemma 4.2.8 and the hypothesis the n_1 columns of $Q_{\bar{n}}$ and the n_1 rows of $P_{\bar{n}}$ on the left of this expression are linearly independent, as are the n_2 columns of $Q_{\bar{n}}$ and the n_2 rows of $P_{\bar{n}}$ on the right. Hence $n_1 = n_2$. □

As above we see that the dimension of a minimal realization is given by

$$n_0 = \operatorname{rank}\{\mathcal{Q}_n[Ne^{At}, Ce^{At}]\mathcal{P}_n[e^{At}N, e^{At}B]\}.$$

for any realization $\{A, N, B, C\}$.

Further results on triangular canonical forms for bilinear systems are discussed by Di Benedetto and Isidori (1978) and for a somewhat different approach to discrete bilinear systems see Pearlman (1978) and Pearlman and Denham (1979). Finally, Frazho (1980) has derived a shift operator description of bilinear systems.

4.2.2 Realization of finite Volterra series

In this section we shall consider the realization of Volterra series which contain only a finite number of nonzero kernels, as discussed by Crouch (1981). In particular we shall be concerned with linear analytic systems of the form

$$\begin{aligned} \dot{x} &= f(x) + \sum_{i=1}^{m} u_i g_i(x), \qquad x(0) = x_0, \quad x \in M \\ y &= h(x) \in \mathbb{R}^q \end{aligned} \tag{4.2.19}$$

defined on an n-dimensional analytic manifold M, where the vector fields $f + \sum_{i=1}^{m} \alpha_i g_i, (\alpha_1, \dots, \alpha_m) \in \mathbb{R}^m$ are assumed to be complete. A simple generalization of the Volterra series for scalar input–output systems developed in Section 4.1.1 shows that such a system has an input–output relation of the form

$$y(t) = w_0(t) + \int_0^t w_1(t, \sigma_1, x_0)(u(\sigma_1))\, d\sigma_1$$

$$+ \int_0^t \int_0^{\sigma_1} w_2(t, \sigma_1, \sigma_2, x_0)(u(\sigma_1))(u(\sigma_2))\, d\sigma_1\, d\sigma_2 + \cdots$$

where the kernels $w_k(t, \sigma_1, \ldots, \sigma_k, x)$ are multilinear maps for each $(t, \sigma_1, \ldots, \sigma_k, x) \in \mathbb{R}^{k+1} \times M$ from $\mathbb{R}^m \times \cdots \times \mathbb{R}^m \to \mathbb{R}^q$. The components of w will be denoted by $w_n^{j_0 j_1 \ldots j_k}(t, \sigma_1, \ldots, \sigma_k, x)$, $1 \leqslant j_0 \leqslant q, 1 \leqslant j_l \leqslant m$, $1 \leqslant l \leqslant k$. Recall from Section 4.1.2 that the kernels are given inductively by

$$w_0(t, x) = h(\gamma(t)x)$$

and

$$\bar{w}_i(t, \sigma_1, \ldots, \sigma_i, x) = \frac{\partial \bar{w}_{i-1}(t, \sigma_1, \ldots, \sigma_{i-1}, \gamma(\sigma_{i-1}, \sigma_i, x))}{\partial x}\, g(x)$$

$$w_i(t, \sigma_1, \ldots, \sigma_i, x) = \bar{w}_i(t, \sigma_1, \ldots, \sigma_i, \gamma(\sigma_i, 0, x)).$$

where we write $\gamma(t, x) = \gamma(t)x$.
Hence,

$$w_i(t, \sigma_1, \ldots, \sigma_i, x) = \frac{\partial}{\partial x}\, w_{i-1}\, (t, \sigma_1, \ldots, \sigma_{i-1}, \gamma(-\sigma_i)x) g(\gamma(\sigma_i)x)$$

by the group property of $\gamma(t)$, and so

$$w_i^{j_0 j_1 \ldots j_i}(t, \sigma_1, \ldots, \sigma_i, x) = g_{j_i}(\gamma(\sigma_i)x)\, \frac{\partial}{\partial x}\, w_{i-1}^{j_0 j_1 \ldots j_{i-1}}(t, \sigma_1, \ldots, \sigma_{i-1}, \gamma(-\sigma_i)x)$$

$$= \gamma(-\sigma_i)_* g_{j_i}(\gamma(\sigma_i)x)(w_{i-1}^{j_0 j_1 \ldots j_{i-1}}(t, \sigma_1, \ldots, \sigma_{i-1}, \ldots)) \tag{4.2.20}$$

regarding g as a vector field and by definition of the differential. This gives a coordinate-free expression for the kernels. Iterating (4.2.20) gives

$$w_i^{j_0 j_1 \ldots j_i}(t, \sigma_1, \ldots, \sigma_i, x)$$
$$= g_{j_i}(\gamma(\sigma_i)x)(g_{j_{i-1}}(\gamma(\sigma_{i-1} - \sigma_i).)(\ldots h_{j_0}(\gamma(t - \sigma_1).)\ldots) \tag{4.2.21}$$

and so

$$w_i(t - s, \sigma_1 - s, \ldots, \sigma_i - s, \gamma(s)x) = w_i(t, \sigma_1, \ldots, \sigma_i, x).$$

We shall be interested in obtaining linear-analytic realizations of finite Volterra series. Recall from Chapter 2 that a realization is minimal if it is reachable and observable, and since the system is analytic, reachability is equivalent to accessibility. Hence we shall seek accessible and observable realizations of finite Volterra series. As we have seen, two such realizations

$$\dot{x}_i = f_i(x_i) + \sum_{j=1}^{m} u_j g_{ij}(x_i), \qquad x_i(0) = x_i^0, \quad x_i \in M_i,$$

$$y_i = h_i(x_i)$$

for $i = 1, 2$, are equivalent in the sense that there exists a (unique) analytic diffeomorphism $\Phi : M_1 \to M_2$ such that

$$\Phi_* f_1 = f_2 \circ \Phi, \quad \Phi_* g_{1j} = g_{2_j} \circ \Phi, \quad h_1 = h_2 \circ \Phi, \quad \Phi(x_1^0) = x_2^0.$$

Next we recall the notion of weak observability (see Chapter 2). Two states $x_0, x_1 \in M$ are indistinguishable if the input–output maps with these as initial states are identical, and we say that a system is weakly observable if for all $x_0 \in M$ there is a neighbourhood U of x_0 such that $x_1 \in U$ is indistinguishable from x_0 implies that $x_0 = x_1$. Then we have seen that an accessible system is weakly observable if and only if

$$T_x^* M = \mathrm{d}\mathscr{H}(x)$$

where $\mathscr{H}$ is the smallest linear subspace of $\mathscr{F}(M)$ containing the functions h_i, $1 \leqslant i \leqslant q$, and closed under Lie differentiation by elements of $\mathscr{L}$, where, for the system (4.2.19), $\mathscr{L}$ denotes the Lie subalgebra of $D(M)$ of vector fields generated by f and $g_1, \ldots, g_m$. Let $\mathscr{I}$ be the ideal in $\mathscr{L}$ generated by $g_1, \ldots, g_m$. Then, again from Chapter 2, we recall that the system is accessible if and only if

$$T_x M = \mathscr{L}(x)$$

and strongly accessible if and only if

$$T_x M = \mathscr{I}(x).$$

Using the simply connected cover of an accessible, weakly observable system (see p. 135), it is not difficult to prove that there is a one-to-one correspondence between the accessible, weakly observable analytic realizations of an input–output map and the covering manifolds of the state space of a minimal realization. (For details, see Crouch (1981).) This implies that strongly accessible, weakly observable realizations and minimal realizations of an input–output map have isomorphic Lie algebras.

Now let $\mathscr{R}$ denote the linear span of the vector fields $(\mathrm{ad}^i f) g_j$ for $i \geqslant 0$ and $1 \leqslant j \leqslant m$. Then $\mathscr{R} \subseteq \mathscr{I}$ and generates $\mathscr{I}$ as a subalgebra of $\mathscr{L}$. An important property of systems with finite Volterra series is that, for any strongly accessible realization with state space M, the Volterra series has the same length at each point of M. More precisely, we have

Lemma 4.2.10

Let $w_i(t, \sigma_1, \ldots, \sigma_i, x)$, $2 \leqslant i < \infty$ be the kernels of a Volterra series and suppose that M is the state space of a strongly accessible realization. Then, if the series has length p at x_0 in the sense that

$$w_{p+1}(t, \sigma_1, \ldots, \sigma_{p+1}, x_0) \equiv 0$$

(in $t, \sigma_1, \ldots, \sigma_{p+1}$) for some $x_0 \in M$, we have

(a) The series has length p for any $x \in M$.
(b) $w_p(t, \sigma_1, \ldots, \sigma_p, x) = w_p(t, \sigma_1, \ldots, \sigma_p)$ is independent of x.

Proof
Since $w_{p+1}(t, \sigma_1, \ldots, \sigma_{p+1}, x_0) \equiv 0$, it follows from the definition of the kernel w_p that

$$L_a w_p(t, \sigma_1, \ldots, \sigma_p, x_0) \equiv 0,$$

for any vector field $a \in \mathscr{R}$, where L_a is the Lie derivative with respect to a. (This is obtained by differentiating w_p repeatedly with respect to σ_{p+1} and putting $\sigma_{p+1} = 0$.) More generally, if $a_1, \ldots, a_i \in \mathscr{R}$ we have

$$L_{a1} L_{a2} \ldots L_{a_i} w_p(t, \sigma_1, \ldots, \sigma_p, x_0) \equiv 0.$$

However, $L_a L_b - L_b L_a = L_{[a,b]}$ and so this implies that

$$L_{b_1} L_{b_2} \ldots L_{b_i} w_p(t, \sigma_1, \ldots, \sigma_p, x_0) \equiv 0$$

for all $b_1, \ldots, b_i \in \mathscr{I}$, since $\mathscr{R}$ generates $\mathscr{I}$.

Since we are considering a strongly accessible realization with state space M, we can find $a_1, \ldots, a_n \in \mathscr{I}$ which span $T_{x_0}M$ at x_0. Let $\phi_i(t)$ be the flow of a_i. Then the function $(s_1, \ldots, s_n) \to \phi_1(s_1) \circ \cdots \circ \phi_n(s_n)x_0$ maps a neighbourhood of $0 \in \mathbb{R}^n$ onto a neighbourhood of x_0. Since we have just seen that all the derivatives of the function

$$(s_1, \ldots, s_n) \to w_p(t, \sigma_1, \ldots, \sigma_p, \gamma_1(s_1) \circ \cdots \circ \gamma_n(s_n)x_0)$$

vanish at $0 \in \mathbb{R}^n$, it follows that all the partial derivatives of the function $x \to w_p(t, \sigma_1, \ldots, \sigma_p, x)$ vanish at x. However, w_p depends analytically on x and so w_p is independent of x, proving (b).

In the same way we can show that $w_{p+i}, i \geqslant 1$, is independent of the state and since each one is identically zero at x_0 part (a) now follows. □

The finiteness of a Volterra series, as might be expected, is reflected in the structure of the Lie algebras $\mathscr{I}$ and $\mathscr{L}$. In fact we have

Theorem 4.2.11
Let $\mathscr{I}$ and $\mathscr{L}$ be associated with a strongly accessible, weakly observable realization of a finite Volterra series of length p as above. Then $\mathscr{I}$ is nilpotent and $\mathscr{L}$ is solvable. Moreover, the descending central series of $\mathscr{I}$ has length $\leqslant p$.

Proof
Since w_{p+1} is identically zero on M, repeated differentiation of (4.2.21) shows that

$$L_{a_1} L_{a_2} \cdots L_{a_{p+1}}(h_k \circ \gamma(t)x) \equiv 0, \qquad 1 \leqslant k \leqslant q$$

for some vector fields $a_1, \ldots, a_{p+1} \in \mathscr{I}$. Differentiating this with respect to t sufficiently many times gives

$$L_{a_1} L_{a_2} \cdots L_{a_{p+1}} L_{b_1} \cdots L_{bj}(h_k) = 0, \qquad j \geqslant 0$$

for any vector fields $b_1, \ldots, b_j \in \mathscr{L}$; i.e.

$$L_{a_1} L_{a_2} \cdots L_{a_{p+1}}(\eta) = 0$$

for all $\eta \in \mathscr{H}$. Hence

$$L_a(\eta) = 0$$

for all $\eta \in \mathscr{H}$ and $a \in \mathscr{I}^{p+1} (= [\mathscr{I}, \mathscr{I}^p])$ and so by weak observability, $\mathscr{I}^{p+1} = \{0\}$.

For solvability, define $\mathscr{L}^{(1)} = \mathscr{L}$, $\mathscr{L}^{(i)} = [\mathscr{L}^{(i-1)}, \mathscr{L}^{(i-1)}]$ (the derived series – see Chapter 1). Then $\mathscr{L}^{(i+1)} \subseteq \mathscr{I}^{(i)}$ and so $\mathscr{L}^{(p+2)} \subseteq \mathscr{I}^{(p+1)} \subseteq \mathscr{I}^{2p} = \{0\}$. □

It is possible for the length of the descending central series to be strictly less than p (the length of the Volterra series) as the system

$$\begin{aligned} \dot{x} &= u, \qquad x(0) = 0, \quad x \in \mathbb{R} \\ y &= x^p, \qquad p \geqslant 1 \end{aligned}$$

shows. (Since $g \equiv 1$, $[\mathscr{I}, \mathscr{I}] = 0$, so the descending central series has length 1. However, y is p-multilinear in u and so the Volterra series has length p.)

Consider again a strongly accessible, weakly observable realization of a finite Volterra series. By strong accessibility we can find vector fields $a_i \in \mathscr{I}$, $1 \leqslant i \leqslant n$, so that $\mathbb{R}a_1(x_0) \oplus \cdots \oplus \mathbb{R}a_n(x_0) = T_{x_0}M$ and so the function

$$(s_1, \ldots, s_n) \to \phi_{a_1}(s_1) \circ \cdots \circ \phi_{a_n}(s_n)x_0$$

is a diffeomorphism between open sets $U \ni x_0$ and $V \ni 0$ (in $\mathbb{R}^n$) which we denote $s \to \Phi(s)x_0$, $s = (s_1, \ldots, s_n)$. Now recall again the Campbell–Hausdorff formula

$$\phi_a(-s)_* b(\phi_a(s)x) = \sum_{i=0}^{\infty} \frac{s!}{i!}((\mathrm{ad}^i a)b)(x) \tag{4.2.22}$$

for any $a, b \in D(M)$. Hence, if the length of the descending central series of $\mathscr{I}$ is p we have, by repeated application of (4.2.22)

$$(\Phi^{-1}(s))_* b(\Phi(s)x_0) = \phi_{a_n}(-s_n)_* \circ \cdots \circ \phi_{a_1}(-s_1)_* b(\phi_{a_1}(s_1) \circ \cdots \circ \phi_{a_n}(s_n)x_0)$$
$$= \sum_{i=1}^{n} p_i(s) b_i(x_0),$$

for any $b \in \mathscr{I}$, where $p_i(s) = p_i(s_1, \ldots, s_n)$ are polynomials of total order $\leqslant p$, and so, on U,

$$b(\Phi(s)x_0) = \sum_{i=1}^{n} p_i(s)\Phi(s)_* b_i(x_0), \qquad b \in \mathscr{I}.$$

Since $s \to \Phi(s)$ is a one-to-one map onto U we see that, on U, any $b \in \mathscr{I}$ can be written as a finite linear combination of vectors at x_0, where the coefficients are polynomials. The linear space of all such combinations is clearly finite-dimensional and so $\mathscr{I}$ is a finite-dimensional Lie algebra on U, and by analyticity also on M. Hence $\mathscr{L}$ is also finite-dimensional.

Note, however, that $\mathscr{L}$ is not necessarily finite-dimensional if the realization is not strongly accessible. For example, the system

$$\dot{x}_1 = x_1$$
$$\dot{x}_2 = e^{x_1} u$$

is of the form $\dot{x} = f(x) + ug(x)$, where

$$f = x_1 \frac{\partial}{\partial x_1}, \qquad g = e^{x_1} \frac{\partial}{\partial x_2},$$

and $\mathscr{L}$ is clearly infinite dimensional.

By using the finite-dimensionality of $\mathscr{L}$ to generate a Lie transformation group on M via the flows of the elements of $\mathscr{L}$ and by using the theory of homogeneous subspaces of nilpotent Lie groups from Matsushima (1951), it is possible to show that the state space of a strongly accessible, observable realization of a finite Volterra series is homeomorphic to $\mathbb{R}^n$, for some n (and hence is simply connected) – details are given in Crouch (1981). Moreover, it can be shown that if $\mathscr{N}$ is the subalgebra of $\mathscr{I}$ consisting of vector fields which vanish at x_0, then $\mathscr{I}$ has a basis $\{a_1, \ldots, a_n, a_{n+1}, \ldots, a_{n+s}\}$, where $\{a_{n+1}, \ldots, a_{n+s}\}$ spans $\mathscr{N}$, such that, if ϕ_i is the flow of a_i, then the map

$$\Phi(t_1, \ldots, t_n) = \phi_1(t_1) \circ \cdots \circ \phi_n(t_n)x_0 \qquad (4.2.23)$$

is a homeomorphism of $\mathbb{R}^n$ onto the state space of a minimal realization (of a finite Volterra series).

Let Σ_1 be a strongly accessible, weakly observable realization of a finite Volterra series with state space M_1. Then we have seen above that M_1 can be chosen to be a covering space of the state space M_2 of a minimal (i.e.

observable, strongly accessible) realization Σ_2. Since M_2 is simply connected by the above remarks, and the universal covering space of M_1 is simply connected, it follows that the covering projection $\pi : M_1 \to M_2$ is a diffeomorphism by the connectedness of M_1. Hence Σ_2 is isomorphic to Σ_1 and so, in this case, weak observability implies observability. If follows that an analytic realization of a finite Volterra series is strongly accessible and observable if and only if

$$T_xM = \mathscr{I}(x), \qquad T_x^* M = \mathrm{d}\mathscr{H}(x)$$

for all $x \in M$.

In order to derive a canonical form for a minimal realization, consider the sequence

$$\mathscr{I} \supseteq \mathscr{I}^2 + \mathscr{N} \supseteq \cdots \supseteq \mathscr{I}^p + \mathscr{N} \supseteq \mathscr{N}$$

of subalgebras of $\mathscr{I}$. Recalling that $\mathscr{R}$ is the linear span of the vector fields $(\mathrm{ad}^i f)g_j, i \geqslant 0, 1 \leqslant j \leqslant m$, we define

$$\mathscr{R}^1 = \mathscr{R}, \qquad \mathscr{R}^i = [\mathscr{R}, \mathscr{R}^{i-1}], \quad i \geqslant 2.$$

Then $\mathscr{R}^i$ is the linear subspace of $\mathscr{I}$ spanned by brackets (of $\mathscr{I}$) of length i. Then

$$\mathscr{I}^i = \mathscr{R}^i + \mathscr{I}^{i+1}, \qquad \mathscr{I}^i + \mathscr{N} = \mathscr{R}^i + (\mathscr{I}^{i+1} + \mathscr{N}).$$

Hence the basis $\{a_1, \ldots, a_{n+s}\}$ introduced above can be assumed to be decomposed in the form

$$\{a_{r_1}, \ldots, a_{r_2-1}\} \cup \{a_{r_2}, \ldots, a_{r_3-1}\} \cup \cdots \cup \{a_{r_p}, \ldots, a_n\} \cup \{a_{n+1} \ldots, a_{n+s}\},$$

where

$$\{a_{r_i}, \ldots, a_{r_{i+1}-1}\} \subseteq \mathscr{R}^i.$$

Lemma 4.2.12
The map $\Phi : \mathbb{R}^n \to M$ given by (4.2.23) is a diffeomorphism and hence $\Phi^{-1} : M \to \mathbb{R}^n$ is a global coordinate system for M.

Proof
The result will follow from the fact that Φ_* is of full rank at each $t \in \mathbb{R}^n$. To prove this note that the map $\Psi(t) : M \to M$ given by

$$\Psi(t)x = \phi_1(t_1) \circ \cdots \circ \phi_n(t_n)x$$

is a diffeomorphism (since the flows are complete). Hence

$$(\Psi(t)_*)^{-1} : T_{\Phi(t)}M \to T_{x_0}M$$

is an isomorphism. However,

$$(\Psi(t)_*)^{-1} \frac{\partial \Phi}{\partial t_i}(t) = \phi_n(-t_n)_* \cdots \phi_i(-t_i)_* a_i(\phi_i(t_i) \circ \cdots \circ \phi_n(t_n)x_0),$$

and so by the Campbell–Hausdorff formula and the fact that $[\mathscr{R}^i, \mathscr{R}^j] \subseteq \mathscr{R}^{i+j}$ we can write

$$(\Psi(t)_*)^{-1} \frac{\partial \Phi}{\partial t_i}(t) = a_i(x_0) + \sum_{j=i+1}^{n} p_j(t_1, \ldots, t_n) a_j(x_0)$$

where the p_j are polynomials in t. Since $\{a_1(x_0), \ldots, a_n(x_0)\}$ spans $T_{x_0}M$, it follows that $\{(\Psi(t)_*)^{-1}(\partial\Phi/\partial t_i)(t),\ 1 \leqslant i \leqslant n\}$ also spans $T_{x_0}M$. □

(Compare this proof with that of Theorem 3.1.9.) Composing Φ^{-1} with the usual coordinate system for $T_{x_0}M$ with basis $a_1(x_0), \ldots, a_n(x_0)$ we see that M is diffeomorphic to $T_{x_0}M$.

Next, by strong accessibility, we have $\mathscr{I}(x_0) = T_{x_0}M$ and so there exists $a_0 \in \mathscr{I}$ such that $(a_0 + f)(x_0) = 0$. Thus, since $\mathrm{ad}(f + a_0) : \mathscr{N} \to \mathscr{N}$ it is easy to see that the map

$$\mathrm{ad}(f + a_0) : \mathscr{I}^{i+1} + \mathscr{N} \to \mathscr{I}^{i+1} + \mathscr{N}$$

is a linear endomorphism for each i. Hence $\mathrm{ad}(f + a_0)$ induces a mapping on

$$(\mathscr{I}^i + \mathscr{N})/(\mathscr{I}^{i+1} + \mathscr{N}) \cong \mathscr{R}^i + (\mathscr{I}^{i+1} + \mathscr{N})/(\mathscr{I}^{i+1} + \mathscr{N}) \cong \mathscr{R}^i(\mathscr{R}^i \cap (\mathscr{I}^{i+1} + \mathscr{N}))$$

If we consider matrix representations of these mappings we obtain the desired canonical form:

Theorem 4.2.13

A strongly accessible, observable realization of dimension n of a finite Volterra series of length p has an isomorphic realization, with state vector $z = (z_1, \ldots, z_p)$, $z_i \in \mathscr{R}^{n_i}$, $\sum_{i=1}^{p} n_i = n$, of the form

$$\begin{bmatrix} \dot z_1 \\ \dot z_2 \\ \vdots \\ \dot z_p \end{bmatrix} = \begin{bmatrix} A_1 z_1 + d_1 \\ A_2 z_2 + d_2(z_1) \\ \vdots \\ A_p z_p + d_p(z_1, \ldots, z_{p-1}) \end{bmatrix} + \sum_{i=1}^{m} u_i \begin{bmatrix} b_{i1} \\ b_{i2}(z_1) \\ \vdots \\ b_{ip}(z_1, \ldots, z_{p-1}) \end{bmatrix} \qquad (4.2.24)$$

$$y = c(z_1, \ldots, z_p)$$

with $z_i(0) = 0$, $1 \leqslant i \leqslant p$, where b_{ij}, d_i and c are vector-valued polynomials in z_i and $n_i = \dim\ \mathscr{R}^i/\mathscr{R}^i \cap (\mathscr{I}^{i+1} + \mathscr{N})$ (which depends only on the input–output map).

Proof

Let $a_0, a_1, \ldots, a_n, \ldots, a_{n+s}$ be as above and let $\phi_1(t), \ldots, \phi_n(t)$ be the flows of $a_1, \ldots, a_n$. Partition x corresponding to the basis of $\mathscr{R}^i$, i.e.

$$x = ((x_1, \ldots, x_{r_2-1}), (x_{r_2}, \ldots, x_{r_3-1}), \ldots, (x_{r_p}, \ldots, x_n))$$

and let z_k be the subvector $(x_{r_k}, \ldots, x_{r_{k+1}-1})$ where $r_1 = 1$, $r_{p+1} = n+1$. We would like to show that any solution $x(t)$ of

$$\dot{x} = f(x) + \sum_{i=1}^{m} u_i g_i(x), \qquad x(0) = x_0 \tag{4.2.25}$$

satisfies

$$x(t) = \phi_1(x_1(t)) \circ \cdots \circ \phi_n(x_n(t))x \tag{4.2.26}$$

i.e.

$$x(t) = \phi_1(z_{11}(t)) \circ \phi_2(z_{12}(t)) \circ \cdots \circ \phi_{r_2-1}(z_{1r_2-1}(t)) \circ \cdots \circ \phi_n(z_{pr_{p+1}-1}(t))x_0$$

which will prove that a system of the form (4.2.24) is isomorphic to one of the form (4.2.25).

To prove (4.2.26) we define, recursively,

$$v_0(t) = x(t)$$

$$v_k(t) = \phi_{s_k}(-x_{s_k}(t)) \circ \cdots \circ \phi_{r_k}(-x_{r_k}(t))v_{k-1}(t), \tag{4.2.27}$$

where $x(t)$ is the solution of (4.2.25) and $s_k = r_{k+1} - 1$. Then we show by induction that if $z_1, \ldots, z_{k-1}$ satisfy equations of the form (4.2.24) then v_{k-1} satisfies a differential equation of the form

$$\dot{v}_{k-1} = (f + a_0)(v_{k-1}) + \sum_{i=r_k}^{n+s} \left[\alpha_i(z_1, \ldots, z_{k-1}) + \sum_{j=1}^{m} u_j \beta_{ij}(z_1, \ldots, z_{k-1})\right] a_i(v_{k-1}) \tag{4.2.28}$$

where the functions α_i and β_{ij} are polynomials in $z_1, \ldots, z_{k-1}$. To show that this is true for $k = 1$ note that since a_0, $g_i \in \mathscr{I}$ we can write

$$-a_0 = \sum_{i=1}^{n+s} \alpha_i a_i, \qquad g_i = \sum_{i=1}^{n+s} \beta_{ij} a_i \tag{4.2.29}$$

and so $v_0 = x$, which is the solution of (4.2.25), can be written in the form of (4.2.28) with $k = 1$. If (4.2.28) is true for $k - 1$, consider v_k, which from (4.2.27) can be written as

$$\begin{aligned}
\dot{v}_k &= \phi_{s_k}(-x_{s_k})_* \cdots \phi_{r_k}(-x_{r_k})_* \dot{v}_{k-1} - \dot{x}_{s_k} a_{s_k}(v_k) \\
&\quad - \sum_{l=r_k}^{s_k-1} \dot{x}_l \phi_{s_k}(-x_{s_k})_* \cdots \phi_{l+1}(-x_{l+1})_* a_l(\phi_l(-x_l) \circ \cdots \circ \phi_{r_k}(-x_{r_k})v_{k-1}) \\
&= \phi_{s_k}(-x_{s_k})_* \cdots \phi_{r_k}(-x_{r_k})_* \dot{v}_{k-1} - \dot{x}_{s_k} a_{s_k}(v_k) \\
&\quad - \sum_{l=r_k}^{s_k-1} \dot{x}_l \phi_{s_k}(-x_{s_k})_* \cdots \phi_{l+1}(-x_{l+1})_* a_l(\phi_{l+1}(x_{l+1}) \circ \cdots \circ \phi_{s_k}(x_{s_k})v_k)
\end{aligned}$$

$$= \phi_{s_k}(-x_{s_k})_* \cdots \phi_{r_k}(-x_{r_k})_* \dot{v}_{k-1} - \dot{x}_{s_k} a_{s_k}(v_k)$$
$$- \dot{x}_{s_k-1} \exp(-x_{s_k} \text{ ad } a_{s_k})(a_{s_k-1})(v_k)$$
$$- \cdots - \dot{x}_{r_k}[\exp(-x_{s_k} \text{ ad } a_{s_k})](\cdots(\exp(-x_{r_k} \text{ ad } a_{r_k}))(a_{r_k})\cdots)(v_k).$$

Since $\mathscr{R}^i \subseteq \mathscr{I}^i$ and $[\mathscr{I}^k, \mathscr{I}^j] \subseteq \mathscr{I}^{k+j}$, it follows that $\dot{v}_k$ can be written in the form

$$\dot{v}_k = \phi_{s_k}(-x_{s_k})_* \cdots \phi_{r_k}(-x_{r_k})_* \dot{v}_{k-1} - \sum_{i=r_k}^{s_k} \dot{x}_i a_i(v_k) + \sum_{i=r_{k+1}}^{n+s} \eta_i(z_k, \dot{z}_k) a_i(v_k) \tag{4.2.30}$$

where η_i is linear in $\dot{z}_k$ and a polynomial in z_k.

Consider next the map ad$(f + a_0)$ restricted to $\mathscr{I}^k + \mathscr{N}$ and let (A_{ij}), $r_k \leqslant i, j \leqslant s_k$ be its matrix representation on $\mathscr{R}^k/(\mathscr{R}^k \cap (\mathscr{I}^{k+1} + \mathscr{N}))$. Then from (4.2.27) we have

$$v_{k-1} = \phi_{r_k}(x_{r_k}) \circ \cdots \circ \phi_{s_k}(x_{s_k}) v_k \tag{4.2.31}$$

and by substituting (4.2.28) into (4.2.30), replacing v_{k-1} by the right-hand side of (4.2.31) and using the Campbell–Hausdorff formula, it is not difficult to see that we obtain an equation of the form

$$\dot{v}_k = (f + a_0)(v_k) + \sum_{i=r_{k+1}}^{n+s} \left[\eta_i(z_k, \dot{z}_k) + \alpha_i'(z_1, \ldots, z_k) + \sum_{j=1}^{m} u_j \beta_{ij}'(z_1, \ldots, z_k)\right] a_i(v_k)$$
$$+ \sum_{i=r_k}^{s_k} \left[-\dot{x}_i + \sum_{j=r_k}^{s_k} A_{ij} x_j + \alpha_i(z_1, \ldots, z_{k-1}) + \sum_{j=1}^{m} u_j \beta_{ij}(z_1, \ldots, z_{k-1})\right] a_i(v_k) \tag{4.2.32}$$

By the induction hypothesis, z_k satisfies an equation of the form in (4.2.24) and so we can set the last bracket to zero and then substitute the resulting expression for z_k back into (4.2.32), giving an equation in v_k of the required form. Setting $z_k(0) = 0$ in (4.2.27) gives $v_k(0) = x_0$. Since the equation (4.2.28) with $k = p + 2$ contains only vector fields which vanish at x_0, it follows that $v_{p+1}(t) \equiv x_0$ and so (4.2.26) now follows from (4.2.31).

Note finally that $y = h(x) = h(\phi_1(x_1(t)) \circ \cdots \circ \phi_n(x_n(t))x_0)$ and so if $\sum_{i=1}^n k_i = p + m$, $m > 0$, we have

$$\left.\frac{\partial^{p+m}}{\partial x_1^{k_1} \cdots \partial x_n^{k_n}}\right|_{x_1 = \cdots = x_n = 0} = L_a^{k_n} L_a^{k_{n-1}} \cdots L_a^{k_1} h(x_0)$$
$$= 0$$

by definition of $\mathscr{H}$. Hence $y = c(z_1, \ldots, z_p)$ is a polynomial in the elements of the vectors z_i. □

The polynomials d_i, b_{ij} and c in the canonical form (4.2.24) have a special structure which can be determined by writing (4.2.24) in the form

$$\dot{z} = F(z) + \sum_{i=1}^{m} u_i G_i(z), \qquad z(0) = 0, \quad z \in \mathbb{R}^n$$
$$y = H(z) \tag{4.2.33}$$

and defining the map $\delta_\lambda : \mathbb{R}^n \to \mathbb{R}^n$ by

$$\delta_\lambda(z) = (\lambda z_1, \dots, \lambda^p z_p), \quad z = (z_1, \dots, z_p), \qquad z_i \in \mathbb{R}^{n_i}.$$

Consider the one-parameter family of bases of $\mathscr{I}$ given by

$$\mathscr{B}_\lambda \triangleq (\lambda a_{r_1}, \dots, \lambda a_{s_1}, \lambda^2 a_{r_2}, \dots, \lambda^p a_{r_p}, \dots, \lambda^p a_{s_p}, a_{n+1}, \dots, a_{n+s}). \tag{4.2.34}$$

Then if we use this basis as in Theorem 4.2.13 we obtain the canonical form

$$\dot{z}_\lambda = F_\lambda(z_\lambda) + \sum_{i=1}^{m} u_i G_{i\lambda}(z_\lambda), \qquad z_\lambda(0) = 0,$$
$$y = H_\lambda(z_\lambda). \tag{4.2.35}$$

in the same way as (4.2.33) was derived. Since δ_λ is a diffeomorphism for $\lambda > 0$ (corresponding to the diffeomorphism on M induced by the flows of the vector fields in the basis (4.2.34)) it follows that the systems (4.2.33) and (4.2.35) are equivalent. Hence

$$F_\lambda = \delta_\lambda^{-1} \circ F \circ \delta_\lambda, \qquad G_{i\lambda} = \delta_\lambda^{-1} \circ G_i \circ \delta_\lambda, \qquad H_\lambda = H \circ \delta_\lambda. \tag{4.2.36}$$

From the construction of the canonical realization in Theorem 4.2.13 using the basis $\mathscr{B}_\lambda$, and in particular (4.2.29), it is clear that we can decompose F and G_i in (4.2.33) in the form

$$F_\lambda = \sum_{i=0}^{p} \frac{F_i}{\lambda^i}, \qquad G_{i\lambda} = \sum_{j=1}^{P} \frac{G_{ij}}{\lambda^j},$$

and from (4.2.36) it follows that

$$\lambda^i F_i \circ \delta_\lambda = \delta_\lambda \circ F_i, \qquad \lambda^i G_{ij} \circ \delta_\lambda = \delta_\lambda \circ G_{ij}. \tag{4.2.37}$$

It follows from (4.2.37) that we may decompose the polynomials in the canonical form (4.2.24) in the forms

$$d_i(z_1, \dots, z_{i-1}) = \sum_{j=0}^{i-1} d_{ij}(z_1, \dots, z_j) + d_{ii}(z_1, \dots, z_{i-1}),$$

$$b_{ki}(z_1, \dots, z_{i-1}) = \sum_{j=0}^{i-1} b_{kij}(z_1, \dots, z_j)$$

where

$$d_{ij}(\lambda z_1, \dots, \lambda^j z_j) = \lambda^j d_{ij}(z_1, \dots, z_j)$$
$$d_{ii}(\lambda z_1, \dots, \lambda^{i-1} z_{i-1}) = \lambda^i d_{ii}(z_1, \dots, z_{i-1}),$$
$$b_{kij}(\lambda z_1, \dots, \lambda^j z_j) = \lambda^j b_{kij}(z_1, \dots, z_j).$$

Of course, any system of the form (4.2.24) has a finite Volterra series of length p and so (4.2.24) is a necessary and sufficient condition for a system to have a Volterra series of length p.

Crouch (1981) also presents some results on homogeneous Volterra series (i.e. those with a single nonzero term). We shall not discuss this further here but we shall consider homogeneous systems which have a bilinear realization in Section 4.5. By lifting the dynamics of a system from M to $M \times \mathbb{R}^N$ for some N, Crouch (1984) also shows how to generalize the above results to systems which do not have finite Volterra series and obtains solvable approximations of the above type.

4.3 CONTROLLABILITY, OBSERVABILITY AND STABILIZABILITY

4.3.1 Controllability of bilinear systems

Consider first bilinear systems of the form

$$\dot{x}(t) = u_1(t)B_1x(t) + \cdots + u_r(t)B_rx(t). \tag{4.3.1}$$

We have seen in Chapter 2 that a condition for accessibility of this system is given by the rank condition on the Lie algebra spanned by the matrices $B_1, \ldots, B_r$. Since 0 is an equilibrium point of (4.3.1) we cannot control from 0 to any other point and so we consider controllability of this system on $\mathbb{R}^n \backslash \{0\} = \mathbb{R}^n_*$. It can then be shown by the methods of Section 4.2 that this system is controllable on $\mathbb{R}^n_*$ if and only if the Lie algebra $\mathfrak{g} \subseteq gl(n; \mathbb{R})$ generated by $\{B_1, \ldots, B_r\}$ is the Lie algebra of a Lie group $G \subseteq GL(n; \mathbb{R})$ which is transitive (see p. 42) on $\mathbb{R}^n_*$ (under the natural linear action on $\mathbb{R}^n$). Hence the controllability of systems of type (4.3.1) can be characterized by the set of all transitive Lie algebras $gl(n; \mathbb{R})$. (A *transitive Lie algebra* is one whose corresponding connected Lie group $G \subseteq GL(n; \mathbb{R})$ is transitive.)

A complete list of transitive Lie algebras has been determined by Boothby, (1975) (see also Boothby and Wilson (1979)). We shall present this list without proof. The interested reader should consult the cited references. To present the list we need some extensions of the classical groups (over $\mathbb{R}$) defined in Chapter 1 to the case of $\mathbb{C}$ and the quarternion field $\mathbb{H}$. (For an extensive treatment of $\mathbb{H}$ see Porteous (1969).) First, $sl(m; \mathbb{C})$ denotes the Lie algebra of $m \times m$ matrices over $\mathbb{C}$ with zero trace. Next, for any linear transformation over $\mathbb{H}^m$, represented by an $m \times m$ matrix A with coefficients in $\mathbb{H}$, there is a corresponding $4m \times 4m$ matrix $A_\mathbb{R}$ with coefficients in $\mathbb{R}$ which operates on $\mathbb{R}^{4m}$ in the obvious way. Let $sl(m; \mathbb{H})$ denote the Lie algebra of $m \times m$ matrices A over $\mathbb{H}$ for which trace $A_\mathbb{R} = 0$. For any element $h = a + ib + jc + kd \in \mathbb{H}$, recall that the transpose $\bar{h}$ of h is defined by $\bar{h} = a - ib - jc - kd$. (This, of course, restricts to $\mathbb{R}$ and $\mathbb{C}$ in the usual way.) For any matrix $A = (a_{ij})$ over $\mathbb{H}$ we define

$$A^* = (\bar{a}_{ij})^{\mathsf{T}}.$$

We have defined $so(m; \mathbb{R})$ to be the subalgebra of $sl(m; \mathbb{R})$ consisting of matrices A for which $A^{\mathsf{T}} = -A$. Similarly we define

$$so(m; \mathbb{C}) = \{A \in sl(m; \mathbb{C}) : A^* = -A\}$$
$$so(m; \mathbb{H}) = \{A \in sl(m; \mathbb{H}) : A^* = -A\}.$$

($so(m; \mathbb{C})$ is often denoted by $su(m; \mathbb{C})$.) If J is the $2m \times 2m$ matrix

$$\begin{bmatrix} 0 & I \\ -I & 0 \end{bmatrix},$$

then $sp(m; \mathbb{C})$ denotes the subalgebra of $sl(2m, \mathbb{C})$ consisting of matrices A such that $JAJ = A^*$, generalizing the real symplectic algebra $sp(m; \mathbb{R})$. To specify the final two types of algebras which we require we need the notion of representation of a Lie algebra. A (matrix) *representation* of a Lie algebra L on $\mathbb{R}^n$ is a map $r : L \to gl(n; \mathbb{R})$ such that r is a homomorphism of vector spaces and

$$r([g_1, g_2]) = [rg_1, rg_2], \text{ for all } g_1, g_2 \in L.$$

Then it can be shown (Kirillov (1976)) that $so(2m+1; \mathbb{R})$ has a representation on $\mathbb{R}^{2m}$ if $m \equiv 0, 3 \bmod 4$. This is called the *spin representation* of $so(2m+1; \mathbb{R})$ and the associated Lie algebra of $2^m \times 2^m$ matrices is denoted by spin$(2m+1; \mathbb{R})$. Finally, if G_2 is the compact simple Lie algebra with Cartan subalgebra of dimension 2, then $g_2(-14)$ denotes the algebra

Table 4.1 Transitive matrix Lie algebras

n	N	*Representative*
m	$m(m-1)/2 + 1$	$so(m; \mathbb{R}) \oplus \mathbb{R}$
$2m$	$m^2 - 1 + \varepsilon\ (\varepsilon = 1, 2)$	$so(m; \mathbb{C}) \oplus \mathfrak{c}$, $\mathfrak{c} \cong \mathbb{R}$ or $\mathbb{C}$
$4m$	$2m^2 + m + \varepsilon\ (\varepsilon = 1, 2)$	$so(m; \mathbb{H}) \oplus \mathfrak{c}$, $\mathfrak{c} \cong \mathbb{R}$ or $\mathbb{C}$
$4m$	$2m^2 + m + 4$	$so(m; \mathbb{H}) \oplus \mathbb{H}$
8	22	spin $(7; \mathbb{R}) \oplus \mathbb{R}$
16	37	spin $(9; \mathbb{R}) \oplus \mathbb{R}$
7	15	$g_{2(-14)} \oplus \mathbb{R}$
m	$m^2 - 1 + \varepsilon\ (\varepsilon = 0, 1)$	$sl(m; \mathbb{R}) \oplus \mathfrak{c}$, $\mathfrak{c} \cong \{0\}$ or $\mathbb{R}$
$2m$	$2(m^2 - 1) + \varepsilon\ (\varepsilon = 0, 1, 2)$	$sl(m; \mathbb{C}) \oplus \mathfrak{c}$, $\mathfrak{c} \cong \{0\}$ or $\mathbb{R}$ or $\mathbb{C}$
$4m$	$4m^2 - 1 + \varepsilon\ (\varepsilon = 0, 1, 2)$	$sl(m; \mathbb{H}) \oplus \mathfrak{c}$, $\mathfrak{c} \cong \{0\}$ or $\mathbb{R}$ or $\mathbb{C}$
$2m$	$2m^2 + m - \varepsilon\ (\varepsilon = 0, 1)$	$sp(m; \mathbb{R}) \oplus \mathfrak{c}$, $\mathfrak{c} \cong \{0\}$ or $\mathbb{R}$
$4m$	$4m^2 + 2m + \varepsilon\ (\varepsilon = 0, 1, 2)$	$sp(m; \mathbb{C}) \oplus \mathfrak{c}$. $\mathfrak{c} \cong \{0\}$ or $\mathbb{R}$ or $\mathbb{C}$
$4m$	$4m^2 + 2 + \varepsilon\ (\varepsilon = 0, 1)$	$sl(m; \mathbb{H}) \oplus so(2; \mathbb{C}) + \mathfrak{c}$, $\mathfrak{c} \simeq \{0\}$ or $\mathbb{R}$

of 7×7 matrices which forms a representation of G_2 on $\mathbb{R}^7$ (see Helgason (1962)).

We can now present the list of transitive matrix Lie algebras given in Table 4.1. In this table, n denotes the dimension of the underlying Euclidean space and N the dimension of the algebra.

An algorithm using only rational operations is given by Boothby and Wilson (1979) for determining whether the Lie algebra of a system is transitive, i.e. belongs to the list in Table 4.1.

We finally show that any controllable bilinear system on $\mathbb{R}^n_*$ of the form (4.3.1) is equivalent to a system determined by two vector fields:

$$\dot{x} = (uB_1 + vB_2)x \tag{4.3.2}$$

(u, v piecewise constant) in the sense that the Lie algebra generated by (4.3.1) is the same as that generated by (4.3.2). To do this we need the following:

Lemma 4.3.1

If $\mathfrak{g}$ is a reductive Lie algebra with centre of dimension $\leqslant k$, then if the semisimple part of $\mathfrak{g}$ is generated by k elements, $\mathfrak{g}$ is also generated by k elements.

Proof

Let $\mathfrak{g} = \mathfrak{s} + \mathfrak{c}$, where $\mathfrak{s}$ is a semisimple ideal and $\mathfrak{c}$ is the centre of $\mathfrak{g}$. Let $X_1, \ldots, X_k$ generate $\mathfrak{s}$. Since $\mathfrak{s} = [\mathfrak{s}, \mathfrak{s}]$, X_i is a linear combination of brackets of $X_1, \ldots, X_k$ for each i. Let $C_1, \ldots, C_k$ span $\mathfrak{g}$ and let $X_i' = X_i + C_i$, $1 \leqslant i \leqslant k$. Then any linear combination of brackets in X_i equals the same combination of brackets in X_i' since C_i is a central element in $\mathfrak{g}$. Hence $X_1', \ldots, X_k'$ also generate $\mathfrak{s}$ and so X_i is a linear combination of brackets in $X_1', \ldots, X_k'$. Hence $X_1', \ldots, X_k'$ generate $\mathfrak{g}$. □

The stated equivalence of systems (4.3.1) and (4.3.2) now follows from the classification of Table 4.1, Lemma 4.3.1 and Kuranishi's theorem (Theorem 1.4.18).

We now consider the more general system

$$\dot{x} = (A + \sum_{k=1}^{m} u_k N_k)x + Bu \tag{4.3.3}$$

where $A = (a_{ij})$ and $N_k = (n_{ijk})$, $1 \leqslant k \leqslant m$ are real $n \times n$ matrices and $B = (b_{ik})$ is a real $n \times m$ matrix. In the case of a linear system

$$\dot{x} = Ax + Bu$$

we have the well-known rank condition for (complete) controllability:

$$\text{rank } [B \quad AB \quad A^2B \quad \ldots \quad A^{n-1}B] = n \tag{4.3.4}$$

This, however, assumes that the control u is unbounded. If u is bounded in some compact set in $\mathbb{R}^n$, then it is known that the system may not be completely controllable even if condition (4.3.4) holds. In contrast, the bilinear system (4.3.3) can be completely controllable for bounded controls, as we shall now see. Thus, let Ω be a compact, connected subset of $\mathbb{R}^m$ containing the origin and let $\mathscr{U}$ be the allowable control set consisting of piecewise continuous functions $u(.) : [0, \infty) \to \Omega$. We also define the reachable set $R(x_0)$ from the initial state x_0 to be the set of all states which can be reached from x_0 in finite time and the incident set $I(x_1)$ to a terminal state x_1 to be the set of all initial states from which x_1 is reachable in finite time.

Following Rink and Mohler (1968), we need to consider the *equilibrium set* of the system (4.3.3) for constant controls. This is the set of all solutions of the equation

$$\left(A + \sum_{k=1}^{m} u_k N_k\right) x + Bu = 0 \tag{4.3.5}$$

for any fixed $u \in \Omega$ for which the matrix $A + \Sigma u_k N_k$ is nonsingular. (Of course, for some u, this equation may have no solution.) In order to derive a sufficient condition for the controllability of (4.3.3) we shall require the following local controllability result of Lee and Markus (1961).

Theorem 4.3.2
Let

$$\dot{x} = f(x, u)$$

be a system where $f: \mathbb{R}^n \times \Omega \to \mathbb{R}^n$ is a C^1 map. Let $x \in \mathbb{R}^n$ and suppose that there exists $u(x) \in \text{interior}(\Omega)$ for which

(a) $f(x, u(x)) = 0$
(b) there exists $v \in \mathbb{R}^m$ such that the vector $([\partial f/\partial u](x, u(x)))v$ does not lie in any invariant subspace of the matrix $(\partial f/\partial x)(x, u(x))$ of dimension $\leqslant (n-1)$.

Then $R(x)$ and $I(x)$ are open connected subsets of $\mathbb{R}^n$. □

We can now prove

Theorem 4.3.3
The bilinear system (4.3.3) is (completely) controllable if the following two conditions hold:

(a) There exist constant controls u^+ and u^- in the interior of Ω such that the eigenvalues of the matrices

$$A + \sum_{k=1}^{m} u_k^+ N_k, \qquad A + \sum_{k=1}^{m} u_k^- N_k$$

have, respectively, positive and negative real parts and such that the equilibrium states $x(u^+)$ and $x(u^-)$ are contained in a connected component of the equilibrium set.

(b) For each x in the equilibrium set with a corresponding control $u(x) \in \Omega$ (i.e. $(A + \sum_{k=1}^{m} u(x)N_k)x + Bu(x) = 0$) there exists $v \in \mathbb{R}^m$ such that the vector

$$g \triangleq Bv - \sum_{l=1}^{m} v_l [N_l (A + \sum_{k=1}^{m} u_k(x)N_k)^{-1} Bu(x)]$$

does not lie in any invariant subspace of dimension $\leqslant n-1$ of the matrix

$$E = A + \sum_{k=1}^{m} u_k(x)N_k.$$

Proof

Since u^- makes the eigenvalues of $A + \sum_{k=1}^{m} u_k^- N_k$ all have negative real parts, the trajectories of the system with the constant control u^- cover $\mathbb{R}^n$ and each trajectory is asymptotic to the unique equilibrium state

$$x(u^-) = -(A + \sum_{k=1}^{m} u_k^- N_k)^{-1} Bu^-.$$

The trajectory starting at any initial state $x_0 \in \mathbb{R}^n$ (with control u^-) reaches any neighbourhood of $x(u^-)$ in finite time.

Similarly with control u^+, all trajectories move away from the unique equilibrium state

$$x(u^+) = -(A + \sum_{k=1}^{m} u_k^+ N_k)^{-1} Bu^+,$$

and fill $\mathbb{R}^n$. The trajectory through any terminal state x_1 (with control u^+) reaches x_1 from any neighbourhood of $x(u^+)$ in finite time.

Now consider any initial state $x_0 \in \mathbb{R}^n$ and terminal state x_1. Since $x(u^+)$ and $x(u^-)$ are contained in a connected component of the equilibrium set, we can join them by a continuous control $u: [t_1, t_2] \to \Omega$ such that $u(t_1) = u^-$, $u(t_2) = u^+$, for some times t_1, t_2. The condition (b) of the theorem clearly implies the condition of Theorem 4.3.2 and so $R(x(u^+))$ is open. However, we can reach x_1 in finite time from any neighbourhood of $x(u^+)$ and so we can reach x_1 from $x(u^-)$. Similarly, since $I(x(u^-))$ is open we can reach x_1 from x_0. □

The condition (a) of Theorem 4.3.3 is satisfied if there exists a continuous function $u: [a, b] \to \Omega$ such that the eigenvalues of $A + \sum_{k=1}^{m} u_k(\tau)N_k$ are all nonzero for each $\tau \in [a, b]$ and the eigenvalues of $A + \sum_{k=1}^{m} u_k(a)N_k$ and $A + \sum_{k=1}^{m} u_k(b)N_k$ are respectively in the open left and right half planes; i.e. we can shift the eigenvalues of the system matrix across the imaginary axis of $\mathbb{C}$ without passing through 0. Condition (b) is more delicate in general. However, if we consider the nth-order system

$$\frac{d^n x}{dt^n} = \sum_{i=1}^{n} \left(a_i + \sum_{k=1}^{m} u_k n_{ik}\right) \frac{d^{(i-1)}x}{dt^{(i-1)}} + \sum_{k=1}^{m} b_k u_k,$$

where $b_i \neq 0$ for some k, and $a_1 \neq 0$, and consider the phase variables $x_1 = x, x_2 = x, \ldots, x_n = x^{(n-1)}$, we obtain the system (4.3.2) where

$$A = \begin{bmatrix} 0 & 1 & 0 & 0 & \cdots & 0 \\ 0 & 0 & 1 & 0 & \cdots & 0 \\ \vdots & \vdots & \vdots & \vdots & & \vdots \\ 0 & 0 & 0 & 0 & \cdots & 1 \\ a_1 & a_2 & a_3 & a_4 & \cdots & a_n \end{bmatrix}, \quad N_k = \begin{bmatrix} 0 & 0 & \cdots & 0 \\ \vdots & \vdots & & \vdots \\ 0 & 0 & \cdots & 0 \\ n_{1k} & n_{2k} & \cdots & n_{nk} \end{bmatrix},$$

$$B = \begin{bmatrix} 0 & \cdots & 0 \\ \vdots & & \vdots \\ 0 & \cdots & 0 \\ b_1 & \cdots & b_m \end{bmatrix}.$$

It follows that the equilibrium state $x(u)$ for a given control u is given by

$$x(u) = \frac{-\sum_{k=1}^{m} b_k u_k}{a_1 + \sum_{k=1}^{m} u_k n_{1k}} \begin{bmatrix} 1 \\ 0 \\ \vdots \\ 0 \end{bmatrix},$$

and the matrix $D = \dfrac{\partial}{\partial u}\left(\left(A + \sum_{k=1}^{m} u_k N_k\right) x + Bu\right)\Bigg|_{x=x(u)}$ is given by

$$D = \frac{-\sum_{k=1}^{m} b_k u_k}{a_1 + \sum_{k=1}^{m} u_k n_{1k}} \begin{bmatrix} 0 & 0 & \cdots & 0 \\ \vdots & & \cdots & \vdots \\ 0 & 0 & \cdots & 0 \\ n_{11} & n_{12} & \cdots & n_{1m} \end{bmatrix} + B.$$

Also we have

$$E = \begin{bmatrix} 0 & 1 & 0 & 0 \ \cdots & 0 \\ 0 & 0 & 1 & 0 \ \cdots & 0 \\ \vdots & \vdots & \vdots & \cdots & \cdots \\ 0 & 0 & 0 & \cdots \ \cdots & 1 \\ a_1 + \sum_{k=1}^{m} u_k n_{1k} & a_2 + \sum_{k=1}^{m} u_k n_{2k} & a_3 + \sum_{k=1}^{m} u_k n_{3k} & \cdots & a_n + \sum_{k=1}^{m} u_k n_{nk} \end{bmatrix}.$$

It is easy to check that the condition (b) of Theorem 4.3.3 (or Theorem 4.3.2) is equivalent to the existence of a vector v such that the vectors

$$Dv, EDv, \ldots, E^{(n-1)}Dv$$

are linearly independent. However, from the forms of D and E this is valid if D is not identically zero. To show that D is not identically zero note first that if $\sum_{k=1}^{m} b_k u_k = 0$ then $D = B \neq 0$. Thus we may assume that $\sum_{k=1}^{m} b_k u_k \neq 0$. Let $v = u$. Then

$$Dv = Du = \begin{bmatrix} 0 \\ 0 \\ \vdots \\ 0 \\ \sum_{k=1}^{m} b_k u_k \left[1 - \dfrac{\sum_{k=1}^{m} n_{1k} u_k}{a_1 + \sum_{k=1}^{m} u_k n_{1k}} \right] \end{bmatrix}$$

which is not the zero vector since $a_1 \neq 0$. Hence we see that the condition (b) of Theorem 4.3.3 is satisfied for all phase-variable systems with $B \neq 0$ and $a_1 \neq 0$.

Further results on bilinear controllability are given by Mohler (1973) and for positive orthant controllability by Boothby (1982).

4.3.2 Observability of bilinear systems

We shall now consider the observability of the bilinear system

$$\begin{aligned} \dot{x}(t) &= (A + u(t)N)x(t), \qquad x(t) \in \mathbb{R}^n \\ y(t) &= c^{\mathrm{T}} x(t) \end{aligned} \tag{4.3.6}$$

with a scalar control u and a scalar measurement y, following Williamson (1977). The multivariable case can be considered similarly. The general system

$$\begin{aligned} \dot{x}(t) &= f(x(t), u(t)) \\ y(t) &= h(x(t)) \end{aligned}$$

for a given control u is said to be *completely observable with respect to u* on the interval $[0, t_1]$ if the initial state $x(0)$ can be determined from a knowledge of $y(t)$ on $[0, t_1]$. (Of course, since the dynamics are deterministic, this implies a knowledge of $x(t)$ for all $t \geqslant 0$.) If the system is completely observable with respect to any admissible control then we say that it is *completely observable*.

Observability of the bilinear system (4.3.6) can be described in terms of that of the linear time-varying system

$$\begin{aligned} \dot{x}(t) &= L(t)x(t), \quad x(t) \in \mathbb{R}^n \\ y(t) &= c^{\mathrm{T}}x(t) \end{aligned} \tag{4.3.7}$$

In fact, we have the following well-known result from linear systems theory (see D'Angelo (1970)):

Theorem 4.3.4
The system (4.3.7) is completely observable on $[0, t_1]$ if $L(t)$ is differentiable almost everywhere $n-2$ times on $[0, t_1]$ and the *observability matrix*

$$\theta = [c \quad \Delta c \quad \dots \quad \Delta^{n-1}c],$$

where Δ is the operator

$$\Delta = L^{\mathrm{T}}(t) + I\frac{\mathrm{d}}{\mathrm{d}t},$$

has rank n almost everywhere on some finite subinterval of $[0, t_1]$. □

This leads immediately to

Corollary 4.3.5
The bilinear system (4.3.6) is completely observable with respect to an $n-2$ times differentiable control u on $[0, t_1]$ if the observability matrix

$$\theta = [c \quad \Delta c \quad \dots \quad \Delta^{n-1}c],$$

where

$$\Delta = (A + u(t)N)^{\mathrm{T}} + I\frac{\mathrm{d}}{\mathrm{d}t},$$

has rank n almost everywhere. □

As we have seen before, observability is related to the existence of canonical forms and, in fact, we have the following theorem.

Theorem 4.3.6
The bilinear system (4.3.6) is completely observable if and only if there exists a constant nonsingular matrix P such that

$$PAP^{-1} = \begin{bmatrix} 0 & 0 & \dots & 0 & a_1 \\ 1 & 0 & \dots & 0 & a_2 \\ \vdots & \vdots & & & \vdots \\ 0 & 0 & \dots & 1 & a_n \end{bmatrix}$$

$$PNP^{-1} = \begin{bmatrix} n_{11} & n_{12} & \dots & n_{1,n-1} & n_{1n} \\ 0 & n_{22} & \dots & n_{2,n-1} & n_{2n} \\ 0 & 0 & \dots & & \\ \vdots & \vdots & & \dots & \dots \\ 0 & 0 & \dots & 0 & n_{nn} \end{bmatrix}$$

and

$$c^{\mathrm{T}} P^{-1} = (0 \quad \dots \quad 0 \quad 1).$$

Proof

If P exists, then with $L = A + u(t)N$ we have

$$\theta^{\mathrm{T}} = \begin{bmatrix} c^{\mathrm{T}} \\ c^{\mathrm{T}} L \\ c^{\mathrm{T}} L^2 + c^{\mathrm{T}} \dot{L} \\ \vdots \end{bmatrix}$$

$$= \begin{bmatrix} c^{\mathrm{T}} P^{-1} \\ c^{\mathrm{T}} P^{-1}(PLP^{-1}) \\ c^{\mathrm{T}} P^{-1}(PLP^{-1})^2 + c^{\mathrm{T}} P^{-1} \overline{(PLP^{-1})}^{\boldsymbol{\cdot}} \\ \vdots \end{bmatrix} P$$

From the structure of $PLP^{-1} = PAP^{-1} + uPNP^{-1}$ it follows that the latter matrix is of the form

$$\begin{bmatrix} 0 & 0 & \dots & 0 & 0 & 1 \\ 0 & 0 & \dots & 0 & 1 & * \\ 0 & 0 & \dots & 1 & * & * \\ \vdots & \vdots & \dots & \vdots & \vdots & \vdots \\ 1 & * & \dots & * & * & * \end{bmatrix} P$$

where the elements $*$ depend on u. Since P is invertible this matrix is of rank n for all u and so the same is true of θ. Observability then follows from Corollary 4.3.5.

Conversely, if the system is completely observable, then it is observable, in particular, with $u(t) \equiv 0$. From the standard linear theory it follows that there exists an invertible matrix P such that

$$PAP^{-1} = \begin{bmatrix} 0 & \dots & 0 & a_1 \\ 1 & \dots & 0 & a_2 \\ \vdots & & \vdots & \vdots \\ 0 & \dots & 1 & a_n \end{bmatrix}, \qquad c^{\mathrm{T}} P^{-1} = [0 \quad 0 \quad \dots \quad 0 \quad 1].$$

If we put $PNP^{-1} = (n_{ij})$, then it suffices to show that $n_{ij} = 0$ for all $i > j$. Assume the contrary, so that $n_{ij} \neq 0$ for some $0 < j < k \leqslant n$. From

$$y = c^{\mathrm{T}} P^{-1} Px = (0 \quad \dots \quad 0 \quad 1) Px,$$

it follows that

$$\begin{bmatrix} y \\ \dot{y} \\ \vdots \\ \\ y^{(n-k-1)} \end{bmatrix} = \begin{bmatrix} 0 & 0 & 0 & \dots & 0 & 1 \\ 0 & 0 & 0 & \dots & 1 & * \\ \vdots & \vdots & \vdots & & \dots & \dots \\ 0 & 1 & * & \dots & * & * \\ 1 & * & * & \dots & * & * \end{bmatrix} P \begin{bmatrix} x_k \\ x_{k+1} \\ \vdots \\ \\ x_n \end{bmatrix},$$

Hence, given the measurements $y, \dot{y}, \dots, y^{(n-1)}$, we can determine uniquely the states $x_k, \dots, x_n$. Let

$$x'(t) = (x_1'(t), \dots, x_n'(t)) \qquad \text{and} \qquad x''(t) = (x_1''(t), \dots, x_n''(t))$$

by any two states associated with these measurements. Then by observability $x' \equiv x''$. However, we shall show that $x_q'(\tau) \not\equiv x_q''(\tau)$ at some time τ and for some $q < k$ follows from the condition $n_{kj} \neq 0$, which is a contradiction.

Since the states $x_k, \dots, x_n$ are determined, we have

$$\Delta x_j \triangleq x_j' - x_j'' = 0 \qquad \text{for } k \leqslant j \leqslant n.$$

Hence, using transformed coordinates, it follows from (4.3.6), with A replaced by PAP^{-1}, etc., that

$$\Delta \dot{x}_j = \Delta x_{j-1} + u \sum_{i=1}^{k-1} n_{ji} \Delta x_i, \qquad 1 \leqslant j \leqslant k. \tag{4.3.8}$$

For $j = k$ this gives

$$(1 + u n_{k,k-1}) \Delta x_{k-1} + u \sum_{i=1}^{k-2} n_{ki} \Delta x_i = 0. \tag{4.3.9}$$

For $j = k - 1$ (4.3.8) implies

$$\Delta \dot{x}_{k-1} = \Delta x_{k-2} + u \sum_{i=1}^{k-1} n_{k-1,i} \Delta x_i. \tag{4.3.10}$$

However, by (4.3.9),

$$\Delta x_{k-1} = -u \sum_{i=1}^{k-1} n_{ki} \Delta x_i$$

and so

$$\Delta \dot{x}_{k-1} = -\frac{\mathrm{d}}{\mathrm{d}t} \left(u \sum_{i=1}^{k-1} n_{ki} \Delta x_i \right). \tag{4.3.11}$$

Combining (4.3.10) and (4.3.11) gives

$$(1 + u n_{k-1,k-2} - \dot{u} n_{k,k-2}) \Delta x_{k-2} + \sum_{\substack{i=1 \\ (i \neq k-2)}}^{k-1} (u n_{ki} - u n_{k-1,i}) \Delta x_i = 0. \tag{4.3.12}$$

Similarly for $j = k - 2, \ldots, 1$ we obtain, together with (4.3.9) and (4.3.12), k equations

$$R\begin{bmatrix}\Delta x_1 \\ \vdots \\ \Delta x_k\end{bmatrix} = 0$$

where $R = (r_{ij})$ is a $k \times k$ matrix whose elements are sums of terms of the form $n_{kj}u^{(i)}$. The element r_{lq}, in particular, contains the additive term $n_{kq}u^{(l)}$ for $q < k$ and no terms with higher u derivatives. Hence if $n_{kq} \neq 0$ we can choose u so that the qth column of R is zero, and reversing the argument we see that we can find two state functions $x'(t)$ and $x''(t)$ such that $\Delta x_q(\tau) = x_q'(\tau) - x_q''(\tau) \neq 0$ at some point τ which both satisfy the equation (4.3.6) and have the same output. □

Using this result Williamson (1977) obtains an observer for a bilinear system associated with a certain biological control problem. Theorem 4.3.6 is generalized to linear analytic systems by Gauthier and Bornard (1981).

4.3.3 Stabilizability of bilinear systems

In the classical theory of linear feedback systems the development of stabilizing controls is very important – very often an unstable plant will be stabilized by an inner loop before the overall controller is designed. In this section we shall therefore examine the existence of stabilizing feedback controls for a bilinear system of the form

$$\dot{x}(t) = Ax(t) + u(t)Bx(t), \quad x \in \mathbb{R}^n. \tag{4.3.13}$$

The first result, although simple, exemplifies the method which has been used very often in the study of the stabilizability of such systems.

Lemma 4.3.7
If A is dissipative in the sense that

$$\langle x, Ax\rangle (= x^{\mathrm{T}}Ax) \leqslant 0$$

for all $x \in \mathbb{R}^n$, then the system (4.3.12) is stabilizable, provided no nontrivial trajectory of the equation

$$\dot{x}(t) = Ax(t)$$

is contained in the set $\{x : \langle x, Bx\rangle = 0\}$.

Proof
Consider the function $V = \langle x, x\rangle > 0$ if $x \neq 0$. Then

$$\begin{aligned}\dot{V} &= 2\langle x, \dot{x}\rangle = 2\langle x, Ax\rangle + 2u\langle x, Bx\rangle \\ &\leqslant -2(\langle x, Bx\rangle)^2\end{aligned}$$

if we choose the control $u = -\langle x, Bx\rangle$. Hence V is a Lyapunov function and the system with this control is stable. The result now follows from LaSalle's invariance principle (Banks (1986a)) since the set on which V is zero contains no complete trajectory of the system. □

Next, following Slemrod (1978), we shall consider the stabilizability of the system (4.3.13) when A satisfies the following condition:

(C) There exists a symmetric, positive definite matrix Q such that

$$QA + A^{\mathrm{T}}Q = 0.$$

By expanding in a series the exponential e^{At} we see that this is equivalent to the Q-norm invariance of the system $\dot{x} = Ax$, i.e. the trajectory $x(t) = e^{At}x_0$ satisfies

$$\langle x(t), Qx(t)\rangle \triangleq \| x(t)\|_Q^2 = \| x_0 \|_Q^2, \qquad t \geqslant 0.$$

Theorem 4.3.8

Let the control u be bounded; $u_1 \leqslant u \leqslant u_2, u_1 < 0, u_2 > 0$. If A satisfies condition (C) and, moreover, there exists a nonempty set $\Omega \subseteq \mathbb{R}^n \backslash \{0\}$ such that

(a) for each $x \in \Omega$,

$$\operatorname{span}\{Ax, Bx, (\operatorname{ad} A)Bx, (\operatorname{ad} A)^2Bx, \ldots, (\operatorname{ad} A)^K Bx\} = \mathbb{R}^n$$

for some integer K.

(b) $\{0\}$ is the only subset of $\Omega^c = \mathbb{R}^n \backslash \Omega$ which is invariant under e^{At}, $-\infty < t < \infty$

then the system (4.3.13) is stabilizable.

Proof

Let $V(x)$ be the function given by

$$V(x) = \tfrac{1}{2}\langle x, Qx\rangle.$$

Then,

$$\begin{aligned}\dot{V} &= \tfrac{1}{2}\langle \dot{x}, Qx\rangle + \tfrac{1}{2}\langle x, Q\dot{x}\rangle \\ &= \tfrac{1}{2}\langle Ax, Qx\rangle + \tfrac{1}{2}\langle x, QAx\rangle + u(t)\langle x, QBx\rangle \\ &= u(t)\langle x, QBx\rangle,\end{aligned}$$

by condition (C). Choosing the control u as the feedback

$$u(x) = \begin{cases} u_1 & \text{if } -\langle x, QBx\rangle < u_1 \\ -\langle x, QBx\rangle & \text{if } u_1 \leqslant -\langle x, QBx\rangle \leqslant u_2 \\ u_2 & \text{if } -\langle x, QBx\rangle > u_2 \end{cases} \tag{4.3.14}$$

we have

$$\dot{V} \leqslant -u^2(x(t)) \leqslant 0,$$

and so the origin is stable. (Note that it is easy to show that equation (4.3.13) with the control (4.3.14) has a unique solution.) To complete the proof we again apply LaSalle's invariance principle which states that the solutions of (4.3.13) with the control (4.3.14) all converge, as $t \to \infty$, to the largest invariant subset S of the set

$$W = \{x \in \mathbb{R}^n : \langle x, QBx \rangle = 0\}.$$

Clearly, for $x_0 \in S$, since the corresponding trajectory $x(t; x_0)$ with control (4.3.14) is also in S (by definition of invariance), we have

$$\langle x(t), QBx(t) \rangle = 0, \qquad \dot{x}(t) = Ax(t), \qquad -\infty < t < \infty$$

i.e.

$$\langle e^{At}x_0, QBe^{At}x_0 \rangle = \langle x_0, Qe^{-At}Be^{At}x_0 \rangle = 0,$$

since

$$e^{A^T t}Q = Qe^{-At}$$

by condition (C). Using the Campbell–Hausdorff formula for $e^{-At}Be^{At}$, we have

$$\left\langle x_0, Q \sum_{k=0}^{\infty} \frac{(-1)^k (\text{ad } A)^k B t^k}{k!} x_0 \right\rangle = 0, \qquad -\infty < t < \infty.$$

Differentiating this with respect to t(at $t = 0$) successively gives

$$\langle x_0, Q(\text{ad } A)^k Bx_0 \rangle = 0, \qquad k = 0, 1, \ldots.$$

Also, by condition (C),

$$\langle x_0, QAx_0 \rangle = \langle A^T Qx_0, x_0 \rangle = -\langle x_0, QAx_0 \rangle = 0.$$

Hence we have shown that the set S is given by

$$S = \{x \in \mathbb{R}^n : \langle x, QAx \rangle = 0, \ \langle x, Q(\text{ad } A)^k Bx \rangle = 0, \ k = 0, 1, \ldots\}$$

From the assumption (a) of the theorem it is clear that $S \cap \Omega = \emptyset$, and so $S \subseteq \Omega^c$. However, by definition, S is invariant under e^{At} and so the result follows from assumption (*b*), since $S = \{0\}$. □

This theorem generalizes a result of Jurdjevic and Quinn (1978). Also, as observed by Ryan and Buckingham (1983), since $S \subseteq \Omega^c$ and $S \subseteq W$, it follows that the condition (b) of the theorem can be replaced by the weaker condition

(b)′ $\{0\}$ is the only subset of $\Omega^c \cap W$ which is invariant under

e^{At}, $-\infty < t < \infty$, where

$$W = \{x \in \mathbb{R}^n : \langle x, QBx \rangle = 0\}.$$

Gutman (1981) obtains stabilizing controls even when the eigenvalues of A have positive real parts and Longchamp (1980) produces controls which determine a sliding mode condition. Finally, more general control laws are derived for bilinear systems by Banks and Morris (1986) which have applications in robotics.

4.4 OPTIMAL CONTROL OF BILINEAR SYSTEMS

4.4.1 Tensor products and the space of formal power series

The optimal control of a linear system

$$\dot{x} = Ax + Bu, \qquad x \in \mathbb{R}^n \tag{4.4.1}$$

subject to the quadratic cost criterion

$$J = \langle x(t_f), Gx(t_f) \rangle + \int_0^{t_f} \{\langle x, Mx \rangle + \langle u, Ru \rangle\} \, dt \tag{4.4.2}$$

is, as is well-known (Banks (1986a)), given by a linear feedback

$$u(t) = -R^{-1}B^T P(t)x(t), \tag{4.4.3}$$

where $P(t)$ satisfies the Riccati equation

$$\begin{aligned} \dot{P} &= -PA - A^{\mathrm{T}}P - M + PBR^{-1}B^{\mathrm{T}}P \\ P(t_f) &= G. \end{aligned} \tag{4.4.4}$$

We shall now generalize this result to the case of a bilinear system

$$\dot{x} = Ax + uBx, \qquad x \in \mathbb{R}^n, u \in \mathbb{R} \tag{4.4.5}$$

subject to the same quadratic cost (4.4.2); see Banks and Yew (1985). Here we use a scalar control for simplicity of exposition – the vector case can be dealt with similarly. It will turn out that we need to consider power series in the state x and so we shall briefly recall some properties of tensors and tensor operators (see also Chapter 1).

We denote the tensor product of i copies of $\mathbb{R}^n$ by $\bigotimes_i \mathbb{R}^n$, and make this space into an inner product space by defining

$$\langle x_1 \otimes \cdots \otimes x_i, y_1 \otimes \cdots \otimes y_i \rangle_{\bigotimes_i \mathbb{R}^n} = \prod_{j=1}^{i} \langle x_j, y_j \rangle = \prod_{j-1}^{i} x_j^T y_j$$

and extending by linearity. Let $\{e_k\}_{1 \leqslant k \leqslant n}$ denote the standard basis of

$\mathbb{R}^n$. Then $\{e_{k_l} \otimes \cdots \otimes e_{k_i}\}$, $1 \leqslant k_j \leqslant n$, $1 \leqslant j \leqslant i$, is an orthonormal basis of $\bigotimes_i \mathbb{R}^n$ and so any tensor $\Xi \in \bigotimes_i \mathbb{R}^n$ can be written in the form

$$\Xi = \sum_{k_1=1}^{n} \cdots \sum_{k_i=1}^{n} \xi_{k_1 \ldots k_i}(e_{k1} \otimes \cdots \otimes e_{k_i}).$$

Since $\bigotimes_i \mathbb{R}^n$ is a Hilbert space we can consider linear operators defined on $\bigotimes_i \mathbb{R}^n$. If $P \in \mathscr{L}(\bigotimes_i \mathbb{R}^n)$, where $\mathscr{L}(X)$ is the space of (bounded) linear operators defined on X, then P has the matrix representation, with respect to the above basis, given by

$$P(e_{k_1} \otimes \ldots \otimes e_{k_i}) = \sum_{\substack{l_j=1 \\ j=1,\ldots,i}}^{n} P_{k_1 \ldots k_i}^{l_1 \ldots l_i} (e_{l_1} \otimes \ldots \otimes e_{l_i}).$$

The adjoint P^* of $P \in \mathscr{L}(\bigotimes_i \mathbb{R}^n)$ is defined as usual by

$$\langle P^*(x_1 \otimes \cdots \otimes x_i), (y_1 \otimes \cdots \otimes y_i)\rangle = \langle (x_1 \otimes \cdots \otimes x_i),\ P(y_1 \otimes \cdots \otimes y_i)\rangle$$

for all $x_j, y_j \in \mathbb{R}^n$. Clearly, P is self-adjoint if

$$P_{k_1 \ldots k_i}^{l_1 \ldots l_i} = P_{l_1 \ldots l_i}^{k_1 \ldots k_i}$$

for all indices $l_1, \ldots, l_i, k_1, \ldots, k_i$. Then we say that P is *symmetric* and this generalizes the notion of a symmetric matrix.

We shall need the following result, which is easy to prove:

Lemma 4.4.1
Let $P \in \mathscr{L}(\bigotimes_i \mathbb{R}^n)$, $Q \in \mathscr{L}(\bigotimes_j \mathbb{R}^n)$. Then

$$[\mathscr{F}_x \langle \textstyle\bigotimes_i x, P \bigotimes_i x\rangle]\, x = 2i \langle \textstyle\bigotimes_i x, P \bigotimes_i x\rangle \tag{4.4.6}$$

if P is symmetric, where $\mathscr{F}_x$ denotes the Frechet derivative with respect to x. Moreover,

$$\langle \textstyle\bigotimes_i x, P \bigotimes_i x\rangle \langle \bigotimes_j x, Q \bigotimes_j x\rangle = \langle \bigotimes_{i+j} x, (P \otimes Q) \bigotimes_{i+j} x\rangle, \tag{4.4.7}$$

where

$$\textstyle\bigotimes_i x = x \otimes \cdots \otimes x \qquad (i \text{ times})$$

and $P \otimes Q$ is defined by

$$(P \otimes Q)(\xi \otimes \eta) = P\xi \otimes Q\eta, \qquad \xi \in \textstyle\bigotimes_i \mathbb{R}^n, \quad \eta \in \bigotimes_j \mathbb{R}^n.$$

Finally, we have

$$\| P \otimes Q \|_{\mathscr{L}(\otimes_{i+j}\mathbb{R}^n)} \leqslant \| P \|_{\mathscr{L}(\otimes_i \mathbb{R}^n)} \| Q \|_{\mathscr{L}(\otimes_j \mathbb{R}^n)}.$$ □

A simple generalization of (4.4.6) shows that

$$\mathscr{F}_x \langle \textstyle\bigotimes_i x, P \bigotimes_i x\rangle Cx = 2\langle \bigotimes_i x, (PC) \bigotimes_i x\rangle \tag{4.4.8}$$

where $C \in \mathscr{L}(\mathbb{R}^n)$ is (represented by) a matrix C_{ij} and PC is the tensor defined by

$$(PC)^{\varkappa_1 \dots \varkappa_i}_{k_1 \dots k_i} = \sum_{j=1}^{i} \left(\sum_{k_j'=1}^{n} P^{\varkappa_1 \dots \varkappa_j \dots \varkappa_i}_{k_1 \dots k_j' \dots k_i} C_{k_j' k_j} \right). \tag{4.4.9}$$

Hence we have

$$\begin{aligned} \| (PC) \bigotimes_i x \|_{\otimes_i \mathbb{R}^n} &= \| P(Cx \otimes x \otimes \dots \otimes x) + P(x \otimes Cx \otimes \cdots \otimes x) \\ &\quad + P(x \otimes \cdots \otimes Cx) \|_{\otimes_i \mathbb{R}^n} \\ &\leqslant i \| P \|_{\mathscr{L}(\otimes_i \mathbb{R}^n)} \| Cx \|_{\mathbb{R}^n} \| x \|^{i-1}_{\mathbb{R}^n}, \end{aligned}$$

and so

$$\| PC \|_{\mathscr{L}(\otimes_i \mathbb{R}^n)} \leqslant i \| P \|_{\mathscr{L}(\otimes_i \mathbb{R}^n)} \| C \|_{\mathscr{L}(\mathbb{R}^n)}. \tag{4.4.10}$$

4.4.2 The formal solution for the optimal control

We shall now consider the optimal control problem (4.4.5) and (4.4.2) in the ring of formal power series $\mathbb{R}^e[[x]]$ in the indeterminate vector $x = (x_1, \dots, x_n)$ which have only even order powers. Thus $\phi \in \mathbb{R}^e[[x]]$ if and only if we can write

$$\phi = \sum_{i=1}^{\infty} \langle \bigotimes_i x, \phi_i \bigotimes_i x \rangle_{\otimes_i \mathbb{R}^n}$$

for some tensor operators $\phi_i \in \mathscr{L}(\bigotimes_i \mathbb{R}^n)$. Convergence of the formal power series solution for the optimal control will be considered in the next section.

The problem (4.4.5) and (4.4.2) will be solved by dynamic programming (see Banks(1986a)). Thus, let $V(t, x)$ be the usual value function. Then the dynamic programming equation for V is

$$\langle x, Mx \rangle + V_t + (\mathscr{F}_x V)Ax + \min_u \, (Ru^2 + (\mathscr{F}_x V)Bxu) = 0 \tag{4.4.11}$$

(Recall that u is a scalar control, so $R \in \mathbb{R}$.) If $c = (\mathscr{F}_x V)Bx$, then

$$Ru^2 + cu = (u + \tfrac{1}{2}R^{-1}c)^2 R - \tfrac{1}{4}c^2 R^{-1}$$

and so the minimum in (4.4.11) is attained when $u = -\frac{1}{2}R^{-1}c$. Thus (4.4.11) becomes

$$V_t + \langle x, Mx \rangle + (\mathscr{F}_x V)Ax - \tfrac{1}{4}\langle (\mathscr{F}_x V)Bx, R^{-1}(\mathscr{F}_x V)Bx \rangle = 0. \tag{4.4.12}$$

To solve this partial differential equation we suppose that, for each t, $V(t, x) \in R^e[[x]]$, i.e.

$$V(t, x) = \sum_{i=1}^{\infty} \langle \bigotimes_i x, P_i(t) \bigotimes_i x \rangle \tag{4.4.13}$$

where $P_i(.) \in C^1[0, t_f; \mathscr{L}(\bigotimes_i \mathbb{R}^n)]$. Thus, substituting (4.4.13) into (4.4.12) we obtain

$$\sum_{i=1}^{\infty} \langle \bigotimes_i x, \dot{P}_i(t) \bigotimes_i x \rangle_{\otimes_i \mathbb{R}^n} + \sum_{i=1}^{\infty} 2 \langle \bigotimes_i x, (P_i A) \bigotimes_i x \rangle_{\otimes_i \mathbb{R}^n} + \langle x, Mx \rangle_{\mathbb{R}^n}$$

$$- \sum_{i=1}^{\infty} \sum_{j=1}^{\infty} R^{-1} \langle \bigotimes_{i+j} x, (P_i B \otimes P_j B) \bigotimes_{i+j} x \rangle_{\otimes_{i+j} \mathbb{R}^n} = 0. \quad (4.4.14)$$

Since

$$V(t, x) = \langle x, Gx \rangle + \int_t^{t_f} \{ \langle x, Mx \rangle + Ru^2 \} \, dt$$

we have

$$V(t_f, x) = \langle x, Gx \rangle = \sum_{i=1}^{\infty} \langle \bigotimes_i x, P_i(t_f) \bigotimes_i x \rangle$$

and so the final conditions for the equations (4.4.14) are

$$\begin{aligned} P_1(t_f) &= G \\ P_i(t_f) &= 0, \qquad i > 1. \end{aligned} \quad (4.4.15)$$

It is now clear why we have not included a zeroth order term or odd order terms in (4.4.13). In fact, it follows from (4.4.14) and (4.4.15) that these terms are identically zero.

Now, equating like powers in (4.4.14) leads to the equations

$$\dot{P}_1(t) + P_1(t)A + A^{\mathrm{T}} P_1(t) + M = 0, \qquad P_1(t_f) = G$$

$$(4.4.16)$$

$$\dot{P}_m(t) + P_m(t)A + A^{\mathrm{T}} P_m(t) - R^{-1} \sum_{\substack{i+j=m \\ i,j \geqslant 1}} \{ P_i B \otimes P_j B \} = 0, \qquad P_m(t_f) = 0,$$

for $m > 1$. Here, $P_m(t)A$ and $A^{\mathrm{T}} P_m(t)$ are defined in a similar way to PC as in (4.4.8). From ($4.4.16_1$) it follows that $P_1(t)$ is symmetric (assuming, of course, that G is symmetric). Hence, by induction, we can write ($4.4.16_m$) in the form

$$\dot{P}_m(t) + P_m(t)A + A^{\mathrm{T}} P_m(t) - \frac{1}{2R} \left\{ \sum_{\substack{i+j=m \\ i,j \geqslant 1}} (B^{\mathrm{T}} P_i \otimes B^{\mathrm{T}} P_j + P_i B \otimes P_j B) \right\} = 0$$

and so $P_m(t)$ is symmetric.

Consider the operators $\mathscr{A}_i$ defined on the Banach space $\mathscr{L}(\bigotimes_i \mathbb{R}^n)$ by

$$\mathscr{A}_i P_i = P_i A, \quad P_i \in \mathscr{L}(\bigotimes_i \mathbb{R}^n), \qquad i \geqslant 1.$$

Then $\mathscr{A}_i$ is clearly a bounded operator and $\| \mathscr{A}_i P_i \| \leqslant i \| P_i \| . \| A \|$ by

(4.4.10). Hence

$$\| \mathscr{A}_i \|_{\mathscr{L}(\mathscr{L}(\bigotimes_i \mathbb{R}^n))} \leqslant i \| A \|_{\mathscr{L}(\mathbb{R}^n)}$$

and so

$$\| e^{\mathscr{A}_i t} \| \leqslant e^{i \| A \| t}. \tag{4.4.17}$$

Thefore we see that the operator $e^{\mathscr{A}_i t} \in \mathscr{L}(\mathscr{L}(\bigotimes_i \mathbb{R}^n))$ is well defined and the solution of ($4.4.16_1$) is given by

$$P_1(t) = e^{\mathscr{A}_1(t_f - t)} G e^{\mathscr{A}_1^*(t_f - t)} + \int_0^{t_f - t} e^{\mathscr{A}_1(t_f - t - s)} M e^{\mathscr{A}_1^*(t_f - t - s)} \, ds \qquad (4.4.18_1)$$

Similarly, from ($4.4.16_m$), we have

$$P_m(t) = -R^{-1} \sum_{\substack{i+j=m \\ i,j \geqslant 1}} \int_0^{t_f - t} e^{\mathscr{A}_m(t_f - t - s)} P_i(t_f - s) B \otimes P_j(t_f - s) B e^{\mathscr{A}_m^*(t_f - t - s)} \, ds. \qquad (4.4.18_m)$$

The optimal control is then given formally by

$$u(t) = -\tfrac{1}{2} R^{-1} (\mathscr{F}_x V) B x = -R^{-1} \sum_{i=1}^{\infty} \langle \otimes_i x, (P_i B) \otimes_i x \rangle. \tag{4.4.19}$$

In the following section we shall consider the convergence of this formal power series.

4.4.3 Convergence of the formal series

Next we shall obtain an expression for the radius of convergence of the power series in (4.4.19) and thus we shall be able to determine for which values of the state this feedback control is well defined. To obtain such an expression we first estimate $P_1(t)$ from ($4.4.18_1$) and (4.4.17) as follows:

$$\begin{aligned} \| P_1(t) \| &\leqslant e^{2\| A \|(t_f - t)} \| G \| + (t_f - t) \| M \| . \sup_{t \in [0, t_f]} e^{2 \| A \| (t_f - t)} \\ &\leqslant (\| G \| + t_f \| M \|) \max\{1, e^{2 \| A \| t_f}\} \\ &\triangleq \alpha. \end{aligned} \tag{4.4.20}$$

Similarly, from ($4.4.18_m$) we have

$$\| P_m(t) \| \leqslant R^{-1} \| B \|^2 \sum_{i+j=m} \int_0^{t_f - t} e^{2 \| A \| m (t_f - t - s)} ij \| P_i(t_f - s) \| . \| P_j(t_f - s) \| \, ds.$$

Define

$$p_m(t) = m e^{2m \| A \| t} \alpha^{-m} (R^{-1} \| B \|^2)^{-m+1} \| P_m(t) \|, \qquad m \geqslant 2. \tag{4.4.21}$$

Then, it follows that

$$p_m(t) \leqslant m \sum_{i+j=m} \int_0^{t_f - t} p_i(t_f - s) p_j(t_f - s)\, ds. \qquad (4.4.22)$$

Lemma 4.4.2
If $p_m(t)$ satisfies (4.4.22), then

$$q_m \leqslant \frac{m}{m-1} \sum_{i+j=m} q_i q_j,$$

where

$$q_k = p_k(t)(t_f - t)^{-k+1}, \qquad k \geqslant 1. \qquad (4.4.23)$$

Proof
By (4.4.22) we have

$$q_m(t_f - t)^{m-1} \leqslant m \sum_{i+j=m} \int_0^{t_f - t} q_i q_j s^{i-1} s^{j-1}\, ds$$

$$= \frac{m}{m-1} \sum_{i+j=m} (t_f - t)^{m-1} q_i q_j. \qquad \square$$

It follows from Lemma 4.4.2 that

$$q_m \leqslant \sum_{i+j=m} q_i q_j, \qquad q_1 = 2.$$

Consider the corresponding equality

$$r_m = \sum_{i+j=m} r_i r_j, \qquad r_1 = 2. \qquad (4.4.24)$$

Then, clearly, $q_m \leqslant r_m$ for all $m \geqslant 1$.

Lemma 4.4.3
The power series $\sum_{i=1}^{\infty} r_i z^{2i}$ has radius of convergence $1/2\sqrt{2}$, where r_i is given by (4.4.24).

Proof
Consider the formal power series $R(z) = \sum_{i=1}^{\infty} r_i z^{2i}$. It is easy to see that the coefficients of this formal series satisfy (4.4.24) if and only if

$$r_1 z^2 + R^2(z) = R(z). \qquad (4.4.25)$$

Hence the formal power series is convergent to an analytic function in some region if and only if the equation (4.4.25) has an analytic solution $R(z)$.

However, (4.4.25) implies that $R(z)$ must satisfy

$$R(z) = \frac{1 \pm (1 - 8z^2)^{1/2}}{2}$$

each branch of which is analytic inside the disc $\{z : |z| < 1/2\sqrt{2}\}$. □

From (4.4.21) and (4.4.23) we have

$$\| P_m(t) \| = \frac{1}{m} e^{-2\|A\| mt} \alpha^m (R^{-1} \|B\|^2)^{m-1} p_m(t)$$

$$= \frac{1}{m} e^{-2\|A\| mt} \alpha^m (R^{-1} \|B\|^2)^{m-1} (t_f - t)^{m-1} q_m$$

and so

$$(\| P_m(t) \|)^{1/2m}$$

$$= \left(\frac{1}{m}\right)^{1/2m} e^{-\|A\| t} \alpha^{\frac{1}{2}} (R^{-1} \|B\|^2)^{\frac{1}{2} - 1/2m} (t_f - t)^{\frac{1}{2} - 1/2m} (q_m)^{1/2m}$$

$$\leqslant \left(\frac{1}{m}\right)^{1/2m} e^{-\|A\| t} \alpha^{\frac{1}{2}} R^{-\frac{1}{2}} \|B\| (t_f - t)^{\frac{1}{2}}. \, 2\sqrt{2}$$

$$\to 2\sqrt{2} e^{-\|A\| t} \left(\frac{\alpha}{R}\right)^{\frac{1}{2}} \|B\| (t_f - t)^{\frac{1}{2}}$$

as $m \to \infty$. Hence, the optimal control (4.4.19) exists as a power series in the state, provided

$$\| x \| < \frac{1}{2\sqrt{2}} \frac{e^{\|A\| t}}{\|B\|} \left(\frac{R}{\alpha (t_f - t)}\right)^{\frac{1}{2}}.$$

We can then prove

Theorem 4.4.4

The bilinear–quadratic regulator problem

$$\dot{x} = Ax + uBx, \qquad x(0) = x_0 \tag{4.4.26}$$

$$J = \langle x(t_f), Gx(t_f) \rangle + \int_0^{t_f} \{ \langle x, Mx \rangle + Ru^2 \} \, dt \tag{4.4.27}$$

has the optimal solution

$$u(t) = -R^{-1} \sum_{i=1}^{\infty} \langle \bigotimes_i x, (P_i B) \bigotimes_i x \rangle \tag{4.4.28}$$

where

$$\dot{P}_1(t) = -P_1(t)A - A^*P_1(t) - M, \qquad P_1(t_f) = G$$

and

$$\dot{P}_m(t) = -P_m(t)A - A^*P_m(t) + R^{-1} \sum_{\substack{i+j=m \\ i,j \geqslant 1}} \{P_iB \otimes P_jB\}, \quad P_m(t_f) = 0, \qquad m \geqslant 2$$

provided

$$\| x_0 \| < \frac{1}{2\sqrt{2}} \frac{1}{\| B \|} \left(\frac{R}{\alpha t_f}\right)^{\frac{1}{2}}. \tag{4.4.29}$$

Proof
We must show that the control u given by (4.4.28) is well defined on $[0, t_f]$, i.e. that the solution of (4.4.26) with this control exists and satisfies

$$\| x(t) \| < \frac{1}{2\sqrt{2}} \frac{e^{\| A \| t}}{\| B \|} \left(\frac{R}{\alpha(t_f - t)}\right)^{\frac{1}{2}} \tag{4.4.30}$$

on this interval. However, this follows from the fact that the optimal cost is

$$J(x_0) = \sum_{i=1}^{\infty} \langle \bigotimes_i x_0, P_i \bigotimes_i x_0 \rangle$$

which exists by (4.4.29), and so the optimal cost from any time t is

$$\sum_{i=1}^{\infty} \langle \bigotimes_i x(t), P_i \bigotimes_i x(t) \rangle \leqslant J(x_0).$$

A simple argument now shows that (4.4.30) holds. □

From Theorem 4.4.4 we see that the optimal control for the bilinear–quadratic regulator problem exists (as a power series) for initial conditions which satisfy (4.4.30), unlike the linear–quadratic regulator problem where the optimal control exists for all initial values. For large initial states we must therefore choose the 'horizon' time t_f to be small; more precisely, we must have

$$t_f < \frac{1}{2\sqrt{2}} \frac{R}{\alpha \| x_0 \|^2 \| B \|}.$$

This is similar to the receding horizon extension of linear–quadratic control where the horizon time is chosen as a function of the state (see Banks, (1986a)). The optimal control of bilinear systems has also been discussed by Banks and Yew (1986) and Tzafestas *et al.* (1984).

4.5 FREQUENCY DOMAIN REPRESENTATION – MULTIDIMENSIONAL LAPLACE TRANSFORMS

4.5.1 The transfer function of a homogeneous system

The importance of the Laplace (or complex frequency) domain approach to linear systems theory is, of course, well known, and is the basis of all 'classical' control and much of the modern multivariable control theory. Any attempt to generalize the notion of frequency response to arbitrary nonlinear systems is confounded by the very requirement of linearity – which is needed in order that a transfer function may be defined. However, each term in a Volterra series is multilinear in u and so we can generalize the single-variable Laplace transform to a multidimensional version.

To be precise we shall derive a transfer function for a stationary homogeneous nonlinear system of degree m, described by the mth term in a Volterra series; see also Mitzel *et al.* (1979). Such a system may be written in the form

$$y(t) = \int_0^\infty \cdots \int_0^\infty w_{\text{tri}}(t_1, \ldots, t_m)u(t-t_1)\cdots u(t-t_m)\,\mathrm{d}t_1\cdots\mathrm{d}t_m, \qquad (4.5.1)$$

where we have added the subscript 'tri' to indicate explicitly that w is defined on the triangular domain $t_1 \geqslant t_2 \geqslant \cdots \geqslant t_m \geqslant 0$. We have seen above that we may also write y in the form

$$y(t) = \int_0^\infty \cdots \int_0^\infty w_{\text{sym}}(t_1, \ldots, t_m)u(t-t_1)\cdots u(t-t_m)\,\mathrm{d}t_1\cdots\mathrm{d}t_m$$

where w_{tri} and w_{sym} are related by

$$w_{\text{tri}}(t_1, \ldots, t_m) = m!\,w_{\text{sym}}(t_1, \ldots, t_m)S(t_1-t_2)S(t_2-t_3)\cdots S(t_{m-1}-t_m),$$

where S is the unit step function. Another convenient representation of y is via the regular kernel w_{reg} given by

$$y(t) = \int_0^\infty \cdots \int_0^\infty w_{\text{reg}}(t_1, \ldots, t_m)u(t-t_1-\cdots-t_m)u(t-t_2-\cdots-t_m)\cdots u(t-t_m)\,\mathrm{d}t_m\cdots\mathrm{d}t_1 \qquad (4.5.2)$$

with

$$w_{\text{reg}}(t_1, \ldots, t_m) = w_{\text{tri}}(t_1+\cdots+t_m, t_2+\cdots+t_m, \ldots, t_{m-1}+t_m, t_m), \quad t_1, \ldots, t_m \geqslant 0 \qquad (4.5.3)$$

$$w_{\text{tri}}(t_1, \ldots, t_m) = w_{\text{reg}}(t_1-t_2, t_2-t_3, \ldots, t_{m-1}-t_m, t_m), \quad t_1 \geqslant \cdots \geqslant t_m \geqslant 0. \qquad (4.5.4)$$

Since the integral on the right of (4.5.1) is an m-dimensional convolution, it makes sense to generalize the linear system transfer function to the *transfer function* of a homogeneous (time-invariant) mth-order system by defining

$$W_{\text{tri}}(s_1, \ldots, s_m) = \int_0^\infty \cdots \int_0^\infty w_{\text{tri}}(t_1, \ldots, t_m) e^{-(s_1 t_1 + \cdots + s_m t_m)} \, dt_1 \cdots dt_m.$$

Similar definitions can be given for $W_{\text{reg}}(s_1, \ldots, s_m)$ and $W_{\text{sym}}(s_1, \ldots, s_m)$. From (4.5.3) and (4.5.4) we have

$$\begin{aligned} W_{\text{reg}}(s_1, \ldots, s_m) &= W_{\text{tri}}(s_1, s_2 - s_1, \ldots, s_m - s_{m-1}) \\ W_{\text{tri}}(s_1, \ldots, s_m) &= W_{\text{reg}}(s_1, s_1 + s_2, \ldots, s_1 + \cdots + s_m). \end{aligned} \tag{4.5.5}$$

As in the single variable case, we say that a rational function

$$R(s_1, \ldots, s_m) = P(s_1, \ldots, s_m)/Q(s_1, \ldots, s_m)$$

is *strictly proper* if the degree of P in s_j is less than that of Q for $1 \leqslant j \leqslant m$. We call R *reduced* if there are no nontrivial common factors between P and Q and *recognizable* if Q can be decomposed in the form

$$Q(s_1, \ldots, s_m) = \prod_{i=1}^{m} Q_i(s_i).$$

4.5.2 Realization theory

We shall be interested in deriving a frequency domain realization theory of a homogeneous system corresponding to the state-space realization theory of bilinear systems given earlier. We have seen that a triangular kernel has a bilinear realization of dimension n if and only if

$$h_{\text{tri}}(t_1, \ldots, t_m) = c e^{A t_m} N e^{A(t_{m-1} - t_m)} N \cdots N e^{A(t_1 - t_2)} b, \qquad t_1 \geqslant t_2 \geqslant \cdots \geqslant t_m \geqslant 0,$$

where c, A, N and b are real matrices of respective dimensions $1 \times n, n \times n, n \times n$ and $n \times 1$. The bilinear realization is given by

$$\begin{aligned} \dot{x}(t) &= Ax(t) + Nx(t)u(t) + bu(t), \qquad x(0) = 0. \\ y(t) &= cx(t). \end{aligned} \tag{4.5.6}$$

Theorem 4.5.1

A homogeneous degree m system has a finite-dimensional bilinear realization if and only if the reduced regular transfer function is a strictly proper, recognizable function.

Proof
If the system has such a realization, then the mth degree regular kernel is given by

$$w_{\text{reg}}(t_1, \dots, t_m) = ce^{At_m}Ne^{At_{m-1}}N \cdots Ne^{At_1}b, \qquad t_1, \dots, t_m \geqslant 0.$$

Hence,

$$W_{\text{reg}}(s_1, \dots, s_m) = c(s_mI - A)^{-1}N(s_{m-1}I - A)^{-1}N \cdots N(s_1I - A)^{-1}b, \tag{4.5.7}$$

which is strictly proper and recognizable.

Conversely, assume that the reduced transfer function W_{reg} is recognizable and strictly proper and so

$$W_{\text{reg}}(s_1, \dots, s_m) = \frac{P(s_1, \dots, s_m)}{Q_1(s_1) \dots Q_m(s_m)}$$

where P and the Q_i may be written in the forms

$$P(s_1, \dots, s_m) = \sum_{i_1=0}^{l_1-1} \cdots \sum_{i_m=0}^{l_m-1} p_{i_1 \dots i_m} s_1^{i_1} \cdots s_m^{i_m}$$

$$Q_k(s_k) = s_k^{l_k} + \sum_{i_k=0}^{l_k-1} q_{k,i_k} s_k^{i_k}, \qquad k = 1, \dots, m$$

As in the time domain, to obtain a realization we require a factorization of W_{reg} as a product of terms in each variable s_k separately. Hence we write

$$P(s_1, \dots, s_m) = S_mS_{m-1} \dots S_1P \tag{4.5.8}$$

where S_k is a matrix which contains only terms in s_k and P is a vector consisting of the coefficients $p_{i_1 \dots i_m}$ of the polynomial $P(s_1, \dots, s_m)$ arranged antilexicographically, i.e.

$$P^{\mathsf{T}} = (p_{0,\dots,0}, p_{1,0,\dots,0}, p_{l_1-1,0,\dots,0}, p_{0,1,0,\dots,0}, p_{1,1,0,\dots,0}, \dots,$$
$$p_{l_1-1,1,0,\dots,0}, \dots, p_{0,l_2-1,\dots,l_m-1}, \dots, p_{l_1-1,l_2-1,\dots,l_m-1}).$$

It is easy to check that we may choose S_k to be given inductively by $S_m = (1, s_m, \dots, s_m^{l_m-1})$, and S_k is the matrix of dimension $(l_m \cdots l_{k+1})x(l_m \cdots l_k)$ with ith row

$$(0_{1x(il_k-l_k)} \qquad (1, s_k, \dots, s_k^{l_k-1}) \qquad 0_{1x(l_m \dots l_k - il_k)}),$$

for $1 \leqslant i \leqslant l_n \cdots l_{k+1}$. Hence, if we define

$$G_1(s_1) = \frac{S_1P}{Q_1(s_1)}, \qquad G_k(s_k) = \frac{S_k}{Q_k(s_k)}, \quad k = 2, \dots, m$$

then we have

$$W_{\text{reg}}(s_1, \dots, s_m) = G_m(s_m)G_{m-1}(s_{m-1}) \cdots G_1(s_1). \tag{4.5.9}$$

Since W_{reg} is strictly proper, each $G_k(s_k)$ has a linear realization

$$G_k(s_k) = C_k(s_k I - A_k)^{-1}B_k.$$

Define the matrices

$$A = \begin{bmatrix} A_1 & 0 & \dots & 0 \\ 0 & A_2 & \dots & 0 \\ \vdots & \vdots & & \vdots \\ 0 & 0 & \dots & A_m \end{bmatrix}, \quad b = \begin{bmatrix} B_1 \\ 0 \\ \vdots \\ 0 \end{bmatrix}$$

$$N = \begin{bmatrix} 0 & 0 & \dots & 0 & 0 \\ B_2C_1 & 0 & \dots & 0 & 0 \\ 0 & B_3C_2 & \dots & 0 & 0 \\ \vdots & \vdots & & \vdots & \vdots \\ 0 & 0 & \dots & B_mC_{m-1} & 0 \end{bmatrix}, \qquad c = [0, \dots, 0, C_m]. \quad (4.5.10)$$

Then it is easy to check that all regular transfer functions of the form (4.5.7) with degree $\neq m$ are zero and

$$\begin{aligned} W_{\text{reg}}(s_1, \dots, s_m) &= C_m(s_m I - A_m)^{-1}B_mC_{m-1}(s_{m-1}I - A_{m-1})^{-1} \\ &\qquad B_{m-1}C_{m-2}\dots B_2C_1(s_1 I - A_1)^{-1}B_1 \\ &= G_m(s_m)\cdots G_1(s_1). \end{aligned}$$
□

The main drawback with the factorization (4.5.9) is that the subsystems $G_k(s_k)$ may be of very high dimension. Hence we shall seek another method of factorization which leads to one which is 'minimal'. To be precise about the meaning of minimal here we first define, as in the linear case, the *order* M_k of $G_k(s_k)$ to be the least common denominator of all the minors of $G_k(s_k)$ in its reduced form. Then we call $M = M_1 + \dots + M_m$ the *order* of the factorization (4.5.9). In order to obtain minimal order factorizations, we need to consider full rank factorizations of a matrix.

Lemma 4.5.2
Let A be an $m \times n$ matrix of rank $r > 0$. Then we can write $A = BC$ where B and C have full rank r.

Proof
Let B be any matrix of order $m \times r$ of full rank, and let $b_1, \dots, b_r$ denote the columns of B. Then we wish to find a matrix C such that $A = BC$. If $a_1, \dots, a_n$ are the columns of A, then we must have

$$a_j = \sum_{i=1}^{r} c_{ij} b_i$$

which specifies C uniquely since the columns $b_1, \dots, b_r$ are linearly independent. Since the rank of B equals r, C must also be of full rank. Hence given any matrix B of rank r we can find a matrix C of full rank r such that $A = BC$. To determine C note that the matrix $(B^{\mathrm{T}}B)^{-1}B^{\mathrm{T}} = B^{\dagger}$ (the generalized inverse of B) acts as a left inverse on B and so it is necessary that

$$B^{\dagger}A = B^{\dagger}BC = C. \qquad \square$$

Hence any $m \times n$ matrix A of rank r has the full-rank factorization

$$A = B(B^{\dagger}A)$$

for any $m \times r$ matrix B of full rank. Similarly, choosing any $r \times n$ matrix C of rank r, we have

$$A = (A({}^{\dagger}C))C$$

where

$${}^{\dagger}C = ((C^{\mathrm{T}})^{\dagger})^{\mathrm{T}}$$

We also require the notion of block transpose of a matrix. If W is an $N \times (mM)$ matrix partitioned in the form (w_{ij}) where w_{ij} is a $1 \times m$ row vector, then we define the *right block transpose* of W to be the $(mN) \times M$ matrix

$$W^{RT(m)} = (w_{ij}^{\mathrm{T}}).$$

Similarly, for an $(nN) \times M$ matrix V of the form (v_{ij}), where v_{ij} is an $n \times 1$ column vector, we define the *left block transpose* of V to be the $N \times (nM)$ matrix

$$V^{LT(n)} = (v_{ij}^{\mathrm{T}}).$$

If P is the column vector in (4.5.8) we define inductively

$$\begin{aligned} &X_1 = P^{LT(l_1)}, \quad V_2 = L_{X_1}, \quad W_1 = R_{X_1} \quad (X_1 = L_{X_1}R_{X_1}) \\ &X_i = V_i^{LT(l_i)}, \quad V_{i+1} = L_{X_i}, \quad W_i = R_{X_i} \quad (X_i = L_{X_i}R_{X_i}), \qquad 2 \leqslant i \leqslant n-1 \end{aligned}$$

for any left–right full-rank factorization $L_{X_i}R_{X_i}$ of X_i. Then (4.5.8) becomes

$$\begin{aligned} S_m S_{m-1} \cdots S_2 S_1 P &= S_m S_{m-1} \cdots S_3 S_2 P^{LT(l_1)} S_1 \\ &= S_m S_{m-1} \cdots S_3 S_2 V_2 W_1 \tilde{S}_1 \\ &= S_m S_{m-1} \cdots S_3 V_2^{LT(l_2)} \tilde{S}_2 W_1 \tilde{S}_1 \\ &= S_m S_{m-1} \cdots S_3 V_3 W_2 \tilde{S}_2 W_1 \tilde{S}_1 \\ &\;\;\vdots \\ &= S_m V_m W_{m-1} \tilde{S}_{m-1} W_{m-2} \tilde{S}_{m-2} \cdots W_1 \tilde{S}_1 \end{aligned}$$

where $\tilde{S}_1 = (1, s_1, \ldots, s_1^{l_1 - 1})^{\mathrm{T}}$ and $\tilde{S}_j$ is the $(r_{j-1}l_j) \times r_{j-1}$ matrix with ith column

$$(0_{1 \times (il_j - l_j)}, (1, s_j, \ldots, s_j^{l_j - 1}), 0_{1 \times (r_{j-1}l_j - il_j)})^{\mathrm{T}}$$

where $r_j = \text{rank } W_j$. Hence we have the factorization

$$H_{\text{reg}}(s_1, \ldots, s_m) = \frac{S_m V_m}{Q_m(s_m)} \frac{W_{m-1}\tilde{S}_{m-1}}{Q_{m-1}(s_{m-1})} \cdots \frac{W_1 \tilde{S}_1}{Q_1(s_1)}$$

$$\triangleq G_m^0(s_m) G_{m-1}^0(s_{m-1}) \cdots G_1^0(s_1). \tag{4.5.11}$$

The next result shows that this factorization is of minimal order.

Theorem 4.5.3

If $H_{\text{reg}}(s_1, \ldots, s_m)$ is a reduced, strictly proper, recognizable, regular transfer function, then the factorization (4.5.11) is minimal.

Proof

Given any factorization

$$H_{\text{reg}}(s_1, \ldots, s_m) = \hat{G}_m(s_m) \hat{G}_{m-1}(s_{m-1}) \cdots \hat{G}_1(s_1) \tag{4.5.12}$$

of the form (4.5.11) we shall show that there exist matrices T_i, $1 \leqslant i \leqslant m-1$, and $\hat{T}_i$, $2 \leqslant i \leqslant m$, such that

$$\begin{aligned} G_1^0(s_1) &= T_1 \hat{G}_1(s_1), \qquad G_m^0(s_m) = \hat{G}_m(s_m) \hat{T}_m, \\ G_j^0(s_j) &= T_j \hat{G}_j(s_j) \hat{T}_j, \qquad 2 \leqslant j \leqslant m-1. \end{aligned} \tag{4.5.13}$$

A standard result of linear systems theory (see Kalman (1965)) then shows that

$$M_j^0 \leqslant \hat{M}_j, \qquad 1 \leqslant j \leqslant m,$$

i.e.

$$M^0 \leqslant \hat{M}$$

where M_j^0 is the order of G_j^0 and $\hat{M}_j$ is the order of $\hat{G}_j$.

Since the factorization (4.5.12) is of the form of that in (4.5.11) we have

$$\hat{G}_m(s_m) = \frac{S_m \hat{V}_m}{Q_m(s_m)}, \qquad \hat{G}_1(s_1) = \frac{\hat{W}_1 \tilde{S}_1}{Q_1(s_1)}$$

$$\hat{G}_j(s_j) = \frac{\hat{W}_j \hat{S}_j}{Q_j(s_j)}, \qquad 2 \leqslant j \leqslant m-1$$

where $\hat{S}_j$ is of the same form as S_j but with possibly different dimension. Since (4.5.11) and (4.5.12) are equal factorizations we have

$$S_m \cdots S_{j+1} V_{j+1} W_j \tilde{S}_j W_{j-1} \tilde{S}_{j-1} \cdots W_1 \tilde{S}_1$$
$$= S_m \hat{V}_m \hat{W}_{m-1} \hat{S}_{m-1} \hat{W}_{m-2} \hat{S}_{m-2} \cdots \hat{W}_1 \tilde{S}_1.$$

$\hat{V}_m \hat{W}_{m-1} \hat{S}_{m-1}$ is a polynomial matrix of degree l_{m-1} and so we can write

$$\hat{V}_m \hat{W}_{m-1} \hat{S}_{m-1} = S_{m-1}(\hat{V}_m \hat{W}_{m-1})^{RT(l_{m-1})}.$$

Thus

$$\begin{aligned} S_m \cdots S_{j+1} V_{j+1} W_j \tilde{S}_j W_{j-1} \tilde{S}_{j-1} \cdots W_1 \tilde{S}_1 &\\ &= S_m S_{m-1} (\hat{V}_m \hat{W}_{m-1})^{RT(l_{m-1})} \hat{W}_{m-2} \hat{S}_{m-2} \cdots \hat{W}_1 \tilde{S}_1 \\ &\ \vdots \\ &= S_m S_{m-1} \ldots S_{j+1} \bar{V}_{j+1} \hat{W}_j \hat{S}_j \cdots \hat{W}_1 \hat{S}_1, \end{aligned}$$

where

$$\bar{V}_{j+1} = (\ldots((\hat{V}_m \hat{W}_{m-1})^{RT(l_{m-1})} \hat{W}_{m-2})^{RT(l_{m-2})} \ldots \hat{W}_{j+1})^{RT(l_{j+1})}.$$

From the structure of S_i it follows that

$$V_{j+1} W_j \tilde{S}_j W_{j-1} \tilde{S}_{j-1} \cdots \tilde{S}_2 W_1 = \bar{V}_{j+1} \hat{W}_j \hat{S}_j \cdots \hat{W}_2 \hat{S}_2 \hat{W}_1.$$

Hence

$$W_j \tilde{S}_j W_{j-1} \tilde{S}_{j-1} \cdots W_2 \tilde{S}_2 = T_j \hat{W}_j \hat{S}_j \hat{W}_{j-1} \cdots \hat{W}_2 \hat{S}_2 \bar{W}_1,$$

where

$$T_j = (V_{j+1})^{\dagger} \bar{V}_{j+1}, \qquad \bar{W}_1 = \hat{W}_1({}^{\dagger}W_1).$$

But $\hat{S}_2 \bar{W}_1 = \bar{W}_1{}^{LT(l_2)} \tilde{S}_2$ and so

$$W_j \tilde{S}_j W_{j-1} \tilde{S}_{j-1} \cdots W_2 = T_j \hat{W}_j \hat{S}_j \cdots \hat{W}_2 \bar{W}_1{}^{LT(l_2)}.$$

Continuing in this way we obtain

$$W_j \tilde{S}_j W_{j-1} = T_j \tilde{W}_j \tilde{S}_j \hat{W}_{j-1} \bar{W}_{j-2}^{LT(l_{j-1})}$$

for an appropriate matrix $\bar{W}_{j-2}$ and so

$$W_j \tilde{S}_j = T_j \hat{W}_j \hat{S}_j \hat{T}_j, \qquad 2 \leqslant j \leqslant m-1,$$

where

$$\hat{T}_j = \hat{W}_{j-1} \bar{W}_{j-2}^{LT(l_{j-1})} ({}^{\dagger}W_{j-1}).$$

Similarly,

$$W_1 \tilde{S}_1 = T_1 \hat{W}_1 \tilde{S}_1, \qquad S_m V_m = S_m \hat{V}_m \hat{T}_m$$

for some matrices T_1, $\hat{T}_m$ and (4.5.13) follows. □

Since (4.5.11) specifies a minimal factorization, it is reasonable to choose a minimal linear realization for each single variable transfer function $G^0(s_j)$ and then construct a bilinear realization as in (4.5.10). The next result shows that such a bilinear realization is, in fact, minimal.

Theorem 4.5.4

If $H_{\text{reg}}(s_1, \ldots, s_m)$ is a reduced, strictly proper, recognizable regular transfer function with a minimal factorization (4.5.11) of order M^0 then $H_{\text{reg}}(s_1, \ldots, s_m)$ has a minimal bilinear realization of dimension M^0.

Proof

A bilinear realization of dimension M^0 can be constructed in the manner just described. Suppose there exists a bilinear realization of dimension $M < M^0$, given by

$$\dot{x}(t) = Ax(t) + Nx(t)u(t) + bu(t)$$
$$y(t) = cx(t).$$

Define the matrices

$$R_1 = [b \quad Ab \quad \ldots \quad A^i b \quad \ldots]$$
$$R_2 = [Nb \quad ANb \quad NAb \quad \ldots \quad A^i NA^j b \quad \ldots]$$
$$R_3 = [N^2 b \quad AN^2 b \quad NANb \quad N^2 Ab \quad \ldots \quad A^i NA^j NA^k b \quad \ldots]$$
$$\vdots$$

and put

$$M_1 = \text{rank } R_1$$

$$M_j = \text{rank}\,[R_1, \ldots, R_j] - \sum_{i=1}^{j-1} M_i, \qquad j = 2, \ldots, m-1$$

and

$$M_m = M - \sum_{i=1}^{m-1} M_j.$$

Introduce the mutually independent subspaces L_i of $\mathbb{R}^M$ by

$$L_1 = \text{column span } [R_1]$$
$$L_j = \{\text{column span}[R_j]\} \cap L_{j-1}^{\perp} \cdots \cap L_1^{\perp}, \qquad j = 2, \ldots, m-1$$
$$\vdots$$
$$L_m = \mathbb{R}^M \cap L_{m-1}^{\perp} \cdots \cap L_1^{\perp},$$

and let $\{\alpha_1, \ldots, \alpha_{M_1}\}, \{\alpha_{M_1+1}, \ldots, \alpha_{M_1+M_2}\}, \ldots$ be bases for the spaces $L_1, L_2, \ldots$.

Then $\mathbb{R}^M = \bigoplus_{i=1}^{m} L_i$ and

$$\gamma \in L_j \Rightarrow A\gamma \in L_1 \oplus \cdots \oplus L_j, \qquad j \leqslant m$$
$$\gamma \in L_j \Rightarrow N\gamma \in L_1 \oplus \cdots \oplus L_{j+1}, \quad j < m.$$

Hence, changing coordinates to $z = [\alpha_1 \quad \dots \quad \alpha_m]x$ gives the bilinear system

$$z(t) = \begin{bmatrix} A_{11} & A_{12} & \dots & A_{1m} \\ 0 & A_{22} & \dots & A_{2m} \\ \vdots & \vdots & & \vdots \\ 0 & 0 & \dots & A_{mm} \end{bmatrix} z(t) + \begin{bmatrix} N_{11} & N_{12} & \dots & N_{1m-1} & N_{1m} \\ N_{21} & N_{22} & \dots & N_{2m-1} & N_{2m} \\ 0 & N_{32} & \dots & N_{3m-1} & N_{3m} \\ \vdots & \vdots & & \vdots & \vdots \\ 0 & 0 & \dots & N_{mm-1} & N_{mm} \end{bmatrix} z(t)u(t)$$

$$+ \begin{bmatrix} b_1 \\ 0 \\ \vdots \\ 0 \end{bmatrix} u(t)$$

$$y(t) = [c_1 \dots c_m] z(t),$$

with matrices A_{ij}, N_{ij} of appropriate sizes. If we compute the triangular kernels of this realization then we see that we must choose

$$A_{ij} = 0, \quad i < j, \quad N_{ij} = 0, \quad i \leqslant j, \quad c_i = 0, \qquad i < m$$

in order that only the kernel of degree m is nonzero. Moreover it is easy to see that this choice does not alter the input–output map.

Finally taking full rank factorizations $N_{j,j-1} = B_j C_{j-1}$, $2 \leqslant j \leqslant m$ gives an M-dimensional realization. However this leads to an Mth order factorization of H_{reg} and this contradicts the minimality of M^0. □

The case of multivariable homogeneous input-output maps with a bilinear realization is discussed by Evans, 1983.

4.5.3 Input–output response

We have considered above a homogeneous Volterra series of the form

$$y(t) = \int_0^\infty \cdots \int_0^\infty h(\tau_1, \dots, \tau_m) u(t - \tau_1) \cdots u(t - \tau_m) \mathrm{d}\tau_1 \cdots \mathrm{d}\tau_m \qquad (4.5.14)$$

and have defined its transfer function $H(s_1, \dots, s_m)$ as the m-variable Laplace transform of h. In the case of the linear system

$$y(t) = \int_0^\infty h(\tau_1) u(t - \tau) \,\mathrm{d}\tau$$

we have the familiar relation

$$Y(s) = H(s)U(s).$$

In order to generalize this to systems of the form (4.5.14) we introduce the function $\bar{y}(t_1, \ldots, t_m)$ defined by

$$\bar{y}(t_1, \ldots, t_m) = \int_0^\infty \cdots \int_0^\infty h(\tau_1, \ldots, \tau_m) u(t_1 - \tau_1) \cdots u(t_m - \tau_m) \, d\tau_1 \cdots d\tau_m.$$

Then, taking m-variable Laplace transforms we have

$$\bar{Y}(s_1, \ldots, s_m) = H(s_1, \ldots, s_m) U(s_1) \ldots U(s_m).$$

Lemma 4.5.5

The following relation holds between $Y(s)$ and $\bar{Y}(s_1, \ldots, s_m)$:

$$Y(s) = \frac{1}{(2\pi i)^{m-1}} \int_{\sigma_1 - i\infty}^{\sigma_1 + i\infty} \cdots \int_{\sigma_{m-1} - i\infty}^{\sigma_{m-1} + i\infty} \bar{Y}(s - w_1 - \cdots - w_{m-1}, w_1, w_2, \ldots, w_{m-1}) \, dw_{m-1} \cdots dw_1.$$

Proof

We shall give a brief heuristic 'proof' – a rigorous proof is simple but tedious. First note that

$$\bar{Y}(s - w_1 - \cdots - w_{m-1}, w_1, w_2, \ldots, w_{m-1}) = \int_0^\infty \cdots \int_0^\infty \int_0^\infty \cdots \int_0^\infty h(\tau_1, \ldots, \tau_m) .$$
$$u(t_1 - \tau_1) \cdots u(t_m - \tau_m) \, d\tau_1 \cdots d\tau_m e^{-st_1 + w_1(t_1 - t_2) + w_2(t_1 - t_3) + \cdots + w_{m-1}(t_1 - t_m)} \, dt_1 \cdots dt_m.$$

However, if we recall the representation

$$\delta(t_1 - t_2) = \frac{1}{2\pi i} \int_{\sigma - i\infty}^{\sigma + i\infty} e^{w(t_1 - t_2)} \, dw$$

and assume that we may interchange the integrals we have

$$\frac{1}{(2\pi i)^{m-1}} \int_{\sigma_1 - i\infty}^{\sigma_1 + i\infty} \cdots \int_{\sigma_{m-1} - i\infty}^{\sigma_{m-1} + i\infty} \bar{Y}(s - w_1 - \cdots - w_{m-1}, w_1, w_2, \ldots, w_{m-1}) \, dw_{m-1} \cdots dw_1$$

$$= \int_0^\infty \cdots \int_0^\infty \int_0^\infty \cdots \int_0^\infty h(\tau_1, \ldots, \tau_m) u(t_1 - \tau_1) \cdots u(t_m - \tau_m) \, d\tau_1 \cdots d\tau_m$$
$$. \delta(t_1 - t_2) \cdots \delta(t_1 - t_m) e^{-st_1 + w_1(t_1 - t_2) + \cdots + w_{m-1}(t_1 - t_m)} \, dt_1 \cdots dt_m$$

$$= \int_0^\infty \int_0^\infty \cdots \int_0^\infty h(\tau_1, \ldots, \tau_m) u(t - \tau_1) \cdots u(t - \tau_m) \, d\tau_1 \cdots d\tau_m \, dt$$

$$= Y(s).$$

□

REMARK We say that $Y(s)$ is obtained from $\bar{Y}(s_1, \ldots, s_m)$ by *association of variables* and is valid for the triangular or symmetric kernel. Note, of course, that

$$y(t) = \bar{y}(t, \ldots, t).$$

Since step and sinusoidal responses are so important in linear systems theory, it is interesting to generalize these notions to bilinear systems with homogeneous input–output maps.

Theorem 4.5.6

If $H_{\text{reg}}(s_1, \ldots, s_m)$ is a strictly proper, recognizable, regular transfer function of a homogeneous system, then if the input to the system is

$$U(s) = \sum_{i=1}^{r} \frac{a_i}{s + \gamma_i},$$

we have

$$Y(s) = \sum_{i_1=1}^{r} \cdots \sum_{i_{m-1}=1}^{r} a_{i_1} \cdots a_{i_{m-1}} H_{\text{reg}}(s + \gamma_{i_1} + \cdots + \gamma_{i_{m-1}}, s + \gamma_{i_2} + \cdots + \gamma_{i_{m-1}}, \ldots, s + \gamma_{i_{m-1}}, s) \,.\, U(s + \gamma_{i_1} + \cdots + \gamma_{i_{m-1}}).$$

In particular, if $U(s) = 1/s$ (i.e. a unit step input) then

$$Y(s) = H_{\text{reg}}(s, s, \ldots, s)/s.$$

Proof

By Lemma 4.5.5 and (4.5.5) we have

$$Y(s) = \frac{1}{(2\pi \mathrm{i})^{m-1}} \int_{\sigma_1 - \mathrm{i}\infty}^{\sigma_1 + \mathrm{i}\infty} \cdots \int_{\sigma_{m-1} - \mathrm{i}\infty}^{\sigma_{m-1} + \mathrm{i}\infty} H_{\text{reg}}\,(s - w_1 - \cdots - w_{m-1}, s - w_2 - \cdots - w_{m-1}, \ldots s - w_{m-1}, s) \,.\, U(s - w_1 - \cdots - w_{m-1}) U(w_1) \cdots U(w_{m-1})\, \mathrm{d}w_{m-1} \cdots \mathrm{d}w_1.$$

By (4.5.11) we have

$$Y(s) = \frac{1}{(2\pi \mathrm{i})^{m-1}} G_m^0(s) \int_{\sigma_1 - \mathrm{i}\infty}^{\sigma_1 + \mathrm{i}\infty} \cdots \int_{\sigma_{m-1} - \mathrm{i}\infty}^{\sigma_{m-1} + \mathrm{i}\infty} G_{m-1}^0(s - w_{m-1}) \cdots G_2^0(s - w_2 - \cdots - w_{m-1}) G_1^0(s - w_1 - \cdots - w_{m-1}) \,.\, U(s - w_1 - \cdots - w_{m-1}) U(w_1) \cdots U(w_{m-1})\, \mathrm{d}w_{m-1} \cdots \mathrm{d}w_1$$

and so by the residue theorem

$$Y(s) = \sum_{i_1=1}^{r} a_{i_1} \frac{1}{(2\pi \mathrm{i})^{m-2}} \int_{\sigma_1 - \mathrm{i}\infty}^{\sigma_1 + \mathrm{i}\infty} \cdots \int_{\sigma_{m-2} - \mathrm{i}\infty}^{\sigma_{m-2} + \mathrm{i}\infty} H_{\text{reg}}(s - w_1 - \cdots - w_{m-2} + \gamma_{i_1}, s - w_2 - \cdots - w_{m-2} + \gamma_{i_1}, \ldots, s) U(s - w_1 - \cdots - w_{m-2} + \gamma_{i_1}) U(w_1) \cdots U(w_{m-2})\, \mathrm{d}w_{m-2} \cdots \mathrm{d}w_1.$$

Repeating this calculation gives the result. □

Further applications of this method of multidimensional Laplace transforms are given in Crum and Heinen (1974), Lubbock (1969) and Tang *et al.* (1983). In the final section of this chapter we shall consider nonrecognizable transfer functions and describe a spectral theory for nonlinear homogeneous systems.

4.6 A SPECTRAL THEORY FOR NONLINEAR SYSTEMS

4.6.1 Inversion of the multidimensional Laplace transform

We have considered above the theory of homogeneous recognizable systems, i.e. those with a transfer function of the form

$$H(s_1, \ldots, s_m) = \frac{P(s_1, \ldots, s_m)}{Q_1(s_1) \cdots Q_m(s_m)}$$

which has an mth order Volterra series realization. In the last section of this chapter we shall consider systems of this type which are not necessarily recognizable. In particular, we shall discuss two aspects of such systems which generalize the corresponding notions for the linear case; namely the inversion of the multidimensional Laplace transform and the root locus (see Banks (1985a)). To simplify the notation transfer functions which depend on only two variables s_1, s_2 will be considered – much of what follows can be generalized to mth order systems.

Let

$$H(s_1, s_2) = \frac{P(s_1, s_2)}{Q(s_1, s_2)} \tag{4.6.1}$$

be the transfer function of a homogeneous second degree system. Although H is not in general recognizable the denominator polynomial $Q(s_1, s_2)$ can be factorized in the form

$$Q(s_1, s_2) = Q_1(s_1, s_2) \cdots Q_k(s_1, s_2), \tag{4.6.2}$$

where each $Q_i(s_1, s_2)$ is irreducible. The simplest method of inverting a rational function of a single variable s is via partial fraction expansion and so it is interesting to generalize this method to rational functions of several variables. For simplicity we assume that the Q_i in (4.6.2) are distinct although the general case can be developed without difficulty (see Banks (1986c)).

Let us first note that the situation is completely different for two or more variables from the single variable case. For it is well known that a rational function

$$H(s) = \frac{p(s)}{q(s)} = \frac{p(s)}{\sum_{i=1}^{m} (s - s_i)},$$

where $\{s_i : 1 \leqslant i \leqslant m\}$ is the set of (distinct) zeros of the polynomial $q(s)$, may be written in the form

$$H(s) = \sum_{i=1}^{m} \frac{a_i}{s - s_i}$$

for some (complex) constants a_i. This is no longer true for transfer functions of the form (4.6.1); for example, we have

$$\frac{1}{s_1^2 - s_2^2} \neq \frac{a_1}{s_1 - s_2} + \frac{a_2}{s_1 + s_2} \tag{4.6.3}$$

even if we allow the a_i to be polynomials. Thus, (4.6.3) holds for all $a_1, a_2 \in \mathbb{C}[s_1, s_2]$ and so the problem is more difficult in this case.

Lemma 4.6.1
If

$$r(s_1, s_2) = \frac{p(s_1, s_2)}{q_1(s_1, s_2) \cdots q_m(s_1, s_2)}$$

where each q_i is irreducible and p and $q_1, \ldots, q_m$ are relatively prime, then we can write

$$r(s_1, s_2) = \sum_{i=1}^{m} \frac{p_i(s_1, s_2)}{q_i(s_1, s_2)} \tag{4.6.4}$$

for some polynomials p_i, $1 \leqslant i \leqslant m$, if and only if

$$(p) \subseteq (q_1') + \cdots + (q_m')$$

where

$$q_i' = \prod_{j \neq i} q_j.$$

(Recall that (p) is the principal ideal generated by p.)

Proof
If $p \in (p) \subseteq (q_1') + \cdots + (q_m')$ then there must exist $p_i \in \mathbb{C}[s_1, s_2]$ such that

$$p = \sum_{i=1}^{m} \left(\prod_{j \neq i} q_j \right) p_i,$$

and so the sufficiency part is true.

For the necessity of the condition, suppose that r can be written in the form of (4.6.4). Then if $t(s_1, s_2) \in (p)$ we have $t = pt_1$ for some polynomial

t_1 and so

$$t = \sum_{i=1}^{m} q_i' \, . \, t_1 p_i \in (q_1') + \cdots + (q_m'). \qquad \square$$

Corollary 4.6.2

Under the assumptions of Lemma 4.6.1, in order that r may be written in the form (4.6.4), it is necessary that‡

$$V(q_i) \cap V(q_j) \subseteq V(p) \qquad \text{for all } i \neq j$$

for any distinct pair of polynomials q_i in the denominator of r.

Moreover, it is sufficient that

$$V(q_i) \cap V(q_j) \subseteq V(p_1) \qquad \text{for all } i \neq j \tag{4.6.5}$$

where $p = p_1^l$ for a suitable polynomial p_1 and positive integer l.

Proof

From Lemma 4.6.1 we have the necessary condition

$$(p) \subseteq (q_1') + \cdots + (q_m').$$

Hence

$$V\left(\sum_{j=1}^{m} (q_j')\right) \subseteq V(p)$$

and

$$V\left(\sum_{j=1}^{m} (q_j')\right) = \bigcap_{j=1}^{m} (V(q_j')) = \bigcap_{j=1}^{m} \left(\bigcup_{i \neq j} V(q_i)\right)$$

$$= \bigcup_{i \neq j} (V(q_i) \cap V(q_j)).$$

For sufficiency, note that

$$\bigcup_{i \neq j} (V(q_i) \cap V(q_j)) \subset V(p_1)$$

implies, as above, that

$$V\left(\sum_{j=1}^{m} (q_j')\right) \subseteq V(p_1).$$

Now for any ideals I_1 and I_2 we have

$$V(I_1) \subseteq V(I_2) \Rightarrow \sqrt{I_2} \subseteq \sqrt{I_1}$$

‡ See Chapter 1 for the definition of $V(p)$.

where we recall that the *radical* $\sqrt{I}$ of an ideal I is defined by $\sqrt{I} = \{\alpha \in \mathbb{C}[s_1, s_2] : \alpha^l \in I \text{ for some positive integer } l\}$.

Thus,

$$(p_1) \subseteq \sqrt{(p_1)} \subseteq \sqrt{\left(\sum_{j=1}^{m} (q_j')\right)}$$

and so

$$p_1 \in \sqrt{\left(\sum_{j=1}^{m} (q')\right)}$$

i.e.

$$p_1^l = p \in \sum_{j=1}^{m} (q_j') \qquad \text{for some } l. \qquad \square$$

It follows that we can write $r = p/(q_1 \cdots q_m)$ as a partial fraction expansion of the form (4.6.4) only if the variety defined by p contains the intersections of pairs of the affine varieties defined by the q_i. Since $V(p) = \emptyset$ if $p = 1$ we see now why (4.6.3) holds. For,

$$V(s_1 - s_2) \cap V(s_1 + s_2) = \{0\}.$$

The significance of the condition (4.6.5) in Corollary 4.6.2 can be understood by considering the example

$$r(s_1, s_2) = \frac{s_2}{(s_1 - s_2^2)(s_1 + s_2^2)}.$$

Note that

$$r(s_1, s_2) \neq \frac{a_1}{s_1 - s_2^2} + \frac{a_2}{s_1 + s_2^2}$$

for all $a_1, a_2 \in \mathbb{C}[s_1, s_2]$, even though $V(s_2) \supseteq V(s_1 - s_2^2) \cap V(s_1 + s_2^2)$. However, we can factorize

$$\frac{s_2^2}{(s_1 - s_2^2)(s_1 + s_2^2)} = \frac{1}{2}\left\{\frac{1}{s_1 - s_2^2}\right\} - \frac{1}{2}\left\{\frac{1}{s_1 + s_2^2}\right\}$$

and so Corollary 4.6.2 applies with $p_1 = s_2$, $p = s_2^2$.

If we can write a rational function $r(s_1, s_2)$ in the form (4.6.4) then we require a formula for the polynomials p_i. Let us first note that, unlike the single variable case, the p_i are not unique. For if

$$r = \frac{p}{q_1 q_2} = \frac{p_1}{q_1} + \frac{p_2}{q_2}$$

then

$$r = \frac{(p_1 + \bar{p}_1 q_1)}{q_1} + \frac{(p_2 - \bar{p}_1 q_2)}{q_2}$$

for any $\bar{p}_1 \in \mathbb{C}[s_1, s_2]$. However, if

$$r = \sum_{i=1}^{m} \frac{p_i}{q_i}, \qquad r = \sum_{i=1}^{m} \frac{\bar{p}_i}{q_i}$$

are two representations of r in partial fractions, then

$$\sum_{i=1}^{m} (p_i - \bar{p}_i) q_i' = 0. \tag{4.6.6}$$

We show that $p_i - \bar{p}_i$ is divisible by q_i. In fact, q_i occurs in all terms in the sum of (4.6.6) except the ith and the q_i are irreducible, proving the assertion. Hence

$$\{p_i - \bar{p}_i\} \in (q_i)$$

and so p_i is determined modulo the ideal (q_i). Thus, p_i is a uniquely determined element of the *coordinate ring* $\mathbb{C}[s_1, s_2]/(q_i)$ of the variety $V(q_i)$ which we denote by $\mathbb{C}_{q_i}[\bar{s}_1, \bar{s}_2]$. Note that

$$\bar{s}_1 = s_1 + (q_i), \qquad \bar{s}_2 = s_2 + (q_i).$$

If follows easily now that the polynomials p_i in (4.6.4) are given by

$$p_i(\bar{s}_1, \bar{s}_2) = \frac{p(\bar{s}_1, \bar{s}_2)}{\prod_{j \neq i = 1}^{m} \{q_j(\bar{s}_1, \bar{s}_2)\}}.$$

For example the function

$$H(s_1, s_2) = \frac{s_1}{(s_1 - s_2^2)(s_1 + 2s_2^2)}$$

clearly has a partial fraction expansion of the form

$$H(s_1, s_2) = \frac{p_1}{s_1 - s_2^2} + \frac{p_2}{s_1 + 2s_2^2}$$

where

$$p_1 = \frac{p}{s_1 + 2s_2^2}\bigg|_{s_1 = s_2^2} = \tfrac{1}{3}, \qquad p_2 = \frac{p}{s_1 - s_2^2}\bigg|_{s_1 = -2s_2^2} = \tfrac{2}{3}.$$

Using the above techniques we can reduce the problem of inverting the general rational function p/q to the problem of inverting the rational function p/q where q is irreducible. We can further reduce the problem to the inversion of $1/q$ where q is irreducible. For, if

$$p(s_1, s_2) = \Sigma a_{ij} s_1^i s_2^j$$

then

$$\mathscr{L}^{-1}_{s_1, s_2}\, p = \sum a_{ij} \delta^{(i)}(t_1) \delta^{(j)}(t_2) \triangleq F(t_1, t_2)$$

where $\delta^{(k)}$ denotes the kth derivative of the delta function. then

$$\mathscr{L}^{-1}_{s_1,s_2}\,(p\,.\,U(s_1)U(s_2)) = F(t_1,t_2)*_{t_1,t_2}(u(t_1)u(t_2)) \triangleq v(t_1,t_2)$$

where $*_{t_1,t_2}$ denotes the double (t_1,t_2)-convolution and u is the input to the system. The input–output relation

$$y(s_1,s_2) = \frac{p(s_1,s_2)}{q(s_1,s_2)}\,U(s_1)U(s_2)$$

is equivalent to the partial differential equation

$$\sum_{i=1}^{m}\sum_{j=1}^{n} b_{ij}\,\frac{\partial^{i+j}y}{\partial t_1^i \partial t_2^j} = v(t_1,t_2) \tag{4.6.7}$$

with zero 'initial' conditions

$$\begin{aligned} \frac{\partial^i y}{\partial t_1^i}(0,t_2) &= 0 \qquad \text{for } 0 \leqslant i \leqslant m-1 \\ \frac{\partial^j y}{\partial t_2^j}(t_1,0) &= 0 \qquad \text{for } 0 \leqslant j \leqslant n-1, \end{aligned} \tag{4.5.8}$$

where

$$q(s_1,s_2) = \sum_{i=1}^{m}\sum_{j=1}^{n} b_{ij}s_1^i s_2^j.$$

If $G(t_1,\xi_1;t_2,\xi_2)$ is the Green's function for the equations (4.6.7) and (4.6.8), i.e.

$$\sum_{i=1}^{m}\sum_{j=1}^{n} b_{ij}\,\frac{\partial^{i+j}}{\partial t_1^i \partial t_2^j}\,G(t_1,\xi_1;t_2,\xi_2) = \delta(t_1-\xi_1)\delta(t_2-\xi_2),$$

then we have

$$y(t_1,t_2) = \iint G(t_1,\xi_1;t_2,\xi_2)v(\xi_1,\xi_2)\,\mathrm{d}\xi_1\,\mathrm{d}\xi_2$$

and we have reduced the inversion problem to finding the Green's function for a differential operator with zero initial conditions and an irreducible characteristic polynomial.

For example, consider the system

$$y(s_1,s_2) = U(s_1)U(s_2)/(s_1 - s_2^2).$$

We must solve the equation

$$\frac{\partial G}{\partial t_1} - \frac{\partial^2 G}{\partial t_2^2} = \delta(t_1-\xi_1)\delta(t_2-\xi_2)$$

for G. It is well known that this equation has the solution

$$G(t_1;t_2) = (4\pi t_1)^{-1/2}\exp[-(4t_1)^{-1}t_2^2]$$

and we have

$$y(t_1, t_2) = G*_{t_1, t_2}u(t_1)u(t_2)$$

$$= \int_0^{t_1}\int_0^{t_2} G(t_1 - \xi_1; t_2 - \xi_2)u(\xi_1)u(\xi_2)\, d\xi_1\, d\xi_2$$

and so

$$y(t) = \int_0^{t}\int_0^{t} G(t - \xi_1; t - \xi_2)u(\xi_1)u(\xi_2)\, d\xi_1\, d\xi_2.$$

4.6.2 Poles and zeros of nonlinear systems

We have defined the transfer function $H(s_1, s_2)$ of a homogeneous degree 2 system in Section 4.5 and it was shown that H is a rational function of the variables s_1 and s_2. Our interest was mainly with recognizable functions H which lead to bilinear realizations. In this section we shall consider a general rational function of the form

$$H(s_1, s_2) = \frac{p(s_1, s_2)}{q(s_1, s_2)} = \frac{p_1^{i_1}(s_1, s_2) \cdots p_m^{i_m}(s_1, s_2)}{q_1^{j_1}(s_1, s_2) \cdots q_n^{j_n}(s_1, s_2)} \tag{4.6.9}$$

where each p_i and q_j is a polynomial in s_1 and s_2. By analogy with the case of linear systems it is natural to make the following definition. Let

$$Z_k = \overline{\{ (s_1, s_2) \in \mathbb{C}^2 : p_k(s_1, s_2) = 0 \}}, \qquad 1 \leqslant k \leqslant m$$

$$P_l = \overline{\{ (s_1, s_2) \in \mathbb{C}^2 : q_l(s_1, s_2) = 0 \}}, \qquad 1 \leqslant l \leqslant n,$$

where the bar denotes projective completion (i.e. closure in $\mathbb{P}^2(\mathbb{C})$). Then we say that $Z_k (1 \leqslant k \leqslant m)$ and $\mathbb{P}_l$ $(1 \leqslant l \leqslant n)$ are, respectively, the *open-loop zeros* and *poles* of the system (4.6.9) with multiplicities i_k and j_l. From Chapter 1 we see that the poles and zeros of a system are topologically irreducible algebraic varieties in $\mathbb{P}^2(\mathbb{C})$ (i.e. Riemann surfaces with finitely many points identified).

As in the linear case we must consider what happens to the poles and zeros at 'infinity'. Recall that 'infinity' in $\mathbb{P}^2(\mathbb{C})$ is a one-dimensional projective space (i.e. $\mathbb{P}^1(\mathbb{C})$) which we add to $\mathbb{C}^2$ to form the completion $\mathbb{P}^2(\mathbb{C})$. We shall determine the number of poles and zeros at infinity by counting the sphere coverings of $\mathbb{P}^1(\mathbb{C})$ generated by $p(s_1, s_2)$ and $q(s_1, s_2)$. As we have seen in Chapter 1, $\mathbb{P}^2(\mathbb{C})$ can be defined as the union of $\mathbb{C}^2$ and the set of points added to each complex 1-space of $\mathbb{C}^2$ to form a $\mathbb{P}^1(\mathbb{C})$, together with the obvious topology. Each 1-subspace of $\mathbb{C}^2$ is given by

$$s_1 - \alpha s_2 = 0, \qquad \alpha \in \mathbb{C}\backslash\{0\}$$

(apart from the s_1-axis) and so the projective 1-space $\mathbb{P}^1(\mathbb{C})$ at infinity in

$\mathbb{P}^2(\mathbb{C})$ is parameterized by $\alpha = s_1/s_2$, apart from the point $s_2 = 0$. It follows that a system of the form (4.6.9) behaves at infinity like the system

$$H'(\alpha, s_2) = \frac{p(\alpha s_2, s_2)}{q(\alpha s_2, s_2)} \triangleq \frac{p'(\alpha, s_2)}{q'(\alpha, s_2)}.$$

If $p_1', \ldots, p_m'$, and $q_1', \ldots, q_n'$, are the irreducible components of p' and q', respectively, then

$$H'(\alpha, s_2) = \frac{p_1'^{i_1'} \cdots p_m'^{i_m'}}{q_1'^{j_1'} \cdots q_n'^{j_n'}}$$

for some positive integers i_k', j_l'. Consider any factor, say p_1'. Projecting onto $\mathbb{C}_\alpha$ along $\mathbb{C}_{s_2}$ we obtain either a trivial point (if $p_1'(\alpha, s_2)$ is independent of s_2) or a sphere covering of $\mathbb{C}_\alpha$ which corresponds to a pole at infinity. The nature of this sphere covering determines the topology of the pole. (Note that the factors p_i' correspond to poles at infinity while the factors q_j' correspond to zeros. For example, if $H = 1/s_1$ then $H' = 1/\alpha s_2$ which gives a sphere covering of the s_2-axis and corresponds to the zero of H at $s_1 = \infty$.)

Consider, for example, the system

$$H(s_1, s_2) = \frac{s_1^2 + s_2}{s_1 s_2} \tag{4.6.10}$$

which has a zero and two poles, all of which are topologically spheres. Then,

$$H'(\alpha, s_2) = \frac{\alpha^2 s_2^2 + s_2}{\alpha s_2 \, . \, s_2} = \frac{(\alpha^2 s_2 + 1)}{\alpha s_2} \tag{4.6.11}$$

and $p'(\alpha, s_2) = \alpha^2 s_2 + 1$ is one-sheeted covering of $\mathbb{C}_\alpha$ which is a sphere, $q_1'(\alpha, s_2) = \alpha$ is trivial and does not represent a zero at infinity and $q_2'(\alpha, s_2) = s_2$ is again a sphere covering of $\mathbb{C}_\alpha$. Hence this system has a pole and a zero at infinity. Note that if we parameterize $\mathbb{P}^1(\mathbb{C})$ at infinity by $\beta = s_2/s_1$ instead of α we obtain

$$H''(s_1, \beta) = \frac{s_1^2 + \beta s_1}{\beta s_1^2} = \frac{s_1 + \beta}{\beta s_1}$$

giving the same result.

4.6.3 Root locus theory

Having defined the poles and zeros of a degree 2 homogeneous system it is natural now to attempt to extend the linear root locus theory to such systems. Let us first recall the basic properties of the root locus for the linear system

$$G(s) = \frac{p(s)}{q(s)} = \frac{\prod_{i=1}^{m} (s - z_i)}{\prod_{j=1}^{n} (s - p_i)}$$

where $n > m$. Then we are interested in the poles of the meromorphic function

$$G_k = \frac{kp}{q + kp}$$

for $k \in [0, \infty]$. Considering G_k on $\mathbb{P}^1(\mathbb{C})$ it is well known that the root locus (i.e. the locus of the roots of $q + kp$ as k varies from 0 to ∞) begins on the open-loop poles and ends on the open-loop zeros ($n - m$ of which are at ∞). The number of poles of a meromorphic function on $\mathbb{P}^1(\mathbb{C})$ is equal to the number of zeros (counting multiplicities) and there are, for almost all $k \in [0, \infty]$, n branches of the root locus. An example with $n = 4$ and $m = 2$ is shown in Fig. 4.1.

Consider now the example

$$H = \frac{s_1^2 + s_2}{s_1 s_2}$$

and the feedback system

$$\frac{kH(s_1, s_2)}{1 + kH(s_1, s_2)}$$

(such a feedback system can be realized, as shown in Banks (1985a)). The root locus is specified by the polynomial

$$s_1 s_2 + k(s_1^2 + s_2) = 0. \tag{4.6.12}$$

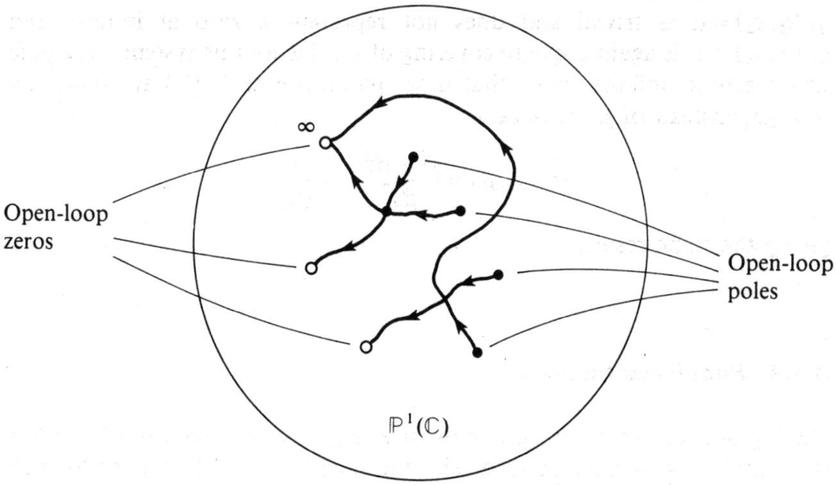

Fig. 4.1 Interpretation of the classical root locus on the Riemann sphere

As in the linear case, simple continuity arguments show that the root locus begins on open-loop poles and ends on open-loop zeros. Each of the polynomial factors in H, i.e. $p = s_1^2 + s_2$, $q_1 = s_1, q_2 = s_2$, is irreducible and defines a sphere in $\mathbb{P}^2(\mathbb{C})$. Moreover these three spheres intersect at 0 (the intersections of p and q_1 and p and q_2 being of multiplicity 2). The polynomial in (4.6.12) for $k > 0$ is irreducible and also defines a sphere in $\mathbb{P}^2(\mathbb{C})$ and so we see that the distinct poles $s_1 = 0$, $s_2 = 0$ at $k = 0$ coalesce to form a sphere as soon as $k > 0$ which remains irreducible for all $k > 0$. Hence the structure of the root locus is significantly different from the linear case. At '∞' in $\mathbb{P}^2(\mathbb{C})$ we have already seen that H has a pole and a zero which 'cancel' so that no branch of the root locus approaches the projective 1-subspace $\mathbb{P}^1(\mathbb{C})$ at infinity.

As a second example, consider the system

$$H(s_1, s_2) = \frac{s_1}{(s_1 + 1)s_2}.$$

This system has two poles and one zero in the finite (affine) space $\mathbb{C}^2$ and, putting $s_1 = \alpha s_2$, we see that

$$H'(\alpha, s_2) = \frac{\alpha s_2}{(\alpha s_2 + 1)s_2}$$

and so it has one zero at infinity. It is easy to show that the root locus starts on the two spheres given by

$$s_1 + 1 = 0, \qquad s_2 = 0$$

and tends to the zero $s_1 = 0$ and the sphere covering of $\mathbb{C}_{s_1}$ as $k \to \infty$.

In general we can now see that the root locus of degree 2 homogeneous systems is a finite collection of connected subsets of $\mathbb{P}^2(\mathbb{C})$ which, in an obvious sense, join open-loop poles to open-loop zeros, as in the linear case, but the number of branches may not equal the number of open-loop poles. let us note finally that much of what we have said above applies to higher-degree systems. However, the main problem in this case is that for $n > 2$ not all projective algebraic varieties of $\mathbb{P}^n(\mathbb{C})$ have been classified topologically. Indeed, a theorem of Markov (1958) shows that the problem of finding a complete set of topological invariants for such varieties is actually insoluble. Hence we will not know, generically, what the poles and zeros 'look like'.

5 DISTRIBUTED NONLINEAR SYSTEMS

5.1 THEORY OF INFINITE-DIMENSIONAL SEMILINEAR DIFFERENTIAL EQUATIONS

5.1.1 Introduction

In this last chapter we shall discuss generalizations of some of the results presented in the previous three chapters to certain infinite-dimensional systems defined by partial differential equations. The setting for the most general nonlinear systems considered above was a differentiable vector bundle over a differentiable manifold which was the state space of the system. We have seen that these systems have global bilinear realizations (of infinite dimension) and that a very wide class of nonlinear systems therefore have finite-dimensional bilinear approximations.

When we consider nonlinear distributed parameter systems, the most general setting should be a differentiable Hilbert (or Banach) bundle over a Hilbert (or Banach) manifold, as discussed by Lang (1972). However, there seems to be little published material on systems theory on such objects, although, of course, the field of global analysis has a large literature (see Palais (1968)). Hence, although some results on the control theory of linear distributed systems on manifolds have been developed (Banks (1984a)) we shall restrict attention in this chapter to semilinear systems of the form

$$\dot{x} = Ax + f(x, u), \tag{5.1.1}$$

where x belongs to a Banach space X, A is a linear operator (to be specified later) on X and f: $X \times U \to X$ is some nonlinear function. In particular, we shall be mainly interested in distributed bilinear systems.

5.1.2 Existence theory

In view of the above remarks we shall begin by discussing the existence theory of equations of the form (5.1.1.). More precisely, we begin by assuming that A is a C^0 semigroup (see Chapter 1) on X and that $f: X \to X$ is independent of u and is a C^k mapping for some $k \geqslant 1$. There are many results on the existence and uniqueness theory of such semilinear equations (Henry (1981) or Tanabe (1979)), but we shall follow Ball *et al.* (1982) since their results are applicable to bilinear systems.

Consider the 'mild form' of the equation

$$\dot{x} = Ax + p(t)B(x(t))$$

given by the integral equation

$$x(t) = T(t)x_0 + \int_0^t T(t-s)p(s)B(x(s))\,\mathrm{d}s, \tag{5.1.2}$$

where $T(t)$ is the semigroup generated by A, defined on the Banach space $(X, \|\,.\,\|)$. Let $B: X \to X$ be a C^k mapping and let $Z(T)$ be any Banach space which is dense and continuously included in $L^1([0, T]; \mathbb{R})$, for some $T > 0$.

Lemma 5.1.1

If $x_0 \in X$ and $p \in Z(T)$, then there exists $t_0 \in (0, T]$ such that (5.1.2) has a unique solution $x \in C([0, t_0]; X)$. Moreover, for any fixed $p_0 \in Z(T)$ the unique solution $x(t; p, x_0)$ exists and is a C^k map of some neighbourhood U of p_0 into $C([0, t_0]; X)$.

Proof

Let F_R be the set

$$F_R = \{x \in C([0, t_0]; X) : \sup \| x(t) - x_0 \| \leqslant R\}$$

and define the map $H_p: F_R \to C([0, t_0]; X)$ by

$$(H_p x)(t) = T(t)x_0 + \int_0^t T(t-s)p(s)B(x(s))\,\mathrm{d}s.$$

Since B is continuous, there exists a constant C and an $R > 0$ such that

$$\| Bx \| \leqslant C \qquad \text{for } \| x - x_0 \| \leqslant R.$$

Also, there exist constants $\omega, M \geqslant 0$ such that $\| T(t) \| \leqslant M\mathrm{e}^{\omega t}$ and so

$$\| (H_p x)(t) - x_0 \| \leqslant \| T(t)x_0 - x_0 \| + M\mathrm{e}^{\omega t_0} C \int_0^{t_0} | p(s) |\,\mathrm{d}s, \qquad 0 \leqslant t \leqslant t_0$$

for $x \in F_R$. Hence

$$\| (H_p x)(t) - x_0 \| \leqslant R$$

for sufficiently small t_0 since $p \in L^1([0, T]; \mathbb{R})$, which implies that H_p maps

F_R into F_R. Moreover, again by the continuity of B, we have

$$\| Bx - By \| \leqslant K \| x - y \|, \qquad \| x - x_0 \|, \quad \| y - y_0 \| \leqslant R$$

for some $K \leqslant 0$. Hence

$$\| (H_p x)(t) - (H_p y)(t) \| \leqslant KMe^{\omega t_0} \int_0^{t_0} | p(s) | \, ds \qquad \sup \| x(t) - y(t) \|$$
$$< \lambda \| x - y \|_{C([0, t_0]; X)}$$

where

$$\lambda = KMe^{\omega t_0} \int_0^{t_0} | p(s) | \, ds < 1$$

if t_0 is sufficiently small. It follows that H_p is a contraction mapping and so (5.1.2) has a unique solution.

If $p_0 \in Z(T)$ is given, choose λ to be independent of p for p in some neighbourhood of p_0. Then H_p is a uniform contraction and is a C^k map of (x, p) on the interior of F_R, and it follows that the limit of H_p^n is C^k as $n \to \infty$. □

Since the map $x(.) \to x(t_0)$ is continuous and linear from $C([0, t_0]; X)$ to X, it is smooth and hence, by the chain rule and the second part of Lemma 5.1.1, it follows that the map $x(t_0; ., x_0) : U \to X$ is C^k. Moreover, by standard arguments we see that the Fréchet derivative $D_p x(t; p_0, x_0)$ of $x(t; p, x_0)$ with respect to p at x_0 is given by

$$D_p x(t; p_0, x_0) . p = \int_0^t T(t-s) p(s) B(x(s; p_0, x_0)) \, ds$$
$$+ \int_0^t T(t-s) p_0(s) DB(x(s; p_0, x_0)) D_p x(s; p_0, x_0) p \, ds,$$

where $DB(x(s; p_0, x_0))$ is the Fréchet derivative of B at $x(s; p_0, x_0)$. Setting $p_0 = 0$ we have

$$D_p x(t; 0, x_0) p = \int_0^t T(t-s) p(s) B(T(s) x_0) \, ds. \qquad (5.1.3)$$

To obtain global solutions on $[0, T]$ we must impose some kind of growth condition on B. A convenient condition for us is a sublinear one, specified by

$$\| B(x) \| \leqslant C + K \| x \| \qquad (5.1.4)$$

for $x \in X$ and some constants C, K. Then we have

Theorem 5.1.2
If B is a C^k mapping which satisfies condition (5.1.4) and A generates a C^0

semigroup, then equation (5.1.2) has a unique solution in $C([0,T];X)$. This solution is a C^k function of $p \in Z(T)$ and $x_0 \in X$.

Proof
By Lemma 5.1.1, the equation (5.1.2) has a local solution on some interval $[0, t_0]$. Suppose that the solution can be extended to the interval $[0, a)$, where $0 \leqslant t < a \leqslant T$. Then

$$\| x(t) \| \leqslant Me^{\omega a}(\| x_0 \| + \int_0^t | p(s) |(C + K \| x(s) \|)\, ds).$$

$$\leqslant Me^{\omega a}(\| x_0 \| + C\int_0^T | p(s) |\, ds) + Me^{\omega a}K \int_0^t | p(s) | \, \| x(s) \| \, ds.$$

Hence, by Gronwall's inequality,

$$\| x(t) \| \leqslant Me^{\omega a}(\| x_0 \| + C \int_0^T | p(s) | \, ds) \exp\left(\int_0^t Me^{\omega a}K | p(s) | \, ds \right).$$

$$\leqslant C_1,$$

for some constant C_1. Hence, if $s, t \in [0, a)$, we have

$$\| x(t) - x(s) \| \leqslant \| T(t)x_0 - T(s)x_0 \| + \int_s^t T(t-\tau)p(\tau)B(x(\tau))\, d\tau$$

$$\leqslant \| T(t)x_0 - T(s)x_0 \| + Me^{\omega a}(C + KC_1) \int_s^t | p(\tau) | \, d\tau.$$

By the strong continuity of T and the fact that $p \in L^1([0,T];\mathbb{R})$ it follows that $\lim_{t \to a^-} x(t)$ exists and so the solution can be continued to a and solutions exist on $[0, T]$. Global uniqueness and C^k differentiability now follow by standard arguments. □

5.1.3 Weak solutions

It can be shown (Tanabe (1979) or Henry (1981)) that if $x_0 \in D(A)$ and $p \in C^1([0,T];\mathbb{R})$, then a mild solution of (5.1.1) (i.e. a solution of (5.1.2)) is also a strict (i.e. differentiable) solution of (5.1.1). We shall also need the concept of weak solution of a differential equation. Consider the more general equation

$$\dot{x}(t) = Ax(t) + f(x(t), t), \tag{5.1.5}$$

where A again generates a C^0 semigroup $T(t)$. Then we say that a function $x \in C([t_0, t_1]; X)$ is a *weak solution* of (5.1.5) on $[t_0, t_1]$ if $x(t_0) = x_0$, $f(x(.),.) \in L^1(t_0, t_1; X)$ and if for each $y \in D(A^*)$ the function $\langle x(t), y \rangle$ is absolutely continuous (i.e. continuous and differentiable almost everywhere) and satisfies the equation

$$\frac{d}{dt}\langle x(t), y\rangle = \langle x(t), A^*y\rangle + \langle f(x(t), t), y\rangle \tag{5.1.6}$$

for almost all $t \in [t_0, t_1]$.

To study weak solutions it is convenient to prove the following lemma.

Lemma 5.1.3
Suppose that A is a closed densely defined linear operator. If $x, y \in X$ satisfy $\langle y, v\rangle = \langle x, A^*v\rangle$ for all $v \in D(A^*)$, then $x \in D(A)$ and $y = Ax$.

Proof
Suppose that $x \notin D(A)$. Consider the graph $G(A)$ of A, which is a closed subset of $X \times X$. Since $(x, y) \notin G(A)$ it follows from the Hahn–Banach theorem that there exists $(v, w) \in (X \times X)^* = X^* \times X^*$ such that

$$\langle (z, Az), (v, w)\rangle = \langle z, v\rangle + \langle Az, w\rangle = 0 \qquad \text{for all } z \in D(A) \tag{5.1.7}$$

and

$$\langle (x, y), (v, w)\rangle = \langle x, v\rangle + \langle y, w\rangle \neq 0. \tag{5.1.8}$$

Then, by (5.1.7), $w \in D(A^*)$ and $v = -A^*w$ (by definition of A^*). Hence, by (5.1.8), $\langle y, w\rangle \neq \langle x, A^*w\rangle$ which is a contradiction. □

We next prove a uniqueness result for a linear system.

Lemma 5.1.4
The equation

$$\dot{x}(t) = Ax(t) + f(t), \qquad t \in (0, \tau], \quad x(0) = x_0 \in X \tag{5.1.9}$$

has at most one weak solution.

Proof
Let $x_1(t)$ and $x_2(t)$ be weak solutions with $x_i(0) = x_0, i = 1, 2$. Put $z = x_1 - x_2$. Then

$$\langle z(t), v\rangle = \left\langle \int_0^t z(s)\,ds, A^*v\right\rangle \qquad \text{for all } v \in D(A^*), \quad t \in [0, \tau].$$

Let $y(t) = \int_0^t z(s)\,ds$. By Lemma 5.1.3, $y(t) \in D(A)$ and $\dot{y}(t) = Ay(t)$. Since A generates a C^0 semigroup, $y(t) = 0$. □

Theorem 5.1.5 (Ball (1977))
The equation (5.1.9) has a unique weak solution if and only if A is the generator of a strongly continuous semigroup $T(t)$ of bounded operators on

X, in which case $x(t)$ is given by

$$x(t) = T(t)x_0 + \int_0^t T(t-s)f(s)\,\mathrm{d}s. \tag{5.1.10}$$

Proof

If A generates a strongly continuous semigroup $T(t)$, then $\| T(t) \| \leqslant M$, $t \in [0, \tau]$, for some $M \geqslant 0$. Let $x \in D(A)$, $v \in D(A^*)$. Then

$$\frac{\mathrm{d}}{\mathrm{d}t} \langle T(t)x, v \rangle = \langle AT(t)x, v \rangle = \langle T(t)x, A^* v \rangle. \tag{5.1.11}$$

Since $D(A)$ is dense in X it follows that the outer equality in (5.1.11) also holds for $x \in X$, $v \in D(A^*)$. Now let $x(t)$ be given by (5.1.10). Then $x \in C([0, \tau]; X)$ and

$$\langle x(t), v \rangle = \langle T(t)x, v \rangle + \int_0^t \langle T(t-s)f(s), v \rangle\,\mathrm{d}s \qquad \text{for all } v \in D(A^*). \tag{5.1.12}$$

If $f \in C([0, \tau]; X)$ it is easy to see that

$$\frac{\mathrm{d}}{\mathrm{d}t} \int_0^t \langle T(t-s)f(s), v \rangle\,\mathrm{d}s = \langle f(t), v \rangle + \int_0^t \langle T(t-s)f(s), A^* v \rangle\,\mathrm{d}s$$

and so by (5.1.12) and (5.1.11),

$$\begin{aligned} \frac{\mathrm{d}}{\mathrm{d}t} \langle x(t), v \rangle &= \langle T(t)x, A^* v \rangle + \int_0^t \langle T(t-s)f(s), A^* v \rangle\,\mathrm{d}s + \langle f(t), v \rangle \\ &= \langle x(t), A^* v \rangle + \langle f(t), v \rangle \end{aligned}$$

by (5.1.10). Hence $x(t)$ is a weak solution. For the case when $f \in L^1([0, \tau]; X)$, let $f_n \in C([0, \tau]; X)$ be a sequence such that $f_n \to f$ in $L^1([0, \tau]; X)$. Then defining

$$x_n(t) = T(t)x + \int_0^t T(t-s)f_n(s)\,\mathrm{d}s,$$

we have

$$\begin{aligned} \sup_{x \in [0, \tau]} \| x_n(t) - x(t) \| &\leqslant M \int_0^\tau \| f_n(s) - f(s) \|\,\mathrm{d}s \\ &\to 0 \end{aligned}$$

as $n \to \infty$. Hence $x_n \to x$ in $C([0, \tau]; X)$. However, we have seen that each $x_n(t)$ is a weak solution and so

$$\langle x_n(t), v \rangle = \langle x_0, v \rangle + \int_0^t [\langle x_n(s), A^* v \rangle + \langle f_n(s), v \rangle]\,\mathrm{d}s \qquad \text{for all } v \in D(A^*)$$

and letting $n \to \infty$ it follows that $x(t)$ is a weak solution of (5.1.9). The uniqueness follows from Lemma 5.1.4.

Conversely, suppose that (5.1.9) has a unique weak solution $x(t) = x(t; x_0)$ for each $x_0 \in X$. Set $T(t)x_0 = x(t; x_0) - x(t; 0)$ for $t \in [0, \tau]$. Extend the definition of $T(t)$ to $[0, \infty)$ by defining

$$T(t)x_0 = T(s)T(\tau)^n x_0 \qquad \text{for } t = n\tau + s, \quad s \in [0, \tau).$$

It is easy to see that $T(t)$ is a strongly continuous semigroup. Let B be the generator of T; we must show that $A = B$. If $x \in D(B)$ and $v \in D(A^*)$ then

$$\frac{\mathrm{d}}{\mathrm{d}t} \langle T(t)x, v\rangle \bigg|_{t=0} = \langle Bx, v\rangle = \langle x, A^* v\rangle$$

since $T(t)x_0 = x(t; x_0) - x(t; 0)$ satisfies

$$\frac{\mathrm{d}}{\mathrm{d}t} \langle T(t)x_0, v\rangle = \langle T(t)x_0, A^* v\rangle \qquad \text{for any } x_0 \in X. \tag{5.1.13}$$

By Lemma 5.1.3 it follows that $x \in D(A)$ and $Bx = Ax$. Hence $D(B) \subseteq D(A)$. To prove that $D(A) \subseteq D(B)$ let $x \in D(A)$ and define

$$y = \int_0^t T(s)x \,\mathrm{d}s, \qquad z = \int_0^t T(s)Ax \,\mathrm{d}s.$$

Then, by (5.1.13),

$$\langle (T(t) - I)x, v\rangle = \langle y, A^* v\rangle, \qquad \text{for all } v \in D(A^*) \tag{5.1.14}$$

and so $y \in D(A)$ by Lemma 5.1.3. Since $D(A^*)$ is dense in X we have

$$T(t)x = x + A\int_0^t T(s)x \,\mathrm{d}s \tag{5.1.15}$$

by (5.1.14). Similarly, $z \in D(A)$ and

$$T(t)Ax = Ax + A\int_0^t T(s)Ax \,\mathrm{d}s. \tag{5.1.16}$$

The function

$$z(t) = \int_0^t T(s)Ax \,\mathrm{d}s - A\int_0^t T(s)x \,\mathrm{d}s$$

belongs to $C([0, \tau]; X)$ by (5.1.15) and $z(0) = 0$. If $v \in D(A^*)$ we have

$$\begin{aligned} \frac{\mathrm{d}}{\mathrm{d}t} \langle z(t), v\rangle &= \frac{\mathrm{d}}{\mathrm{d}t}\left\langle \int_0^t T(s)Ax \,\mathrm{d}s, v\right\rangle - \frac{\mathrm{d}}{\mathrm{d}t}\left\langle A\int_0^t T(s)x \,\mathrm{d}s, v\right\rangle \\ &= \langle T(t)Ax, v\rangle - \langle T(t)x, A^* v\rangle \\ &= \left\langle Ax + A\int_0^t T(s)Ax \,\mathrm{d}s, v\right\rangle - \left\langle x + A\int_0^t T(s)x \,\mathrm{d}s, A^* v\right\rangle \end{aligned}$$

by (5.1.15) and (5.1.16). Hence

$$\frac{\mathrm{d}}{\mathrm{d}t}\langle z(t), v\rangle = \langle z(t), A^*v\rangle, \qquad t \in [0, \tau].$$

By uniqueness of weak solutions, $z(t) = 0$. Hence

$$\int_0^t T(s)Ax\,\mathrm{d}s = A\int_0^t T(s)x\,\mathrm{d}s, \qquad t \in [0, \tau],$$

and so, by (5.1.15),

$$\lim_{t \to 0+} \frac{1}{t}[T(t)x - x] = \lim_{t \to 0+} \frac{1}{t}\int_0^t T(s)Ax\,\mathrm{d}s = Ax.$$

Hence $x \in D(B)$. □

In the same way we can prove a similar result for the nonlinear equation (5.1.5).

Theorem 5.1.6
A function $x\colon [t_0, t_1] \to X$ is a weak solution of equation (5.1.5) on $[t_0, t_1]$ if and only if $f(x(.),.) \in L^1([t_0, t_1]; X)$ and x satisfies the integral equation

$$x(t) = T(t - t_0)x_0 + \int_{t_0}^t T(t - s)f(u(s), s)\,\mathrm{d}s$$

for all $t \in [t_0, t_1]$. □

Hence, to show the existence of weak solutions we can prove that a mild solution $x(t)$ exists for which $f(x(t), t) \in L^1([t_0, t_1]; X)$. In order to do this we impose the following condition on f:

(C) $f\colon X \times \mathbb{R} \to X$ is locally Lipschitz in x and continuous in t; i.e.

$$\| f(x, t) - f(y, t) \| \leqslant \alpha(\| x \|, \| y \|, t) \| x - y \|,$$

for some continuous function α.

Using this condition and the techniques of ordinary differential equations (see Coddington and Levinson (1955) and Segal (1963) we can prove the following theorem.

Theorem 5.1.7
If $f\colon X \times \mathbb{R} \to X$ satisfies condition (C) then, for any $x_0 \in X$, (5.1.5) has a unique weak solution x defined on a maximal interval $[t_0, t_{max})$, $t_{max} > t_0$ such that $x \in C([t_0, t_{max}); X)$. Moreover x is continuous in x_0 in the sense that if $x_n \in C([t_0, t_1]; X)$ are weak solutions such that $x_n(0) \to x_0$ as $n \to \infty$, then $x_n \to x$ in $C([t_0, t_1]; X)$. If $t_{max} < \infty$, then

$$\lim_{t \to t_{max}^-} \| x(t) \| = \infty.$$ □

We shall also require a similar continuity result with respect to the initial condition, but in the weak topology. Actually, we require only the notion of weak convergence and so we do not need to define the weak topology generally. Thus we recall that a sequence x_n in a Banach space X *converges weakly* to $x \in X$ and we write $x_n \rightharpoonup x$ if

$$\langle x_n - x, v \rangle \to 0 \qquad \text{for all } v \in X^*.$$

We assume throughout that X is a reflective Banach space.

Lemma 5.1.8

If $T(t)$ is a strongly continuous semigroup of linear operators, then the map $(t, x) \to T(t)x$ is jointly sequentially weakly continuous on $\mathbb{R}^+ \times X$, i.e.

$$t_n \to t, \qquad x_n \rightharpoonup x \Rightarrow T(t_n)x_n \rightharpoonup T(t)x.$$

Proof

By a well-known result (Dunford and Schwartz (1958)) the map $T(t): X \to X$, for each $t \in \mathbb{R}^+$, is sequentially continuous with respect to the weak topology on X. Let $t_n \to t$ in $\mathbb{R}^+$ and $x_n \rightharpoonup x$ in X. Since any weakly convergent sequence is strongly bounded (Yosida (1974)) it follows from the inequality

$$\| T(t_n)x_n \| \leqslant \| T(t_n) \| \; \| x_n \|$$

that $T(t_n)x_n$ is strongly bounded. A bounded set in a reflective Banach space is weakly compact and so there exist subsequences $\{t_\mu\}$ and $\{x_\mu\}$ such that $T(t_\mu)x_\mu \rightharpoonup y$ for some $y \in X$. We have shown above that

$$\langle T(t_\mu)x_\mu - x_\mu, v \rangle = \int_0^{t_\mu} \langle \mathrm{T}(s)x_\mu, A^* v \rangle \, \mathrm{d}s$$

for each $v \in D(A^*)$ and so, letting $\mu \to \infty$,

$$\begin{aligned} \langle y - x, v \rangle &= \int_0^t \langle T(s)x, A^* v \rangle \, \mathrm{d}s \\ &= \int_0^t \frac{\mathrm{d}}{\mathrm{d}s} \langle T(s)x, v \rangle \, \mathrm{d}s \\ &= \langle T(t)x - x, v \rangle. \end{aligned}$$

Hence

$$\langle y, v \rangle = \langle T(t)x, v \rangle \qquad \text{for all } v \in D(A^*) \tag{5.1.17}$$

and so $y = T(t)x$. It then follows that the whole sequence $T(t_n)x_n$ converges weakly to $T(t)x$, for, if not, there would exist another subsequence converging to $z \neq T(t)x$ which would contradict the above argument leading to (5.1.7). □

We now specialize to a separable Hilbert space H. Then the weak topology on H has a countable basis at 0 and so is metrizable. We denote by H_w the space H with the metrized weak topology.

Theorem 5.1.9

Let $f: H_w \to H_w$ be sequentially continuous and suppose that the equation

$$\begin{aligned} \dot{x}(t) &= Ax(t) + f(x(t)) \\ x(0) &= x_0 \end{aligned} \tag{5.1.18}$$

has a unique weak solution $x(t; x_0)$ on $[0, T]$ for each $x_0 \in H$. Furthermore suppose that

$$\| x(t; x_0) \| \leqslant K \qquad \text{for all } t \in [0, T], \quad x_0 \in B$$

where B is any bounded subset of H, and K is some constant (depending on B). Then $x_{0n} \rightharpoonup x_0$ implies that $x(t; x_{0n}) \rightharpoonup x(t; x_0)$ for all $t \in [0, T]$.

Proof

Set $x_n(t) = x(t; x_{0n})$, $x(t) = x(t; x_0)$. Since the sequence $\{x_{0n}\}$ is bounded it follows by assumption that $\{x_n(t)\}$ is also bounded for all n and $t \in [0, T]$. Since f maps bounded sets to bounded sets we have

$$\| f(x_n(t)) \| \leqslant \text{constant} \qquad \text{for all } n \quad \text{and all } t \in [0, T].$$

We prove that the sequence $\{x_n(t)\}$ is equicontinuous and uniformly bounded in $C([0, T]; H_w)$. First, we have for $t_r \to t+$ and $w \in H$,

$$\begin{aligned} | \langle x_n(t_r) - x_n(t), w \rangle | &\leqslant | \langle (T(t_r) - T(t))x_{0n}, w \rangle | \\ &\quad + \int_0^t | \langle (T(t_r - \tau) - T(t - \tau))f(x_n(\tau)), w \rangle | \, d\tau \\ &\quad + \int_t^{t_r} | \langle T(t_r - \tau)f(x_n(\tau)), w \rangle | \, d\tau \\ &\leqslant C_1 a_r + C_2 | t_r - t | \end{aligned} \tag{5.1.19}$$

for some constants C_1, C_2 where

$$a_r = \sup_{\substack{\| v \| \leqslant 1 \\ 0 \leqslant s \leqslant t}} | \langle (T(t - s) - T(t_r - s))v, w \rangle |.$$

We show that $a_r \to 0$ as $r \to \infty$. If this does not hold, then by the weak compactness of bounded subsets of H we can find sequences $\{v_\mu\}$ and $\{s_\mu\}$ such that $v_\mu \rightharpoonup v$, $s_\mu \to s$ (for some v and some $s \in [0, t]$), $t_\mu \to t$ and an $\varepsilon > 0$ such that

$$| \langle (T(t - s_\mu) - T(t_\mu - s_\mu))v_\mu, w \rangle | \geqslant \varepsilon.$$

Lemma 5.1.8 now gives a contradiction. Hence by (5.1.19), $\langle x_n(t_r) - x_n(t), w\rangle \to 0$ uniformly as $r \to \infty$. A similar argument shows that $\langle x_n(t_r) - x_n(t), w\rangle \to 0$ uniformly as $r \to \infty$ for a sequence $\{t_r\}$ with $t_r \to t-$. Hence $\{x_n(t)\}$ is equicontinuous in $C([0, T]; H_w)$. The uniform boundedness of $\{x_n(t)\}$ in $C([0, T]; H_w)$ follows directly from the assumption of the theorem.

The Arzela–Ascoli theorem (Theorem 1.6.7) shows that there exists an element $\bar{x} \in C([0, T]; H_w)$ and a subsequence $\{x_\nu(t)\}$ of $\{x_n(t)\}$ such that $x_\nu(t) \to \bar{x}(t)$ uniformly on $[0, T]$ as $\nu \to \infty$. Since x_ν is also a mild solution we have

$$\langle x_\nu(t), w\rangle = \langle T(t)x_{0\nu}, w\rangle + \int_0^t \langle T(t-s)f(x_\nu(s)), w\rangle \, ds,$$

for any $w \in H$. Hence, by the sequential weak continuity of f and the dominated convergence theorem, it follows that

$$\langle \bar{x}(t), w\rangle = \langle T(t)x_0, w\rangle + \int_0^t \langle T(t-s)f((\bar{x}(s)), w\rangle \, ds \qquad \text{for all } w \in H$$

whence

$$\bar{x}(t) = T(t)x_0 + \int_0^t T(t-s)f(\bar{x}(s)) \, ds.$$

By uniqueness of weak solutions, $x = \bar{x}$ on $[0, T]$. As in the proof of Lemma 5.1.8 we now see that the whole sequence $x_n(t) \rightharpoonup x(t)$. □

5.2 THE NONLINEAR VARIATION OF CONSTANTS FORMULA AND DISTRIBUTED BILINEAR SYSTEMS

5.2.1 Nonlinear perturbations

In Section 4.1.2 we derived the Volterra series expansion for a linear analytic system by using, essentially, the nonlinear variation of constants formula for the perturbed nonlinear system. We shall now generalize this formula to certain distributed systems as in Banks (1984b). Let us first recall the nonlinear variation of constants formula for finite-dimensional differential equations (Alekseev (1961) and Brauer (1966)).

Consider a nonlinear differential equation

$$\dot{x} = f(x, t), \qquad x(t_0) = x_0 \tag{5.2.1}$$

and suppose that conditions are imposed on f (e.g. the Carathéodory conditions) which guarantee the global existence and uniqueness of solutions $x(t; t_0, x_0)$ which are differentiable with respect to x_0. Now consider a nonlinear perturbation of the equation (5.2.1) given by the equation

$$\dot{y} = f(y, t) + g(y, t). \tag{5.2.2}$$

Then the solutions $x(t; t_0, x_0)$, $y(t; t_0, x_0)$ (with the same initial condition) of (5.2.1) and (5.2.2) are related by

$$y(t; t_0, x_0) - x(t; t_0, x_0) = \int_{t_0}^{t} \Phi(t; s, y(s; t_0, x_0)) g(y(s; t_0, x_0), s) \, \mathrm{d}s \tag{5.2.3}$$

where

$$\Phi(t; t_0, x_0) \triangleq \frac{\partial}{\partial x_0} [x(t; t_0, x_0)]$$

is the fundamental matrix solution of the 'variational system'

$$\dot{Z} = f_x[x(t; t_0, x_0), t] Z. \tag{5.2.4}$$

5.2.2 The Fréchet derivative

We shall use the Fréchet derivative of a function $f: E \to F$ between different Banach spaces E and F and so we shall recall the definition and give a simple example here. In fact, the Fréchet derivative of f at $x \in E$ (if it exists) is an operator $B(x) \in \mathscr{L}(E, F)$ such that

$$\lim_{\| p \|_E \to 0} \frac{\| f(x+p) - f(x) - B(x)p \|_F}{\| p \|_E} = 0.$$

As usual, we shall denote the Fréchet derivative of f at x_0 by $\mathscr{F}f(x_0)$. If $f: E_1 \times E_2 \to F$ is a function of two variables $(x, y) \in E_1 \times E_2$, then we shall denote the partial Fréchet derivatives at (x_0, y_0) by $\mathscr{F}_x f(x_0, y_0)$, and $\mathscr{F}_y f(x_0, y_0)$.

Example

Consider the function $f: H_0^{1,4}(0, 1) \to L^2(0. 1)$ defined by

$$f(\phi) = \left(\frac{\mathrm{d}\phi}{\mathrm{d}x}\right)^2, \quad \phi \in H_0^{1,4}.$$

Denote by $\mathscr{D}(0, 1)$ the space of infinitely differentiable functions with compact support in the interval $[0, 1]$. For $\phi, p \in \mathscr{D}(0, 1)$, we have

$$\left\| \left(\frac{\mathrm{d}}{\mathrm{d}x}(\phi + p)\right)^2 - \left(\frac{\mathrm{d}\phi}{\mathrm{d}x}\right)^2 - 2\frac{d\phi}{\mathrm{d}x}\frac{\mathrm{d}p}{\mathrm{d}x} \right\|_{L^2} \Big/ \| p \|_{H_0^{1,4}}$$

$$= \left\| \left(\frac{\mathrm{d}p}{\mathrm{d}x}\right)^2 \right\|_{L_2} \Big/ \| p \|_{H_0^{1,4}}$$

$$= c \| p \|_{H_0^{1,4}}$$

for some constant c. Since $\mathscr{D}(0,1)$ is dense in $H_0^{1,4}(0,1)$ we have

$$\mathscr{F}f(\phi) = 2\frac{\mathrm{d}\phi}{\mathrm{d}x}\frac{\mathrm{d}}{\mathrm{d}x} : H_0^{1,4}(0,1) \to L^2(0,1).$$

(Note that when the domain of definition Ω in the spaces $H^p(\Omega)$, etc., is clear from the context, we shall simply write H^p, etc.)

5.2.3 Nonlinear semigroups and linear evolution operators

We shall need some elementary results about nonlinear semigroups (see Barbu (1976)). First recall that if C is a closed subset of a Hilbert space H, then a *semigroup of type* ω *on* C is a function $S: \mathbb{R}^+ \times C \to C$ such that

(a) $S(t+s)x = S(t)S(s)x, \quad$ for all $x \in C, \quad t, s \geqslant 0,$
(b) $S(0)x = x, \quad$ for all $x \in C$
(c) $S(t)x$ is continuous in $t \geqslant 0, \quad$ for all $x \in C$
(d) $\| S(t)x - S(t)y \| \leqslant \mathrm{e}^{\omega t} \| x - y \|, \quad$ for all $t \geqslant 0, \quad x, y \in C$

where $S(t)x \triangleq S(t, x)$.

Lemma 5.2.1
Let S be a semigroup of type ω on C and $x \in D_C(A)$, where

$$D_C(A) = \{ x \in C : \liminf_{t \to 0+} \| S(t)x - x \| / t < \infty \}.$$

Then $S(t)x$ is Lipschitz continuous on every interval $[0, T], T > 0$. □

We shall also need the concept of evolution operator, which generalizes the transition matrix of a finite-dimensional system, just as the semigroup generalizes the matrix exponential. An evolution operator on a Hilbert space H is a map $U: \Delta(T) \to \mathscr{B}(H)$, where $\Delta(T) = \{ (t, \tau) : 0 \leqslant \tau \leqslant t \leqslant T \}$, such that

(a) $U(t, \tau) = U(t, s)U(s, \tau), \quad 0 \leqslant \tau \leqslant s \leqslant t \leqslant T$
(b) $U(\tau, \tau) = I, \quad 0 \leqslant \tau \leqslant T$
(c) U is strongly continuous in t and τ
(d) U is strongly differentiable with respect to t.

Then we have the following well-known result of Tanabe (see Yosida (1974) or Friedman (1969)):

Theorem 5.2.2
Let $A(t)$ ($t \in [0, T]$) be a set of linear operators in H with the following properties

(a) The domain $D(A(t))$ of $A(t)$ is independent of t and dense in H.
(b) For each $t \in [0, T]$, the resolvent $R(\lambda; A(t))$ of $A(t)$ exists for all λ

with Re $\lambda \geqslant 0$ and

$$\| R(\lambda; A(t)) \| \leqslant \frac{c}{|\lambda| + 1} \qquad \text{when Re } \lambda \geqslant 0$$

for some constant $c > 0$.

(c) For any $t, s, \tau \in [0, T]$,

$$\| (A(t) - A(\tau))A^{-1}(s) \| \leqslant C | t - \tau |^{\alpha} \qquad \text{for some } \alpha \in (0, 1],$$

where C is a constant independent of t, s, τ.

Then A generates a unique evolution operator $U(t, \tau)$, i.e.

$$\frac{\partial U}{\partial t}(t, \tau) = A(t)U(t, \tau) \qquad \text{when } \tau < t \leqslant T$$

$$U(\tau, \tau) = I,$$

where the derivative of U with respect to t is taken in the strong topology. □

5.2.4 The variational equation

We have stated above that if $x(t; t_0, x_0)$ is the solution of the equation

$$\dot{x} = f(x, t), \qquad x(t_0) = x_0$$

then $\Phi(t; t_0, x_0) = \partial/\partial x_0 [x(t; t_0, x_0)]$ is the fundamental solution of the variational system

$$\dot{Z} = f_x[x(t; t_0, x_0), t] Z.$$

We shall now generalize this result to the case of a nonlinear (autonomous) differential equation

$$\dot{x} = Ax, \qquad x(0) = x_0 \in D(A),$$

where A is an operator with domain $D(A)$, dense in H, which is a linear subspace of H.

Theorem 5.2.3

Let $A : D(A) \rightarrow H$ be a nonlinear operator which generates a semigroup $S(t)$ on $D_C(A) \subseteq D(A)$, and suppose that $\mathscr{F}A(x)$ ($\in \mathscr{L}(D(A), H)$‡) exists uniformly for each $x \in D_C(A)$. Assume that $\mathscr{F}A$ satisfies the following conditions along a solution $S(t)x$ of $\dot{x} = Ax$;

(a) For each $t \in [0, T]$, the resolvent $R(\lambda; \mathscr{F}A(S(t)x))$ exists for all λ such

‡ $D(A)$ is the set $D(A)$ with the graph norm $(\| x \|^2 + \| Ax \|^2)^{1/2}$, $x \in D(A)$.

that $\operatorname{Re}\lambda \geqslant 0$ and

$$\| R(\lambda; \mathscr{F}A(S(t)x)) \| \leqslant \frac{c}{|\lambda| + 1} \text{ when } \operatorname{Re}\lambda \geqslant 0$$

for some $c > 0$.

(b) For any $t, s, \tau \in [0, T]$,

$$\| (\mathscr{F}A(S(t)x) - \mathscr{F}A(S(\tau)x))\mathscr{F}A^{-1}(S(s)x) \| \leqslant C|t - \tau|.$$

(Note that $D(\mathscr{F}A(S(t)x)) = D(A)$ is independent of t and dense in H.)

Then the semigroup $S(t)x$ is differentiable with respect to x (as a map from H into H) and $\mathscr{F}_{x_0}S(t)x_0 = U(t, 0)$, where U is the evolution operator which is the fundamental solution of the equation

$$\dot{Y} = \mathscr{F}_x A(S(t)x_0)Y.$$

Proof

Let $x = x_0 + h \in D(A)$ and set

$$M(t; x_0, h) = S(t)x - S(t)x_0.$$

Since the semigroup $S(t)x$ satisfies the equation $\dot{x} = Ax$, we have

$$\frac{\mathrm{d}M}{\mathrm{d}t}(t; x_0, h) = AS(t)x - AS(t)x_0, \qquad t \geqslant 0$$

$$= (\mathscr{F}_x A(S(t)x_0) + \Gamma)M(t; x_0, h) \tag{5.2.5}$$

where

$$\frac{\| \Gamma \|_{\mathscr{L}(D(A), H)}}{\| h \|_{D(A)}} \to 0 \qquad \text{as } \| h \| \to 0$$

uniformly for $t \geqslant 0$.

Hence, if $\bar{M} = M/\| h \|$, then

$$\frac{\mathrm{d}\bar{M}}{\mathrm{d}t}(t; x_0, h) = (\mathscr{F}_{x_0} A(S(\mathrm{t})x_0) + \Gamma)\bar{M}(t; x_0, h). \tag{5.2.6}$$

Now consider the system

$$\frac{\mathrm{d}Y}{\mathrm{d}t} = \mathscr{F}_x A(S(t)x_0)Y, \qquad t \geqslant 0. \tag{5.2.7}$$

It follows from Theorem 5.2.2 and the conditions on $\mathscr{F}A(x)$ specified in the theorem that this equation defines a unique evolution operator $U(t, s; x_0)$ such that the solution of (5.2.7) is given by

$$Y(t) = U(t, 0; x_0)$$

$$Y(0) = I_H.$$

Hence $M_1 \triangleq Y(t)(x-x_0)-\bar{M}(t;x_0,h)$ satisfies the equation

$$\frac{\mathrm{d}M_1}{\mathrm{d}t}=\mathscr{F}_xA(S(t)x_0)M_1-\Gamma\bar{M}(t;x_0,h), \qquad t\geqslant 0,$$

i.e.

$$M_1=-\int_0^t U(t,s;x_0)\Gamma\bar{M}(s;x_0,h)\,\mathrm{d}s,$$

and so

$$\|M_1\|\leqslant t\sup\left(\|U(t,s;x_0)\|\frac{\|\Gamma\|}{\|h\|}\|M(t;x_0,h)\|\right)$$
$$\to 0 \text{ as } \|h\|\to 0,$$

for all $t\in[0,T]$. Thus, $\lim_{\|h\|\to 0}M(t;x_0,h)$ exists and equals $U(t,0;x_0)(x-x_0)$. □

Example

To illustrate the above theorem consider the nonlinear diffusion equation

$$\frac{\partial\phi}{\partial t}=\frac{\partial^2\phi}{\partial x^2}+\lambda\phi-\rho\phi^2, \qquad \rho,\lambda\geqslant 0, \quad \lambda<\pi^2, \quad x\in[0,1]. \quad (5.2.8)$$

Define

$$C=\{\phi\in H_0^1(0,1):\phi(x)\geqslant 0, \qquad \text{for all } x\in[0,1]\}.$$

It is well-known (Henry (1981)) that the equation (5.2.8) has a unique solution in C for any initial value in C. Morover, defining the operator A by

$$A\phi=\frac{\mathrm{d}^2\phi}{\mathrm{d}x^2}+\lambda\phi-\rho\phi^2, \qquad \phi\in D(A)=H^2(0,1)\cap H_0^1(0,1)$$

we have, for any ϕ_1, $\phi_2\in C\cap D(A)$,

$$\begin{aligned}\langle A\phi_1-A\phi_2,\phi_1-\phi_2\rangle_{L^2(0,1)} &= -\|\phi'\|_{L^2}^2+\lambda\|\phi\|_{L^2}^2-\rho\int_0^1(\phi_1+\phi_2)\phi^2\,\mathrm{d}x\\ &\leqslant -\|\phi'\|_{L^2}^2+\lambda\|\phi\|_{L^2}^2\\ &\leqslant(\lambda-\pi^2)\|\phi\|_{L^2}^2,\end{aligned}$$

where $\phi\triangleq\phi_1-\phi_2$, $\phi'=\mathrm{d}\phi/\mathrm{d}x$, and we have used the inequality

$$\pi^2\|\phi\|_{L^2}^2\leqslant\|\phi'\|_{L^2}^2, \qquad \phi\in H_0^1(0,1).$$

A is therefore dissipative on $D_C(A)\triangleq C\cap D(A)$ (the dissipative domain of A), i.e.

$$\langle A\phi_1-A\phi_2,\phi_1-\phi_2\rangle_{L^2(0,1)}\leqslant 0, \qquad \phi_1,\phi_2\in D_C(A).$$

Hence A generates a semigroup $S(t)$ of type $-(\pi^2-\lambda)$ on $\bar{C}$ (the closure in $L^2(0,1)$).

Consider the derivative $\mathscr{F}A(S(t)\phi)$ of A at $S(t)\phi$ for any $\phi \in D_C(A)$. The first two terms in A are linear so we consider the term $f(\phi) = \phi^2$, for which

$$\frac{\| f(\phi_1 + \phi) - f(\phi_1) - 2\phi_1\phi \|_{L^2}}{\| \phi \|_{D(A)}} \leqslant c \frac{\| \phi^2 \|_{L^2}}{\| \phi \|_{L^4}} = c \| \phi \|_{L^4}$$

(since $D(A) \subseteq L^4$) for a constant c. Thus, $\mathscr{F}A(S(t)\phi)$ exists uniformly in t for any $\phi \in D_C(A)$ and is given by

$$(\mathscr{F}A(S(t)\phi))\psi(x) = \frac{\mathrm{d}^2\psi}{\mathrm{d}x^2}(x) + \lambda\psi(x) - 2\rho(S(t)\phi)(x)\psi(x)$$
$$\text{for } \psi \in D(A). \qquad (5.2.9)$$

To prove that the conditions of Theorem 5.2.3 are satisfied by $\mathscr{F}A$ recall the Sobolev imbedding theorem (Adams (1975))

$$D_C(A) \subseteq H^2(0, 1) \cap C^1(0, 1). \qquad (5.2.10)$$

If $\phi \in D_C(A)$ then $S(t)\phi \in D(A) \subseteq H^2 \cap C^1$ and so $(S(t)\phi)(x)$ is continuous. It follows that the operator $\mathscr{F}A(S(t)\phi)$ is, for each fixed $t \in [0, T]$ and $\phi \in D_C(A)$, a uniformly strong elliptic operator on $D(A)$ with domain independent of t. Moreover, since $S(t)$ is defined on C we have $(S(t)\phi)(x) \geqslant 0$ and so $\mathscr{F}A(S(t)\phi)$ is also dissipative, for each $t \in [0, T]$; in fact, we can show as above that

$$\langle \mathscr{F}A(S(t)\phi_1) - \mathscr{F}A(S(t)\phi_2), \phi_1 - \phi_2 \rangle_{L^2} \leqslant (\lambda - \pi^2) \| \phi^2 \|_{L^2}. \qquad (5.2.11)$$

To verify condition (b) of the theorem, note that from the classical theory of elliptic operators (Friedman (1969)), it follows that

$$\| \{\mathscr{F}A(S(t)\phi)\}^{-1}\psi \|_{D(A)} \leqslant c \| \psi \|_{L^2}, \qquad \psi \in L^2,$$

for some constant $c \geqslant 0$. Hence, by (5.2.10), we have

$$\| \{\mathscr{F}A(S(t)\phi)\}^{-1}\psi \|_{C(0,1)} \leqslant \bar{c} \| \psi \|_{L^2},$$

for a new constant $\bar{c}$. Since the first two terms in (5.2.9) are constant with respect to t we need consider only the third term, which satisfies

$$\begin{aligned} &\| ((S(t)\phi) - (S(\tau)\phi)) \, . \, \{\mathscr{F}A(S(s)\phi)\}^{-1}\psi \|_{L^2} \\ &\qquad \leqslant \| S(t)\phi - S(\tau)\phi \|_{L^2} \| \{\mathscr{F}A(S(s)\phi)\}^{-1}\psi \|_{C(0,1)} \\ &\qquad \leqslant c_2 |t - \tau| \, \| \psi \|_{L^2}, \end{aligned}$$

by Lemma 5.2.1. □

5.2.5. The nonlinear variation of constants formula

In this section we consider the equation

$$\dot{x} = Ax, \qquad (5.2.12)$$

where A is a nonlinear operator which satisfies the conditions of Theorem 5.2.3, and the nonlinear perturbation

$$\dot{y} = Ay + B(y, t) \tag{5.2.13}$$

where B is a nonlinear map from $D(B) \times \mathbb{R}^+$ into H for some subset $D(B)$ of H. We shall assume that these equations have unique solutions through some point $(x_0, t_0) \in H \times \mathbb{R}^+$ which we denote by $x(t; t_0, x_0)$ and $y(t; t_0, y_0)$, respectively. If A generates a semigroup $S(t)$ then we have $x(t; t_0, x_0) = S(t - t_0)x_0$. Since A satisfies the conditions of Theorem 5.2.3 we can write

$$\Phi(t, 0; x_0) = \mathscr{F}_{x_0}[x(t; 0, x_0)]$$

where $\Phi(t, s; x_0)$ is the fundamental solution of the equation

$$\dot{\Psi} = \mathscr{F}_x A(S(t)x_0)\Psi.$$

The finite-dimensional nonlinear variation of constants formula now generalizes directly (see Brauer (1966)) and we have

Theorem 5.2.4
If A satisfies the conditions of Theorem 5.2.3 on $D_C(A)$, then for all $t \geqslant t_0$ such that $x(t; t_0, x_0)$, $y(t; t_0, x_0) \in D_C(A) \cap D(B)$ we have

$$y(t; t_0, x_0) - x(t; t_0, x_0) = \int_{t_0}^{t} \Phi(t, s; y(s; t_0, x_0))B(y(s; t_0, x_0), s)\, \mathrm{d}s. \tag{5.2.14}$$

□

5.2.6 An application to stability theory

We first prove a generalization of a well-known result on finite-dimensional non-autonomous differential equations.

Theorem 5.2.5
Consider the linear differential equation

$$\dot{x} = A(t)x$$

where, for each t, $A(t)$ is a strongly elliptic operator on a region Ω $(\subseteq \mathbb{R}^n)$ (independent of t). If the order of $A(t)$ is $2p$ and we have

$$\left.\frac{\partial^i}{\partial n^i}\right|_{\Gamma} x = 0, \quad \Gamma = \partial\Omega, \qquad 1 \leqslant i \leqslant p,$$

then there exists a function $f(t)$ such that

$$\| x(t) \|_{L^2(\Omega)} \leqslant \| x(t_0) \|_{L^2(\Omega)} \exp\left(\int_{t_0}^{t} f(s)\, \mathrm{d}s\right), \qquad t \geqslant t_0$$

Proof
Since $A(t)$ is strongly elliptic with zero boundary conditions, so is $A(t)+A^*(t)$, which is self-adjoint. $A(t)+A^*(t)$ has compact resolvent (Dunford and Schwartz (1963)) and so has a sequence of real eigenvalues $\{\lambda_n(t)\}$ with finite multiplicity and a corresponding complete orthonormal sequence of eigenvectors $\{\phi_n(t)\}$ such that, if

$$x=\sum_{i=1}^{\infty}\langle x,\phi_i(t)\rangle\phi_i(t),$$

then

$$(A(t)+A^*(t))x=\sum_{i=1}^{\infty}\lambda_i(t)\langle x,\phi_i(t)\rangle\phi_i(t),$$

where λ_i is counted according to its multiplicity. It follows that, since we can order the $\lambda_i(t)$ so that

$$\cdots\leqslant\lambda_n(t)\leqslant\cdots\leqslant\lambda_1(t)\leqslant\lambda(t)$$

where $\lambda(t)$ is independent of n, we have

$$\begin{aligned}\lambda(A(t))&\triangleq\sup_{x\neq 0}\frac{\langle x,(A+A^*)x\rangle_{L^2(\Omega)}}{\|x\|^2}\\&=\frac{\sum_{i=1}^{\infty}\lambda_i(t)\langle x,\phi_i(t)\rangle^2}{\sum_{i=1}^{\infty}\langle x,\phi_i(t)\rangle^2}\\&\leqslant\lambda(t).\end{aligned}$$

Hence,

$$\begin{aligned}\frac{\mathrm{d}}{\mathrm{d}t}\|x(t)\|^2_{L^2(\Omega)}&=\frac{\mathrm{d}}{\mathrm{d}t}\langle x(t),x(t)\rangle\\&=\langle x,A(t)x\rangle+\langle A(t)x,x\rangle\\&=\langle x,(A(t)+A^*(t))x\rangle\\&\leqslant\lambda(t)\|x\|^2_{L^2(\Omega)},\end{aligned}$$

i.e.
$$\|x(t)\|^2_{L^2(\Omega)}\leqslant\|x(0)\|^2_{L^2(\Omega)}\exp\left(\int_0^t\lambda(s)\,\mathrm{d}s\right).$$ □

Corollary 5.2.6
If $A(t)$ satisfies the conditions of Theorem 5.2.5 and, moreover, $A(t)$ is $-\omega(t)$-dissipative on a real Hilbert space H (i.e.

$$\langle A(t)x,x\rangle\leqslant-\omega(t)\|x\|^2,\qquad x\in D(A))$$

and $D(A^*) = D(A)$, then we have

$$\| x(t) \|^2_{L^2(\Omega)} \leqslant \| x(0) \|^2_{L^2(\Omega)} \exp\left(- \int_0^t 2\omega(s)\, ds \right).$$

Proof
Since

$$\langle A(t)x, x \rangle = \langle x, A^*(t)x \rangle = \langle A^*(t)x, x \rangle,$$

we have

$$\langle A(t) + A^*(t)x, x \rangle \leqslant -2\omega(t) \| x \|^2.$$

The largest eigenvalue of $A(t) + A^*(t)$ is given by

$$\lambda(t) = \sup \frac{\langle (A(t) + A^*(t))x, x \rangle}{\| x \|^2} \leqslant -2\omega(t),$$

and the result follows from Theorem 5.2.5. □

Example
Consider again the system discussed above, namely

$$\frac{\partial \phi}{\partial t} = \frac{\partial^2 \phi}{\partial x^2} + \lambda \phi - \rho \phi^2, \qquad \rho, \lambda \geqslant 0, \quad \lambda < \pi^2, \quad x \in [0, 1].$$

If A is defined by

$$A\phi = \frac{d^2 \phi}{dx^2} + \lambda \phi - \rho \phi^2, \qquad \phi \in D_C(A)$$

then

$$(\mathscr{F}_\phi A(S(t)\phi)\psi)(x) = \frac{d^2 \psi}{dx^2} + \lambda \psi(x) - 2\rho (S(t)\phi)(x)\psi(x)$$

and $\mathscr{F}_\phi(S(t)\phi)$ satisfies

$$\dot{\Psi} = \mathscr{F}_\phi A(S(t)\phi)\Psi.$$

Since, as we have seen above, $\mathscr{F}_\phi A(S(t)\phi)$ is dissipative on $D(A)$ and so by Corollary 5.2.6 applied to the linear operator $\mathscr{F}_\phi(S(t)\phi)$, we have

$$\begin{aligned} \| \mathscr{F}_\phi(S(t)\phi) \|_{L^2(0,1)} &\triangleq \| \Phi(t; 0, \phi) \|_{L^2(0,1)} \\ &\leqslant \exp(-2(\pi^2 - \lambda)t). \end{aligned}$$

The next result provides an explicit Lipschitz constant for the unperturbed equation.

Lemma 5.2.7
Under the conditions of Theorem 5.2.3 and Corollary 5.2.6 with $A(t)$

replaced by $\mathscr{F}_\phi A(S(t)\phi)$ we have, for any $x_0, y_0 \in H$,

$$\| x(t; y_0) - x(t; x_0) \| \leqslant \| y_0 - x_0 \|_H \exp\left(- \int_0^t 2\omega(s)\, ds \right)$$

where $x(t; x_0)$ is the solution of the nonlinear equation

$$\dot{x} = Ax, \qquad x(0) = x_0.$$

Proof
Let $\zeta: [0, 1] \to H$ be the function defined by

$$\zeta(\lambda) = x_0 + \lambda(y_0 - x_0), \qquad 0 \leqslant \lambda \leqslant 1.$$

Then

$$\frac{d}{d\lambda}[x(t; \zeta(\lambda))] = \Phi(t; 0, \zeta(\lambda))\zeta'(\lambda) = \Phi(t; 0, \zeta(\lambda))(y_0 - x_0)$$

and

$$x(t; y_0) - x(t; x_0) = \left[\int_0^1 \Phi(t; 0, \zeta(\lambda))\, d\lambda \right] (y_0 - x_0).$$

Hence,

$$\| x(t; y_0) - x(t; x_0) \| \leqslant \max_{0 \leqslant \lambda \leqslant 1} \| \Phi(t; 0, \zeta(\lambda)) | \,.\, \| y_0 - x_0 \|,$$

and the result follows from Corollary 5.2.6. □

Using the nonlinear variation of constants formula we can now obtain a bound on the difference between solutions of the perturbed and unperturbed equations. In fact we have

Theorem 5.2.8
If the nonlinear operator A satisfies the assumptions of Lemma 5.2.7 and $g(t, y): \mathbb{R}^+ \times H \to H$ is bounded, then the solutions $y(t; y_0)$, $x(t; x_0)$ of the equations

$$\dot{x} = Ax, \qquad x(0) = x_0$$

and

$$\dot{y} = Ay + g(t, y), \qquad y(0) = y_0$$

satisfy the inequality

$$\| y(t; y_0) - x(t; x_0) \| \leqslant \| y_0 - x_0 \| \exp\left(\int_0^t - 2\omega(s)\, ds \right)$$

$$+ \int_0^t \exp\left(\int_s^t - 2\omega(\tau)\, d\tau \right) \| g(s; y(s; y_0)) \|\, ds. \quad (5.2.15)$$

Proof
This follows directly from Theorem 5.2.4 and Lemma 5.2.7. □

We can now prove a stability theorem for the perturbed system.

Theorem 5.2.9
Let the conditions of Theorem 5.2.3 hold for the system

$$\dot{x} = Ax$$

where $A0 = 0$ and suppose that the assumptions of Corollary 5.2.6 hold with $A(t) = \mathscr{F}_x A(S(t)x_0)$. Then the perturbed system

$$\dot{y} = Ay + g(t, y),$$

where $g(t, y) = o(\| y \|_H)$ as $\| y \| \to 0$ uniformly in t, is asymptotically stable at $y = 0$.

Proof
If $\varepsilon > 0$ there exists $\delta > 0$ such that $\| g(t, y) \| < \varepsilon \| y \|$ if $\| y \| < \delta$. By Theorem 5.2.8 with $x_0 = 0$ we obtain

$$\| y(t; y_0) \|$$

$$\leqslant \| y_0 \| \exp\left(\int_0^t -2\omega(\tau)\,\mathrm{d}\tau\right) + \int_0^t \exp\left(\int_s^t -2\omega(\tau)\,\mathrm{d}\tau\right) \| g(s, y(s; y_0)) \|\,\mathrm{d}s$$

$$\leqslant y_0 \| \exp\left(\int_0^t -2\omega(\tau)\,\mathrm{d}\tau\right) + \varepsilon \int_0^t \exp\left(\int_s^t -2\omega(\tau)\,\mathrm{d}\tau\right) \| y(s; y_0) \|\,\mathrm{d}s$$

if $\| y(t; y_0) \| < \delta$. The result now follows from an application of Gronwall's inequality, which gives

$$\| y(t; y_0) \| \leqslant \| y_0 \| \exp\left(\int_0^t -2\omega(\tau)\,\mathrm{d}\tau\right) \mathrm{e}^{\varepsilon t}.$$ □

Returning to the above example it follows from this theorem that the system

$$\frac{\partial \phi}{\partial t} = \frac{\partial^2 \phi}{\partial x^2} + \lambda\phi - \rho\phi^2 + g(\phi, t)$$

with $g(\phi, t) = o(\| \phi \|_{L^2(0,1)})$ as $\| \phi \|_{L^2(0,1)} \to 0$ uniformly in t is asymptotically stable at the origin.

5.2.7 An application to controllability

In this section we shall consider a system of the form

$$\dot{x} = Ax + Bu + \Psi(x, u, t), \qquad x = x_0 \qquad (5.2.16)$$

where A is a nonlinear operator defined in a Hilbert space H, B is a linear operator defined on the control space U with values in a Hilbert space V and the controls are assumed to be bounded by k:

$$u(t) \in U_k = \{u : \| u \| \leqslant k\} \subseteq U,$$

for each $t \geqslant 0$. Moreover, we assume that $B \in \mathscr{L}(U,V)$ and $\Psi : H \times U \times \mathbb{R}^+ \to V$ satisfies the inequality

$$\| \Psi(x,u,t) \|_V \leqslant \alpha + \beta \| x \|_H \qquad \text{for all } u \in U_k$$

for some constants α, β and for $\| x \|_H \leqslant \gamma$. Finally we assume that H is densely embedded in V.

We would like to show that the state x of the system (5.2.16) is bounded when the control is constrained in U_k. It is important to consider the effect of bounded controls on the reachable set since if it can be shown that the latter is bounded it will follow that the system is not controllable (over certain time intervals). Thus, consider (5.2.16) as a perturbation of the system

$$\dot{y} = Ay, \qquad y = x_0$$

where A satisfies the above conditions. Let

$$A(t) = \mathscr{F}_y A(S(t)x_0)$$

where A generates the semigroup $S(t)$, and consider the linear (nonautonomous) equation

$$\dot{\xi}(t) = A(t)\xi(t), \qquad \xi(t) \in D(A(t)), \quad t \in [0, T].$$

Suppose that there exists a linear operator $\bar{A}$ which is independent of t such that

$$\| (A(t) - \bar{A})\xi \|_H \leqslant c_1 \| \xi \|_H$$

for all $\xi \in H$ and some constant c_1, and such that $\bar{A}$ generates a semigroup $T(t)$ which is 'smoothing' in the sense that

$$\| T(t)\xi \|_H \leqslant \frac{c_2}{t^\alpha} e^{-at} \| \xi \|_V$$

for each $\xi \in V$ and some constants c_2, $a > 0$, and $0 < \alpha < 1$. Then,

$$\begin{aligned} \dot{\xi} &= A(t)\xi \\ &= \bar{A}\xi + (A(t) - \bar{A})\xi \end{aligned}$$

and so

$$\xi(t) = T(t-s)\xi(s) + \int_0^t T(t-\tau)(A(\tau) - \bar{A})\xi \, d\tau.$$

Hence,

$$\| \xi(t) \|_H \leqslant \frac{c_2}{(t-s)^{\alpha}} \mathrm{e}^{-\alpha(t-s)} \| \xi(s) \|_V + \int_s^t \mathrm{e}^{-\alpha(t-\tau)} c_1 \| \xi \|_H \, \mathrm{d}\tau$$

and by a generalization of Gronwall's lemma (Henry (1981)) we have

$$\| \xi(t) \|_H \leqslant \frac{c}{(t-s)^{\alpha}} \mathrm{e}^{-\alpha(t-s)} \| \xi(s) \|_V$$

for a new constant $c = c(c_1, c_2, \alpha, T)$.

If $\Phi(t, s; x_0)$ is the fundamental solution of the variational equation

$$\dot{Z} = \mathscr{F}_y A(S(t)x_0) Z$$

we have therefore proved

Lemma 5.2.10
Under the above assumptions, Φ satisfies the inequality

$$\| \Phi(t, s; x_0) \|_{\mathscr{L}(V,H)} \leqslant \frac{c}{(t-s)^{\alpha}} \mathrm{e}^{-a(t-s)}$$

$$\triangleq g(t-s), \qquad \text{say.} \qquad \square$$

It follows from the nonlinear variation of constants formula that

$$\| x(t; t_0, x_0) \| \leqslant \mathrm{e}^{-a(t-t_0)} \| x_0 \|_H$$

$$+ \int_{t_0}^t \frac{c}{(t-\tau)^{\alpha}} \mathrm{e}^{-a(t-\tau)} \{ k \| B \|_{\mathscr{L}(U,V)} + \alpha + \beta \| x(\tau; t_0, x_0) \|_H \} \, \mathrm{d}\tau$$

provided $\| x(\tau; t_0, x_0) \|_H \leqslant \gamma$.

We can now determine a bound on the state as follows:

Theorem 5.2.11
Suppose that, for $t_0 \leqslant t \leqslant t_0 + \delta, \delta > 0$, we have

$$1 - \beta \int_{t_0}^t g(t-s) \, \mathrm{d}s > 0$$

and that

$$\left\{ \sup_{t \in [t_0, t_0+\delta]} \left(1 - \beta \int_{t_0}^t g(t-s) \, \mathrm{d}s \right) \right\}^{-1} \times$$

$$\left\{ \sup_{t \in [t_0, t_0+\delta]} \left(\mathrm{e}^{-a(t-t_0)} \| x_0 \|_H + (k \| B \|_{\mathscr{L}(U,V)} + \alpha) \int_{t_0}^t g(t-s) \, \mathrm{d}s \right) \right\} < \gamma. \tag{5.2.17}$$

Then, if $\| x_0 \|_H < \gamma$, we have

$$\| x(t; t_0, x_0) \|_H < \gamma \qquad \text{for } t \in [t_0, t_0 + \delta].$$

Proof
Fix x_0 such that $\| x_0 \| < \gamma$ and choose $\varepsilon > 0$ so that

$$\text{LHS (5.2.17)} \leqslant \gamma - \varepsilon, \; \| x_0 \| \leqslant \gamma - \varepsilon.$$

Suppose there exists $t \in [t_0, t_0 + \delta]$ such that $\| x(t; t_0, x_0) \| > \gamma - \varepsilon$, and put

$$\tau = \inf \{ t \in [t_0, t_0 + \delta] : \| x(t; t_0, x_0) \| > \gamma - \varepsilon \}.$$

By continuity, and since $\| x_0 \| \leqslant \gamma - \varepsilon$,

$$\| x(\tau; t_0, x_0) \| = \gamma - \varepsilon$$

and, again by continuity, there is an interval $[\tau, \tau + \delta_1] \subseteq [t_0, t_0 + \delta]$ such that

$$\| x(t; t_0, x_0) \| \leqslant \gamma - \varepsilon/2, \qquad t \in [\tau, \tau + \delta_1].$$

Hence,

$$\| x(t; t_0, x_0) \|_H \leqslant \mathrm{e}^{-a(t - t_0)} \| x_0 \|_H + \left(k \|B\|_{\mathscr{L}(U, V)} + \alpha + \beta \sup_{t_1 \in [t_0, \tau + \delta_1]} \| x(t_1; t_0, x_0) \| \right) \int_{t_0}^{\tau + \delta_1} g(t - s) \, \mathrm{d}s,$$

for $t \in [t_0, \tau + \delta_1]$. It follows from (5.2.17) that

$$\sup_{t \in [t_0, \tau + \delta_1]} \| x(t; t_0, x_0) \| \leqslant \gamma - \varepsilon$$

which contradicts the definition of τ. □

Since

$$\int_0^t g(t - s) \, \mathrm{d}s \leqslant \int_0^\infty g(t - s) \, \mathrm{d}s \leqslant c \left(\frac{1}{1 - \alpha} + \frac{\mathrm{e}^{-a}}{a} \right)$$

we have

Corollary 5.2.12
Assume that the time T is chosen so that

$$1 - \beta c \left(\frac{1}{1 - \alpha} + \frac{\mathrm{e}^{-a}}{a} \right) > 0$$

and

$$\eta \triangleq \gamma - c \left(\frac{1}{1 - \alpha} + \frac{\mathrm{e}^{-a}}{a} \right) (\gamma \beta + k \| B \|_{\mathscr{L}(U, V)} + \alpha) > 0.$$

Then if $\| x_0 \|_H < \eta$ we have

$$\| x(t; t_0, x_0) \|_H < \gamma \qquad \text{for } t \in [t_0, T].$$ □

Hence the system (5.2.16) with bounded controls for which the conditions of Corollary 5.2.12 hold is not (approximately‡) controllable on $[t_0, T]$, for the controlled state must be less than γ for sufficiently small initial conditions.

Example

We shall return finally to the example

$$\frac{\partial \phi}{\partial t} = \frac{\partial^2 \phi}{\partial x^2} + \lambda\phi - \rho\phi^2, \qquad \lambda \geqslant 0, \quad \lambda < \pi^2, \quad \phi(0,t) = \phi(1,t) = 0 \quad (5.2.18)$$

and consider the nonlinear control system

$$\frac{\partial \phi}{\partial t} = \frac{\partial^2 \phi}{\partial x^2} + \lambda\phi - \rho\phi^2 + \delta(x - x_1)u^2$$

(i.e. point control at $x = x_1$). As before, let $S(t)$ be the semigroup generated by the system (5.2.18) and $T(t)$ the linear semigroup generated by

$$\bar{A}\phi = \frac{\partial^2 \phi}{\partial x^2} + \lambda\phi.$$

Then (Banks (1983)),

$$\| T(t)\phi \|_{L^2(0,1)} \leqslant \frac{c_2}{t^{1/4 + 1/2\varepsilon}} \mathrm{e}^{-(\pi^2 - \lambda)t} \| \phi \|_{H^{-1/2-\varepsilon}(0,1)}$$

for any $\varepsilon > 0$. We have also shown above that

$$A(t)\psi \triangleq (\mathscr{F}A(s(t)\phi_0))\psi = \frac{\mathrm{d}^2\phi}{\mathrm{d}x^2} + \lambda\phi - 2\rho(S(t)\phi_0)\psi$$

and so

$$\| (A(t) - \bar{A})\xi \| = 2\rho \| S(t)\phi_0 \|_{C(0,1)} \| \xi \|_{L^2(0,1)} \triangleq c_1 \| \xi \|_{L^2(0,1)}$$

for $\phi_0 \in D_C(A)$. Let

$$V = H^{-1/2-\varepsilon}(0,1), \qquad U = \mathbb{R}.$$

Then

$$\psi(x,u,t) \triangleq \delta(x - x_1)u^2 : U \to V$$

satisfies

$$\| \psi(x,u,t) \|_V \leqslant \| \delta \|_{H^{-1/2-\varepsilon}(0,1)} k^2 \triangleq \alpha,$$

if $\| u \| \leqslant k$. (Here, $\beta = 0$.) Let c $(= c(c_1, c_2, \alpha, T))$ be the constant determined above and let

$$\bar{\varepsilon} = (\tfrac{3}{4} - \tfrac{1}{2}\varepsilon)^{-1} + \frac{\mathrm{e}^{-(\pi^2 - \lambda)}}{(\pi^2 - \lambda)}.$$

‡ See Banks (1983) for a definition and properties of approximate controllability.

Then, by Corollary 5.2.12, it follows that $\| \phi(t) \| \leqslant \gamma$, $t \in [0, T]$ if

$$\eta = \gamma - c\alpha\bar{\varepsilon} > 0 \qquad \text{and} \qquad \| \phi_0 \|_{L^2(0,1)} \leqslant \eta.$$

5.2.8 A generalization (Banks 1985b)

In the previous sections we have derived and applied the nonlinear variation of constants formula for nonlinear perturbations of nonlinear differential equations. However, the methods developed above apply most easily when the unperturbed system operator is a linear differential operator plus a polynomial in the state. The results proved previously do not apply to systems which are perturbations of, for example, the nonlinear heat equation

$$\frac{\partial \phi}{\partial t} = \varkappa(\phi)\frac{\partial^2 \phi}{\partial x^2}, \qquad 0 \leqslant x \leqslant 1 \tag{5.2.19}$$

where the coefficient $\varkappa$ depends on the temperature. In this section we shall generalize the above results to cover this situation. Firstly we must generalize Theorem 5.2.3 so that we can justify the use of the variational equation of (5.2.19). We shall need the following lemma, whose proof is elementary.

Lemma 5.2.13
Let

$$I_n(t) = \int_0^t \frac{s^{n/2}}{(t-s)^{1/2}}\,\mathrm{d}s.$$

Then

$$I_n(t) = \begin{cases} 2\left(\dfrac{n(n-2)(n-4)\cdots 2}{(n+1)(n-1)\cdots 3.1}\right)t^{(n+1)/2} & (n \text{ even}) \\[2ex] \dfrac{\pi}{2}\left(\dfrac{n(n-2)(n-4)\cdots 1}{(n+1)(n-1)\cdots 4.2}\right)t^{(n+1)/2} & (n \text{ odd}). \end{cases} \qquad \square$$

We define $\gamma_n \triangleq I_n(t)t^{(n+1)/2}$.

Theorem 5.2.14
Suppose that the linear operator $A(t)$ is given by

$$A(t)\phi = A\phi + B(t)\phi$$

where A is linear and $D(A) = D(A(t)) = H^2(0,1) \cap H_0^1(0,1)$ for each t and

$$B(.) \in L^\infty(\mathbb{R}^+; \mathscr{L}(L^2, L^1)).$$

Assume also that A generates a semigroup $T(t)$ such that

$$\| T(t) \|_{\mathscr{L}(L^2)} \leqslant M\mathrm{e}^{-\alpha t}, \ \| T(t) \|_{\mathscr{L}(L^{1+\varepsilon}, L^2)} \leqslant M\mathrm{e}^{-\alpha t} t^{-1/2},$$

for some $\varepsilon, \alpha, M > 0$. Then the equation

$$U(t, \tau)\phi = T(t)\phi + \int_\tau^t T(t-s)B(s)U(s, \tau)\phi \, \mathrm{d}s \qquad (5.2.20)$$

is soluble for $U(t, \tau)\phi$ for all $\phi \in L^2(0, 1)$.

Proof
For simplicity we take $\tau = 0$; the general case can be proved similarly. Define

$$U_0(t, 0)\phi = T(t)\phi,$$

$$U_n(t, 0)\phi = \int_0^t T(t-s)B(s)U_{n-1}(s, 0)\phi \, \mathrm{d}s, \ n \geqslant 1.$$

Then,

$$\| U_0(t, 0)\phi \|_{L^2} \leqslant M\mathrm{e}^{-\alpha t} \| \phi \|_{L^2},$$

$$\| U_0(t, 0)\phi \|_{L^2} \leqslant M\mathrm{e}^{-\alpha t} t^{-1/2} \| \phi \|_{L^{1+\varepsilon}}.$$

Hence,

$$\begin{aligned}
\| U_1(t, 0)\phi \|_{L^2} &\leqslant \int_0^t \| T(t-s)B(s)U_0(s, 0)\phi \|_{L^2} \, \mathrm{d}s \\
&\leqslant \int_0^t \| T(t-s) \|_{\mathscr{L}(L^{1+\varepsilon}, L^2)} \| B(s) \|_{\mathscr{L}(L^2, L^{1+\varepsilon})} \| U_0(s, 0)\phi \|_{L^2} \, \mathrm{d}s \\
&\leqslant \int_0^t \frac{M\mathrm{e}^{-\alpha(t-s)}}{(t-s)^{1/2}} \beta M \mathrm{e}^{-\alpha s} \| \phi \|_{L^2} \, \mathrm{d}s \\
&\leqslant \mathrm{e}^{-\alpha t} \beta M^2 \int_0^t \frac{\mathrm{d}s}{(t-s)^{1/2}} \| \phi \|_{L^2} \\
&= 2\mathrm{e}^{-\alpha t} \beta M^2 t^2 \| \phi \|_{L^2} \qquad \text{(by Lemma 5.2.3)},
\end{aligned}$$

where $\beta = \| B(\,.\,) \|_{L^\infty(\mathscr{L}(L^2, L^{1+\varepsilon}))}$. Similarly,

$$\begin{aligned}
\| U_2(t, 0)\phi \|_{L^2} &\leqslant \int_0^t \| T(t-s)B(s)U_1(s, 0)\phi \|_{L^2} \, \mathrm{d}s \\
&\leqslant \int_0^t \frac{M\mathrm{e}^{-\alpha(t-s)}}{(t-s)^{1/2}} \beta 2\mathrm{e}^{-\alpha s} \beta M^2 s^{1/2} \| \phi \|_{L^2} \, \mathrm{d}s \\
&= 2M^3 \beta^2 \mathrm{e}^{-\alpha t} \int_0^t \frac{s^{1/2}}{(t-s)^{1/2}} ds \| \phi \|_{L^2} \\
&= 2M^3 \beta^2 \mathrm{e}^{-\alpha t} \gamma_1 t \| \phi \|_{L^2}.
\end{aligned}$$

Generally,

$$\| U_m(t,0)\phi \|_{L^2} \leqslant 2M^{m+1}\beta^m e^{-\alpha t}(\gamma_1 \cdots \gamma_{m-1}) t^{1/2m} \| \phi \|_{L^2}.$$

It is easy to check that

$$\gamma_1 \cdots \gamma_{m-1} \leqslant \begin{cases} (\frac{1}{2}\pi)^{1/2(m-1)}/[\frac{1}{2}(m-1)]! & (m \text{ odd}), \\ \pi^{-1/2}(\frac{1}{2}\pi)^{1/2m}/(\frac{1}{2}m)! & (m \text{ even}). \end{cases}$$

Now define

$$U(t,0)\phi = \sum_{m=0}^{\infty} U_m(t,0)\phi$$

for those $\phi \in L^2$ for which the right-hand side exists. Then

$$\begin{aligned} &\| U(t,0)\phi \|_{L^2} \\ &\leqslant \sum_{m=0}^{\infty} \| U_m(t,0)\phi \|_{L^2} \\ &\leqslant M e^{-\alpha t} \| \phi \|_{L^2} + \sum_{\substack{m=1 \\ m \text{ odd}}}^{\infty} 2e^{-\alpha t}\beta^m M^{m+1} \left(\frac{\pi}{2}\right)^{(m-1)/2} \frac{t^{m/2}}{(\frac{1}{2}(m-1))!} \| \phi \|_{L^2} \\ &\qquad + \sum_{\substack{m=2 \\ m \text{ odd}}}^{\infty} \frac{2}{\pi^{1/2}} e^{-\alpha t}\beta^m M^{m+1} \left(\frac{\pi}{2}\right)^{m/2} \frac{t^{m/2}}{(\frac{1}{2}m)!} \| \phi \|_{L^2} \\ &= M e^{-\alpha t}[1 - 2\pi^{-1/2} + 2t^{1/2}\beta M \exp(\tfrac{1}{2}M^2\beta^2\pi t) \\ &\qquad + 2\pi^{-1/2} \exp(\tfrac{1}{2}M^2\beta^2\pi t)] \| \phi \|_{L^2} \\ &= M e^{-\alpha t}[1 - 2\pi^{-1/2} + 2(t^{1/2}\beta M + \pi^{-1/2})\exp(\tfrac{1}{2}M^2\beta^2\pi t)] \| \phi \|_{L^2}. \end{aligned}$$

Thus, $U(t,0)$ is a bounded operator for each $t \geqslant 0$, and the result follows. □

A simple extension to the case where A is time-dependent can be proved:

Corollary 5.2.15
Let

$$A(t)\phi = A_1(t)\phi + B(t)\phi$$

where $B(.)$ is as in Theorem 5.2.14 and $A_1(t)$ generates an evolution operator $V(t,s)$ such that

$$\| V(t,s) \|_{\mathscr{L}(L^2)} \leqslant M e^{-\alpha(t-s)}, \qquad \| V(t,s) \|_{\mathscr{L}(L^{1+\epsilon}, L^2)} \leqslant M e^{-\alpha(t-s)}(t-s)^{-1/2},$$

for some $\varepsilon > 0$. Then the equation

$$U(t,\tau)\phi = V(t,\tau)\phi + \int_\tau^t V(t,s)B(s)U(s,\tau)\phi\,\mathrm{d}s$$

is soluble for $U(t,\tau)\phi$ for all $\phi \in L^2$. □

Example
Consider the nonlinear heat equation (5.2.19), i.e.

$$\frac{\partial\phi}{\partial t} = N\phi \triangleq \varkappa(\phi)\frac{\partial^2\phi}{\partial x^2}, \qquad 0 \leqslant x \leqslant 1,$$

where $\varkappa$ satisfies the condition

$$\left|\frac{\varkappa(v) - \varkappa(w)}{v - w}\right| \leqslant K_d < \infty, \quad \varkappa(v) \geqslant K_m, \quad \text{for all } v, w \in \mathbb{R}^+,$$

for some constants K_d, K_m. Then, if $\phi_1, \phi_2 \in H_0^1(0,1) \cap H_0^2(0,1) \cap C^2(0,1)$, we have

$$\left\langle \varkappa(\phi_1)\frac{\partial^2\phi_1}{\partial x^2} - \varkappa(\phi_2)\frac{\partial^2\phi_2}{\partial x^2}, \phi_1 - \phi_2 \right\rangle_{L^2}$$

$$= \int_0^1 \varkappa(\phi_1)\left(\frac{\partial^2\phi_1}{\partial x^2} - \frac{\partial^2\phi_2}{\partial x^2}\right)(\phi_1 - \phi_2)\,\mathrm{d}x$$

$$+ \int_0^1 [\varkappa(\phi_1) - \varkappa(\phi_2)]\left(\frac{\partial^2\phi_2}{\partial x^2}\right)(\phi_1 - \phi_2)\,\mathrm{d}x$$

$$= -\int_0^1 \varkappa^1(\phi_1)\frac{\partial\phi_1}{\partial x}\left(\frac{\partial\phi_1}{\partial x} - \frac{\partial\phi_2}{\partial x}\right)(\phi_1 - \phi_2)\,\mathrm{d}x$$

$$- \int_0^1 \varkappa(\phi_1)\left(\frac{\partial\phi_1}{\partial x} - \frac{\partial\phi_2}{\partial x}\right)^2\mathrm{d}x + \int_0^1 \frac{\varkappa(\phi_1) - \varkappa(\phi_2)}{\phi_1 - \phi_2}\left(\frac{\partial^2\phi_2}{\partial x^2}\right)(\phi_1 - \phi_2)^2\,\mathrm{d}x$$

$$\leqslant K_d \max_{x \in [0,1]}\left|\frac{\partial\phi_1}{\partial x}\right| \left\|\frac{\partial\phi_1}{\partial x} - \frac{\partial\phi_2}{\partial x}\right\|_{L^2} \|\phi_1 - \phi_2\|_{L^2}^2$$

$$- K_m\left\|\frac{\partial\phi_1}{\partial x} - \frac{\partial\phi_2}{\partial x}\right\|_{L^2}^2 + K_d \max_{x \in [0,1]}\left|\frac{\partial^2\phi_2}{\partial x^2}\right| \|\phi_1 - \phi_2\|_{L^2}^2$$

$$\leqslant \frac{K_d}{\pi}\left(\max_{x \in [0,1]}\left|\frac{\partial\phi_1}{\partial x}\right| + \frac{1}{\pi}\max_{x \in [0,1]}\left|\frac{\partial^2\phi_2}{\partial x^2}\right|\right)\left\|\frac{\partial\phi_1}{\partial x} - \frac{\partial\phi_2}{\partial x}\right\|_{L^2}^2$$

$$- K_m\left\|\frac{\partial\phi_1}{\partial x} - \frac{\partial\phi_2}{\partial x}\right\|_{L^2}^2$$

where we have again used the inequality

$$\pi^2 \| \phi_1 - \phi_2 \|_{L^2}^2 \leqslant \left\| \frac{\partial \phi_1}{\partial x} - \frac{\partial \phi_2}{\partial x} \right\|_{L^2}^2.$$

Now let $C_\varepsilon^2(0, 1)$ denote the set of all $\phi \in C^2(0, 1)$ such that

$$K_m - \frac{K_d}{\pi}\left(\max_{x \in [0,1]} \left| \frac{\partial \phi}{\partial x} \right| + \frac{1}{\pi} \max_{x \in [0,1]} \left| \frac{\partial^2 \phi}{\partial x^2} \right| \right) \geqslant \varepsilon > 0.$$

Then

$$\left\langle \varkappa(\phi_1) \frac{\partial^2 \phi_1}{\partial x^2} - \varkappa(\phi_2) \frac{\partial^2 \phi_2}{\partial x^2}, \phi_1 - \phi_2 \right\rangle_{L^2} \leqslant -\varepsilon\pi^2 \| \phi_1 - \phi_2 \|_{L^2}^2$$

for $\phi_1, \phi_2 \in H_0^1(0, 1) \cap H_0^2(0, 1) \cap C_\varepsilon^2(0, 1)$. It follows that N is a dissipative operator on $D_C(N) \subseteq H_0^1(0, 1) \cap H^2(0, 1) \cap C_\varepsilon^2(0, 1)$ and so N generates a nonlinear contraction semigroup $S(t)$ on $\overline{D_C(N)}$ (see Barbu (1976)) which satisfies the inequality

$$\| S(t)\phi \| \leqslant e^{-\varepsilon\pi^2} \| \phi \|, \qquad \phi \in \overline{D_c(N)}.$$

Consider now the Fréchet derivative of N;

$$\mathscr{F}N(\psi)(\phi) = \varkappa(\psi) \frac{\partial^2 \phi}{\partial x^2} + \varkappa^1(\psi) \frac{\partial^2 \psi}{\partial x^2} \phi. \qquad \phi \in D_C(N).$$

Define the linear operator $A(t)$ by

$$A(t)\phi = \varkappa(\psi(t)) \frac{\partial^2 \phi}{\partial x^2} + \varkappa'(\psi(t)) \frac{\partial^2 \psi(t)}{\partial x^2} \phi,$$

where $\psi(t) \in D_C(N)$ for $t \geqslant 0$. Now, $A \triangleq \partial^2/\partial x^2$ satisfies the conditions of Theorem 5.2.14 since (Banks (1983))

$$\| T(t) \|_{\mathscr{L}(H^{-1}, L^2)} \leqslant M_1 e^{-\alpha t} t^{-1/2}$$

for some $M_1 > 0$, where $T(t)$ is the semigroup generated by A. (Note that $H^1 \subseteq L^{(1+\varepsilon)/\varepsilon}$, by the Sobolev embedding theorem and so, by duality, $L^{1+\varepsilon} \subseteq H^{-1}$.) It is then easy to see that $A_1(t) \triangleq \varkappa(\psi(t))\partial^2/\partial x^2$, with $\psi(t) = S(t)\psi(0)$ for some $\psi(0) \in D_C(N)$, satisfies the conditions of Corollary 5.2.15. Also, if

$$B(t)\phi \triangleq \varkappa'(\psi(t)) \frac{\partial^2 \psi(t)}{\partial x^2} \phi$$

then

$$\| B(t)\phi \|_{L^{1+\varepsilon}} \leqslant K_d \left\| \frac{\partial^2 \psi(t)}{\partial x^2} \right\|_{L^{2\hat{\varepsilon}}} \| \phi \|_{L^2},$$

where $\bar{\varepsilon} = (1+\varepsilon)/(1-\varepsilon)$. However, $\psi(t) = S(t)\psi(0) \in \overline{D_C(N)}$ for all t and so

$$\left\| \frac{\partial^2}{\partial x^2} \psi(t) \right\|_{L^{2\bar{\varepsilon}}} \in L^\infty(0, \infty).$$

Hence, $B(.) \in L^\infty(\mathbb{R}^+, \mathscr{L}(L^2, L^1))$.

Now returning to the general system

$$\dot{\phi} = N\phi, \tag{5.2.21}$$

the following generalization of Theorem 5.2.3 is a simple consequence of Corollary 5.2.15.

Theorem 5.2.16

Let $N: D(N) \to H$ be a nonlinear operator which generates a contraction semigroup $S(t)\phi$ on $D_C(N)$, and suppose that $\mathscr{F}N(\phi)$ $(\in \mathscr{L}[D(N), H])$ exists uniformly for each $\phi \in D_C(N)$. Assume that $\mathscr{F}N[S(t)\phi]$ satisfies the conditions of Corollary 5.2.15, so that it is of the form

$$\mathscr{F}N[S(t)\phi] = A_1(t) + B(t).$$

Then $S(t)\phi$ is differentiable with respect to ϕ and satisfies the relation

$$\mathscr{F}_{\phi_0} S(t)\phi_0 = U(t, 0)$$

where U is the operator which is the solution of the equation

$$U(t,\tau)\phi = V(t,\tau) + \int_\tau^t V(t,s)B(s)U(s,\tau)\phi \, \mathrm{d}s \tag{5.2.22}$$

and V is the evolution operator generated by $A_1(t)$. □

To derive the nonlinear variation of constants formula we require the next lemma.

Lemma 5.2.17

Let

$$\Phi(t, \phi_0) = \mathscr{F}_{\phi_0}[S(t)\phi_0] \in \mathscr{L}[D(N), H].$$

Then

$$\frac{\partial}{\partial t_0} S(t - t_0)\phi_0 = -\Phi(t - t_0, \phi_0)N_0 \quad \text{almost everywhere.} \tag{5.2.23}$$

Proof

We have

$$\frac{\mathrm{d}^+}{\mathrm{d}t} S(t - t_0)\phi_0 = NS(t - t_0)\phi_0$$

where $\mathrm{d}^+/\mathrm{d}t$ denotes the right derivative. However, $NS(t - t_0)\phi_0$ is differentiable from the right with respect to t_0, since

$$\frac{\partial^+}{\partial t_0} NS(t-t_0)\phi_0 = \mathscr{F}N[S(t-t_0)\phi_0]\,\frac{\partial^+ S(t-t_0)\phi_0}{\partial t_0},$$

and so $(\mathrm{d}^+/\mathrm{d}t)S(t-t_0)\phi_0$ is also differentiable from the right with respect to t_0 and

$$\begin{aligned}\frac{\partial^+}{\partial t_0}\frac{\mathrm{d}^+}{\mathrm{d}t} S(t-t_0)\phi_0 &= \frac{\mathrm{d}^+}{\mathrm{d}t}\frac{\partial^+}{\partial t_0} S(t-t_0)\phi_0 \\ &= \mathscr{F}N[S(t-t_0)\phi_0]\,\frac{\partial^+}{\partial t_0} S(t-t_0)\phi_0 \\ &= A_1(t-t_0)\,\frac{\partial^+}{\partial t_0} S(t-t_0)\phi_0 + B(t-t_0)\,\frac{\partial^+}{\partial t_0} S(t-t_0)\phi_0.\end{aligned}$$

Thus,

$$\frac{\partial^+}{\partial t_0} S(t-t_0)\phi_0 = -V(t-t_0,0)N\phi_0 + \int_{t_0}^{t} V(t-t_0, s-t_0)B(s-t_0)\,\frac{\partial^+}{\partial t_0} S(s-t_0)\phi_0\,\mathrm{d}s$$

since $(\partial^+/\partial t_0)S(t-t_0)\phi_0 = -N\phi_0$, when $t=t_0$. (5.2.23) now follows from the uniqueness of solutions of (5.2.22). □

Consider now the perturbed equation

$$\dot{\psi} = N\psi + M(t)\psi \tag{5.2.24}$$

and suppose that the solutions $\phi(t;\phi_0)$ of (5.2.21) and $\psi(t;\psi_0)$ of (5.2.24) satisfy

$$\phi_0, \psi_0 \in C \Rightarrow \phi(t;\phi_0), \psi(t;\psi_0) \in C$$

for some subset $C \subseteq H$. Then using Lemma 5.2.17 we can prove the following result in the same way as the finite-dimensional case (Brauer (1966)).

Theorem 5.2.18
With the above assumptions,

$$\psi(t;\phi_0) - \phi(t;\phi_0) = \int_0^t \Phi(t-s;\psi(s;\phi_0))M(s)\psi(s;\phi_0)\,\mathrm{d}s,$$

for each $\phi_0 \in C$. □

Example
Consider the perturbation of the nonlinear heat equation given by

$$\frac{\partial\psi}{\partial t} = \varkappa(\psi)\frac{\partial^2\psi}{\partial x^2} - \psi^2, \qquad 0 \leqslant x \leqslant 1, \quad \psi(0) = \psi_0, \tag{5.2.25}$$

for $t \in [0, T]$. We show that this equation has a solution by considering the following sequence of approximations. Let I_{nj} be the interval

$$I_{nj} = \left[(j-1)\frac{T}{n}, \frac{jT}{n} \right], \qquad 1 \leqslant j \leqslant n,$$

and consider the system defined on I_{nj} by

$$\frac{\partial \psi_{nj}(x,t)}{\partial t} = \varkappa\left[\psi_{nj}\left(x, (j-1)\frac{T}{n}\right)\right] \frac{\partial^2 \psi_{nj}(x,t)}{\partial x^2} - \psi_{nj}(x,t) \quad (5.2.26)$$

with initial condition

$$\psi_{nj}\left(x, (j-1)\frac{T}{n}\right) = \psi_{n,j-1}\left(x, (j-1)\frac{T}{n}\right), \qquad j > 1$$

and $\psi_{n1}(x, 0) = \psi_0(x)$. Then each of the equations (5.2.26) has the form

$$\frac{\partial \psi_{nj}(x,t)}{\partial t} = A_{nj}\psi_{nj}(x,t) - \psi_{nj}^2(x,t),$$

where each operator A_{nj} is time invariant and strongly elliptic with

$$D(A_{nj}) = H_0^1(0,1) \cap H^2(0,1), \qquad n \geqslant 1, \quad 1 \leqslant j \leqslant n.$$

If $T_{nj}(t)$ denotes the semigroup generated by A_{nj}, then it is well known (Henry (1981)) that the equations (5.2.26) have unique solutions (for $\psi_0 \in D(A_{nj})$) given by

$$\psi_{nj}(x,t) = T_{nj}(t)\psi_{nj}\left(x, (j-1)\frac{T}{n}\right) - \int_{(j-1)T/n}^{t} T_{nj}(t-s)\psi_{nj}^2(x,s)\, ds, \quad (5.2.27)$$

for $t \in I_{nj}$. Now, as above, we have

$$\langle A_{nj}\psi_{nj}^1 - A_{nj}\psi_{nj}^2, \psi_{nj}^1 - \psi_{nj}^2 \rangle_{L^2(0,1)}$$

$$\leqslant \left[\frac{K_d}{\pi} \max_{x \in [0,1]} \left| \frac{\partial}{\partial x} \psi_{n,j-1}\left(x, (j-1)\frac{T}{n}\right) \right| - K_m \right] \left\| \frac{\partial \psi_{nj}^1}{\partial x} - \frac{\partial \psi_{nj}^2}{\partial x} \right\|_{L^2}^2$$

$$\leqslant -\varepsilon \pi^2 \|\psi_{nj}^1 - \psi_{nj}^2\|,$$

for $\psi_{nj}^1, \psi_{nj}^2 \in H_0^1(0,1) \cap H^2(0,1) \cap C_\varepsilon^2(0,1) \cap C_p(0,1) \subseteq D_C(A_{nj})$ where ε is as before and

$$C_p(0,1) = \{\psi \in H_0^1(0,1) : \psi(x) \geqslant 0 \text{ for all } x \in [0,1]\}.$$

Hence $D_C(A_{nj})$ is independent of n and j and (5.2.27) defines a stable dynamical system in $C([0,\infty); D_C(A_{nj}))$. We write $D_C(A) = D_C(A_{nj})$.

Now define the sequence $\{\Psi_n(x,t)\}$ by

$$\Psi_n(x,t) = \psi_{nj}(x,t), \qquad t \in I_{nj}, \quad 1 \leqslant j \leqslant n, \quad n \geqslant 1.$$

This sequence is uniformly bounded and equicontinuous in $C([0,T]; L^2)$

and so, by the Arzela–Ascoli theorem $\{\Psi_n(x, t);\ n \in N\}$ is precompact. Hence the sequence has a limit in $C([0, T]; L^2(0, 1))$, which is a solution of (5.2.25). The solutions $\phi(t; \phi_0)$ and $\psi(t; \phi_0)$ of the nonlinear heat equation $\partial\phi/\partial t = \varkappa(\phi)\partial^2\phi/\partial x^2$ and the perturbation (5.2.24) are therefore related by

$$\psi(t; \phi_0) - \phi(t; \phi_0) = -\int_0^t \Phi(t-s; \psi(s; \phi_0))\psi^2(s; \phi_0)\, \mathrm{d}s$$

where

$$\Phi(t; \xi) = \mathscr{F}_\xi[\phi(t; \xi)].$$

5.3 STABILIZABILITY

5.3.1 Dynamical systems

In this section we shall discuss the stabilizability of certain distributed parameter systems. Firstly, we require some elementary notions from the theory of general dynamical systems. Let H be a real Hilbert space and let $T(t)$ be a semigroup of (nonlinear) operators defined on H. Then, if $\phi \in H$ we define the *positive orbit through* ϕ to be the set

$$\mathscr{O}^+(\phi) = \{T(t)\phi : t \geqslant 0\}.$$

The *ω-limit set* of ϕ is the set

$$\omega(\phi) = \{\psi \in H : T(t_n)\phi \to \psi \text{ as } n \to \infty \text{ for some sequence } t_n \to \infty\}.$$

Similarly if we replace strong convergence by weak convergence in the definition of $\omega(\phi)$ we obtain the *weak ω-limit set* $\omega_w(\phi)$ of ϕ. Finally we say that a subset C of H is *positively invariant* (respectively, *invariant*) if $T(t)C \subseteq C$ for all $t \in \mathbb{R}^+$ (respectively, $T(t)C = C$ for all $t \in \mathbb{R}^+$).

The following result is standard; see Henry (1981), Ball and Slemrod (1979a) or Dafermos (1972).

Theorem 5.3.1

(a) If $\mathscr{O}^+(\phi)$ is precompact, then $\omega(\phi)$ is nonempty and invariant.

(b) If $T(t)$ is sequentially weakly continuous (i.e. $\phi_n \rightharpoonup \phi$ implies $T(t)\phi_n \rightharpoonup T(t)\phi$) and $\mathscr{O}^+(\phi)$ is bounded, then $\omega_w(\phi)$ is nonempty and invariant. □

5.3.2 Stabilizability of dissipative systems

Consider now the nonlinear equation

$$\dot{x}(t) = Ax(t) + f(x(t)), \qquad x(0) = x_0, \tag{5.3.1}$$

where A generates a C^0 semigroup $T(t)$ of contractions on H (which implies that $\langle Ax, x\rangle \leqslant 0$ for $x \in D(A)$) and $f: H \to H$ satisfies the condition

(C) (a) f is locally Lipschitz
(b) $x_n \rightharpoonup x \Rightarrow f(x_n) \rightharpoonup f(x)$,
(c) $\langle f(x), x\rangle \leqslant 0$ for all $x \in H$.

Theorem 5.3.2
If f satisfies condition (C), then the equation (5.3.1) has a unique weak solution $x(t; x_0)$ which exists for all $t \geqslant 0$, for any $x_0 \in H$. Moreover, $S(t)x_0 \triangleq x(t; x_0)$ is a semigroup on H and $\omega_w(x_0)$ is a nonempty invariant set for each $x_0 \in H$. Let $\bar{\omega}$ be the set

$$\bar{\omega} = \{x \in H: \langle S(t)x, f(S(t)x)\rangle = 0 \text{ for all } t \in \mathbb{R}^+\}.$$

Then

$$\omega_w(x_0) \subseteq \bar{\omega}$$

for each $x_0 \in H$. (Thus, if $\bar{\omega} = \{0\}$, then $x(t; x_0) \rightharpoonup 0$ as $t \to \infty$.)

Proof
We have proved in Section 5.1 that such an equation has a unique weak solution $x(t) = x(t; x_0)$ for each $x_0 \in H$ on some interval $[0, \varepsilon]$. Let $F(t) = f(x(t))$ and choose a sequence $\{F_n\} \subseteq C^1([0, \varepsilon]; X)$ such that $F_n \to F$ in $C([0, \varepsilon]; X)$. Moreover, choose a sequence x_{n0} such that $x_{n0} \to x_0$ in H and define

$$x_n(t) = T(t)x_{n0} + \int_0^t T(t-s)F_n(s)\,ds.$$

Then $x_n(t) \in D(A)$, $x_n \in C^1([0, \varepsilon]; X)$ and so

$$\dot{x}_n(t) = Ax_n(t) + F_n(t), \qquad t \in [0, \varepsilon].$$

Hence,

$$\|x_n(t)\|^2 - \|x_{n0}\|^2 = 2\int_0^t \langle Ax_n(s) + F_n(s), x_n(s)\rangle\,ds$$

$$\leqslant 2\int_0^t \langle F_n(s), x_n(s)\rangle\,ds \tag{5.3.2}$$

since A is dissipative. Now, if $z_n(t) = x_n(t) - x(t)$, then

$$\|z_n(t)\| \leqslant Me^{\omega t}\|z_n(0)\| + \int_0^t Me^{\omega(t-s)}\|F_n(s) - F(s)\|\,ds$$

and $x_n \to x$ in $C([0, \varepsilon]; X)$. Therefore, from (5.3.2) we have

$$\|x(t)\|^2 \leqslant \|x_0\|^2 + 2\int_0^t \langle f(x(s)), x(s)\rangle\,ds$$

$$\leqslant \|x_0\|^2$$

by condition (C)(c).

It follows that $x(t; x_0)$ exists for all $t \geqslant 0$. Moreover, by Theorem 5.1.7, $x(t;.): H \to H$ is sequentially weakly continuous. Thus $S(t)$ is a semigroup and $\omega_w(x_0)$ is nonempty and invariant by Theorem 5.3.1. If $x \in \omega_w(x_0)$ then let $\{t_n\}$ be a sequence such that $S(t_n)x_0 \rightharpoonup x$ as $n \to \infty$. Then, as above,

$$\lim_{n \to \infty} \int_{t_n}^{t_n + t} \langle f(S(s)x_0), S(s)x_0 \rangle \, ds =$$

$$\lim_{n \to \infty} \int_0^t \langle f(S(s)S(t_n)x_0), S(s)S(t_n)x_0 \rangle \, ds = 0$$

for all $t \geqslant 0$. Now, by condition (C)(b) and Theorem 5.1.9 we have

$$\lim_{n \to \infty} \langle f(S(s)S(t_n)x_0, S(s)S(t_n)x_0 \rangle = \langle f(S(s)x), S(s)x \rangle$$

for each $s \in [0, t]$ and so, by the dominated convergence theorem,

$$\int_0^t \langle f(S(s)x, S(s)x \rangle \, ds = 0.$$

Finally, by the continuity of f we have

$$\langle f(S(t)x), S(t)x \rangle = 0 \qquad \text{for all } t \geqslant 0. \qquad \square$$

We shall now specialize the discussion and consider a system of the form

$$\dot{x}(t) = Ax(t) + u(t)B(x(t)), \qquad x(0) = x_0 \tag{5.3.3}$$

where A generates a C^0 contraction semigroup $T(t)$, so that $\langle Ax, x \rangle \leqslant 0$ for all $x \in D(A)$, B is a (nonlinear) operator from H into H, and u is a scalar control. We shall say that the system (5.3.3) is stabilizable (or weakly stabilizable) if there exists a continuous function $g: H \to \mathbb{R}$ such that, with the feedback control $u(t) = g(x(t))$, the system (5.3.3) has a weak solution $x(t; x_0)$ and also

(a) $\{0\}$ is a stable equilibrium point of (5.3.3)
(b) $x(t; x_0) \to 0$ (or $x(t; x_0) \rightharpoonup 0$) as $t \to \infty$ for all $x_0 \in H$.

Theorem 5.3.3 (Ball and Slemrod, 1979a)
If $B: H_w \to H$ is sequentially continuous and

$$\langle T(t)x, B(T(t)x) \rangle = 0 \qquad \text{for all } t \in \mathbb{R}^+ \Rightarrow x = 0 \tag{5.3.4}$$

then the system (5.3.3) is weakly stabilizable.

Proof
Choose the control

$$u(x) = -\langle x, B(x) \rangle, \tag{5.3.5}$$

and set $f(x) = -\langle x, B(x)\rangle B(x)$. We shall apply Theorem 5.3.2 and so we must verify condition (C) for f. In fact, f is clearly locally Lipschitz since it maps bounded sets to bounded sets. Next, if $x_n \rightharpoonup x$ then $B(x_n) \to B(x)$, by assumption and so $f(x_n) \to f(x)$. Finally,

$$\langle f(x), x\rangle = -(\langle x, B(x)\rangle)^2 \leqslant 0$$

and so by Theorem 5.3.2, if $x_0 \in H$ and $x \in \omega_w(x_0)$, then

$$\langle S(t)x, f(S(t)x)\rangle = 0 \qquad \text{for all } t \geqslant 0 \tag{5.3.6}$$

and so $\langle S(t)x, B(S(t)x)\rangle = 0$. Hence

$$f(S(t)x) = -\langle S(t)x, B(S(t)x)\rangle B(S(t))x = 0 \qquad \text{for all } t \geqslant 0.$$

Now, with the control (5.3.5), the system (5.3.3) becomes

$$\dot{x}(t) = Ax(t) - B(x(t))\langle x(t), B(x(t))\rangle = Ax(t) - f(x(t))$$

and so,

$$x(t) = T(t)x - \int_0^t T(t-s)f(x(s))\, \mathrm{d}s.$$

However, $x(t) = S(t)x$ and since we have just proved that $f(S(t)x) = 0$ we have

$$x(t) = T(t)x.$$

Hence, by (5.3.6) and assumption (5.3.4), we have $x = 0$. □

This result can be applied to hyperbolic systems as we shall now show. Let

$$\ddot{y} + Py + u(t)y = 0 \tag{5.3.7}$$

be an abstract wave equation where P is a positive self-adjoint, densely defined linear operator on a real Hilbert space V such that P^{-1} exists and is compact. If W is the domain of $P^{1/2}$, i.e. $W = D(P^{1/2})$, then W is a Hilbert space with the inner product

$$\langle w_1, w_2\rangle = \langle P^{1/2}w_1, P^{1/2}w_2\rangle_V.$$

Let $H = W \times V$ be the Hilbert space with inner product

$$\langle (w_1, v_1), (w_2, v_2)\rangle_H = \langle w_1, w_2\rangle_W + \langle v_1, v_2\rangle_V$$

and define the operators

$$A = \begin{bmatrix} 0 & I \\ -P & 0 \end{bmatrix}, \qquad B = \begin{bmatrix} 0 & 0 \\ -I & 0 \end{bmatrix},$$

where A has domain $D(A) = D(P) \times W$. Note that the injection $W \subsetneq H$ is compact (Adams (1975)) and so $B: H \to H$ is a compact operator.

Theorem 5.3.4

The system (5.3.7) is weakly stabilizable if and only if all the eigenvalues λ_m of P are simple.

Proof

If the eigenvalues $\{\lambda_m\}$ are simple, then let $\{\phi_m\}$ be the corresponding sequence of orthonormal eigenfunctions. We must show that condition (5.3.4) of Theorem 5.3.3 holds. It is easy to show that the semigroup $T(t)$ is given by

$$T(t)x = \sum_{m=1}^{\infty} \begin{bmatrix} c_m \cos\sqrt{(\lambda_m)}t + d_m \sin\sqrt{(\lambda_m)}t \\ -\sqrt{\lambda_m} c_m \sin\sqrt{(\lambda_m)}t + \sqrt{\lambda_m}\, d_m \cos\sqrt{(\lambda_m)}t \end{bmatrix} \phi_m,$$

where

$$x = \sum_{m=1}^{\infty} \begin{pmatrix} c_m \\ \sqrt{(\lambda_m)} d_m \end{pmatrix} \phi_m.$$

Hence,

$$\langle T(t)x, BT(t)x \rangle_H = \sum_{m=1}^{\infty} \sqrt{(\lambda_m)}\,[\tfrac{1}{2}(c_m^2 - d_m^2)\sin 2\sqrt{(\lambda_m)}t - c_m d_m \cos 2\sqrt{(\lambda_m)}t],$$

and the right-hand side is zero if and only if $c_m = d_m = 0$, i.e. $x = 0$.

Conversely, if P has an eigenvalue λ with at least two linearly independent eigenfunctions ϕ, ψ, then let

$$\xi = \begin{bmatrix} a \\ b \end{bmatrix} \phi + \begin{bmatrix} c \\ d \end{bmatrix} \psi$$

with $ad \neq bc$. Consider the solution y of (5.3.3) with initial condition $(y(0), \dot{y}(0)) = \xi$. This solution is clearly of the form

$$y(t) = w_1(t) + w_2(t)\psi.$$

Substituting this into (5.3.3) we see that w_1, w_2 satisfy the ordinary differential equations

$$\begin{aligned} \ddot{w}_1 + \lambda w_1 + u(t) w_1 &= 0, \qquad w_1(0) = a,\ \dot{w}_1(0) = b \\ \ddot{w}_2 + \lambda w_2 + u(t) w_2 &= 0, \qquad w_2(0) = c,\ \dot{w}_2(0) = d \end{aligned}$$

i.e. $\ddot{w}_1/w_1 = \ddot{w}_2/w_2$. Hence $\dot{w}_1(t)w_2(t) - w_1(t)\dot{w}_2(t) = \text{constant} = cb - ad \neq 0$. Now, $(y(t), \dot{y}(t)) \rightharpoonup (0,0)$ if and only if $w_1(t), w_2(t) \to 0$, and so

$$(y(t), \dot{y}(t)) \not\rightharpoonup (0,0). \qquad \square$$

We present two examples from Ball and Slemrod (1979a).

Examples

1. If $\Omega \subseteq \mathbb{R}^n$ is a bounded open set, consider the controlled wave equation on Ω:

$$y_{tt}(t,x) - \Delta y(t,x) + u(t)y(t,x) = 0, \qquad x \in \Omega, \quad t \in \mathbb{R}^+ \quad (5.3.8)$$

$$y(t,x)\Big|_{\partial\Omega} = 0.$$

Then we define

$$V = L^2(\Omega), \qquad D(P) = \left\{ v \in V: -\Delta v \in V, v\Big|_{\partial\Omega} = 0 \right\}, \qquad P = -\Delta.$$

Then $W = H_0^1(\Omega)$ and so (5.3.8) defines a dynamical system on $H_0^1(\Omega) \times L^2(\Omega)$ which is weakly stabilizable if and only if the eigenvalues of P are simple.

2. The equation of a flexible beam with distributed control is given by

$$y_{tt} + y_{xxxx} + u(t)y = 0, \qquad x \in [0,1], \quad t \geqslant 0,$$

with boundary conditions

$$y = y_x = 0 \qquad \text{at } x = 0, 1 \qquad (5.3.9)$$

if the beam is clamped at the ends, or

$$y = y_{xx} = 0 \qquad \text{at } x = 0, 1 \qquad (5.3.10)$$

if the beam has simple supports at the ends. The Hilbert space V is defined again to be $L^2(\Omega)$ and the operator $P = \mathrm{d}^4/\mathrm{d}x^4$ with domain

$$D(P) = \{ y \in V: y_{xxxx} \in V, y \text{ satisfies (5.3.9) or (5.3.10)} \}$$

Then,

$$W = H_0^2(0,1) \qquad \text{(in case (5.3.9)}$$

and

$$W = H^2(0,1) \cap H_0^1(0,1) \qquad \text{(in case (5.3.10))}.$$

Hence, the system is stabilizable on $H = W \times L^2(0,1)$ since it is easy to see that the eigenvalues are simple in either case.

Note, however, that if we consider the system

$$y_{tt} + y_{xxxx} + u(t)y_{xx} = 0, \qquad x \in [0,1],$$

then

$$B = \begin{bmatrix} 0 & 0 \\ -\mathrm{d}^2/\mathrm{d}x^2 & 0 \end{bmatrix}$$

is bounded from H into H (where $H = W \times L^2(0,1)$ as in the last example), but B is not compact and so $B: H_w \to H$ is not sequentially continuous. Hence the above theory does not apply. Using the theory of nonharmonic Fourier series, Ball and Slemrod (1979b) develop results which do apply in this case.

5.3.3 Non-dissipative systems

In the above discussion, we have treated systems of the form (5.3.3) for which A is dissipative, i.e.

$$\langle Ax, x\rangle \leqslant 0 \qquad \text{for all } x \in D(A). \tag{5.3.11}$$

The intuitive idea for stabilization of the system is then simple. We differentiate $V = \| x \|^2$ formally along the trajectories of the system to obtain

$$\begin{aligned} \dot{V} &= 2\langle Ax, x\rangle + 2u\langle x, Bx\rangle \\ &\leqslant 2u\langle x, Bx\rangle \end{aligned}$$

and we simply choose u to make $\dot{V}$ negative, e.g. $u = -\langle x, Bx\rangle$. If (5.3.11) does not hold, then we cannot use this argument. To obviate this difficulty we shall consider the bilinear system

$$\dot{x} = Ax + uBx, \qquad x(0) = x_0 \tag{5.3.12}$$

(precise conditions on the operators A and B will be imposed later) and again differentiate $V = \| x \|^2$ formally, to obtain

$$\begin{aligned} \dot{V} &= \langle x, \dot{x}\rangle + \langle \dot{x}, x\rangle \\ &= \langle x, Ax + uBx\rangle + \langle Ax + uBx, x\rangle \\ &= \langle (A + A^*)x, x\rangle + u\langle (B + B^*)x, x\rangle. \end{aligned}$$

Then we define the control

$$\begin{aligned} u &= -\frac{-\langle (A + A^*)x, x\rangle - 1}{\langle (B + B^*)x, x\rangle}, && \text{if } \langle (B + B^*)x, x\rangle \neq 0 \\ u &= 0, && \text{otherwise} \end{aligned} \tag{5.3.13}$$

(see Banks and Morris (1986)). It follows that

$$\dot{V} = -1$$

and so we have stabilizability in finite time. Note, however, that if A is an unbounded operator, then (5.3.13) is only defined for $x \in D(A)$. To overcome this difficulty we shall use the spectral theorem (Theorem 1.6.2) for a self-adjoint operator and assume that A splits into an unbounded stable part and a bounded unstable part.

To be more specific we shall assume that A is a densely defined self-adjoint operator on a Hilbert space H (if A is not self-adjoint we consider $A+A^*$). By the spectral theorem we can find a family of projections $P_A(\lambda)$, $\lambda \in (-\infty, \infty)$ which has the properties of Theorem 1.6.2. Similarly, we shall denote by $P_B(\lambda)$ the spectral projections corresponding to the self-adjoint operator $B+B^*$, where we assume that $B \in \mathscr{L}(H)$. Moreover, we shall assume that $B+B^*$ is non-negative in the sense that the (real) spectrum of $B+B^*$ is in the interval $[\alpha, \beta] \subseteq \mathbb{R}$, $0 \leqslant \alpha, \beta \leqslant \infty$. Now suppose that the spectrum of A is separated, so that

$$\sigma(A) \subseteq (-\infty, -\varepsilon) \cup [\gamma, \delta]$$

where $\varepsilon > 0$, $0 < \gamma \leqslant \delta < \infty$. Then we can write

$$Ah = \int_{-\infty}^{-\varepsilon} \lambda \, dP_A(\lambda)h + \int_{\gamma-0}^{\delta} \lambda \, dP_A(\lambda)h, \qquad h \in D(A). \tag{5.3.14}$$

Note that $P_A(\lambda)$ is constant on $[-\varepsilon, \gamma)$ (with a jump at γ). Let P_ε^A denote this constant projection. Similarly we write

$$(B+B^*)h + \int_{\alpha-0}^{\beta} \lambda \, dP_B(\lambda)h, \qquad h \in H$$

and we define $P_\alpha^B = P_B(\alpha)$. We shall assume that

$$I - P_\varepsilon^A \leqslant P_\alpha^B. \tag{5.3.15}$$

The denominator of (5.3.13) is given by

$$\begin{aligned} \langle (B+B^*)x, x \rangle &= \int_{\alpha-0}^{\beta} \lambda d\langle P_B(\lambda)x, x\rangle \\ &\geqslant \langle P_\alpha^B x, x\rangle, \end{aligned} \tag{5.3.16}$$

and we consider the following orthogonal splitting of the state:

$$x = x_1 + x_2,$$

where

$$x_1 = \bar{P}_\varepsilon^A x \triangleq (I - P_\varepsilon^A)x, \qquad x_2 = P_\varepsilon^A x.$$

Since the integrals in the expression

$$Ax = \int_{-\infty}^{-\varepsilon} \lambda \, dP_A(\lambda)x + \int_{\gamma-0}^{\delta} \lambda dP_A(\lambda)x, \qquad x \in D(A)$$

are given by Riemann–Stieltjes sums, a simple limit argument shows that

$$P_\varepsilon^A Ax = \int_{-\infty}^{-\varepsilon} \lambda dP_A(\lambda)x, \qquad x \in D(A)$$

and so

$$\bar{P}_\varepsilon^A Ax = \int_{\gamma-0}^{\delta} \lambda dP_A(\lambda)x, \qquad x \in D(A).$$

Thus the original equation (5.3.12) splits into the equations

$$\begin{aligned} \dot{x}_1 &= \bar{P}_\varepsilon^A Ax + u\bar{P}_\varepsilon^A Bx \\ \dot{x}_2 &= P_\varepsilon^A Ax + uP_\varepsilon^A Bx. \end{aligned} \tag{5.3.17}$$

We now choose the control u to be given by (5.3.13) on $R(\bar{P}_\varepsilon^A)$ (the range of $\bar{P}_\varepsilon^A$) and to be zero on $R(P_\varepsilon^A)$. Before giving a rigorous proof of stabilizability of the system (5.3.12) we make the following informal observations. We see that x approaches the subspace

$$\{x: \langle (B+B^*)x, x\rangle = 0\} \tag{5.3.18}$$

in finite time. By (5.3.15), we have

$$\langle (B+B^*)x, x\rangle \geqslant \| P_\alpha^B x\|^2 \geqslant \| \bar{P}_\varepsilon^A x\|^2.$$

Hence on the subspace determined by (5.3.18) we have $x_1 = 0$ and so equations (5.3.17) become

$$\begin{aligned} \dot{x}_1 &= 0 \\ \dot{x}_2 &= P_\varepsilon^A Ax = AP_\varepsilon^A x = Ax_2. \end{aligned}$$

The latter equation has the alternative expression

$$\dot{x}_2 = \int_{-\infty}^{-\varepsilon} \lambda \mathrm{d}P_A(\lambda)x_2. \tag{5.3.19}$$

We shall assume that the operator $P_\varepsilon^A A$ generates a stable semigroup. It will then be shown below that the system (5.3.10) is stabilizable. To justify this we must prove that the system

$$\begin{aligned} \dot{x} &= Ax - \left\{\frac{2\langle Ax, x\rangle + 1}{\langle (B+B^*)x, x\rangle}\right\} Bx, &&\text{if } \langle (B+B^*)x, x\rangle \neq 0 \\ \dot{x} &= Ax, &&\text{if } \langle (B+B^*)x, x\rangle = 0 \end{aligned}$$

has a solution. However, to prove that this system has a solution is not particularly easy because of the term $\langle Ax, x\rangle$ in the control, and so we shall define the control to be given by

$$\begin{aligned} u &= -\left[\frac{2\langle \bar{P}^A Ax, x\rangle + 1}{\langle (B+B^*)x, x\rangle}\right], &&\text{for } \langle (B+B^*)x, x\rangle \neq 0 \\ &= 0, &&\text{otherwise} \end{aligned}$$

rather than by (5.3.13). Here,

$$\bar{P}_\varepsilon^A Ax = \int_{\gamma-0}^{\delta} \lambda \mathrm{d}P_A(\lambda)x, \qquad x \in D(A),$$

and the right-hand side defines a bounded operator and so is, in fact, valid for all $x \in H$. Then we obtain

$$\dot{V} = \frac{\mathrm{d}}{\mathrm{d}t} \langle x, x \rangle = 2\left\langle \int_{-\infty}^{-\varepsilon} \lambda \mathrm{d} P_A(\lambda) x, x \right\rangle - 1 \leqslant -1$$

and the subspace $\{x : \langle (B + B^*)x, x \rangle = 0\}$ is still attracting in finite time.

Theorem 5.3.5
If B is a bounded operator and A is a self-adjoint operator which generates a semigroup $T(t)$ on H, which is stable when restricted to the subspace $P_\varepsilon^A H$, then the system

$$\begin{aligned} \dot{x} &= Ax - \left\{ \frac{2\langle \bar{P}_\varepsilon^A Ax, x \rangle + 1}{\langle (B + B^*)x, x \rangle} \right\} Bx, \quad \langle (B + B^*)x, x \rangle \neq 0 \\ \dot{x} &= Ax, \qquad\qquad\qquad\qquad\qquad \langle (B + B^*)x, x \rangle = 0 \\ x(0) &= x_0 \end{aligned} \tag{5.3.20}$$

has a weak solution which converges to 0 as $t \to \infty$.

Proof
If $x_0 \in M$ where $M = \{x \in H : \langle (B + B^*)x, x \rangle = 0\}$, then the result is trivial by (5.3.19).

Consider the case where $x_0 \notin M$. If $x(t) \notin M$ for $t \in [0, \tau]$, then we may write (5.3.20) in the mild form:

$$x(t) = T(t)x_0 - \int_0^t T(t - s) \left\{ \frac{2\langle \bar{P}_\varepsilon^A Ax(s), x(s) \rangle + 1}{\langle (B + B^*)x(s), x(s) \rangle} \right\} Bx(s) \, \mathrm{d}s. \tag{5.3.21}$$

Using the limit argument in the proof of Theorem 5.3.2 it is easy to see that

$$\| x(t) \|^2 \leqslant \| x_0 \|^2 - t \tag{5.3.22}$$

and so $x \to M$ in finite time provided the solution exists. Note that the inequality (5.3.22) is valid only as long as $x(t) \notin M$, so it does not imply that $\| x \| \to 0$, although it does imply that $x \to M$ as we shall see.

Since $\bar{P}_\varepsilon^A A$ is a bounded operator, and since $\| x(t) \|^2$ is decreasing,

$$f(x(\,.\,)) \triangleq \frac{\langle \bar{P}_\varepsilon^A Ax, x \rangle + 1}{\langle (B + B^*)x, x \rangle} \tag{5.3.23}$$

is in $L^1[0, \tau]$ and so a function $x(\,.\,)$ is a weak solution of (5.3.20) on $[0, \tau]$ if and only if it is a mild solution of (5.3.23).

Now, it is easy to check that, for $x \notin M$, f is locally Lipschitz and so (5.3.21) has a solution on $[0, \tau]$ provided $x \notin M$. Moreover, $x \in C([0, \tau]; H)$. From (5.3.22) it follows that there is a minimal time τ_m, say, such that $x \in M$ when $t = \tau_m$ and $x \in C([0, t]; H)$ for any time $t < \tau_m$.

Let $\{t_n\} \to \tau_m$. Then, by (5.3.22), $x(t_i) \rightharpoonup x_m$ (possibly for some subsequence of $\{t_n\}$), where $x_m \in M$. If we define $x(\tau_m) = x_m$, the result follows. □

Remarks

(a) If A is not self-adjoint, then we replace A by $A + A^*$

(b) Note that the weak solution in Theorem 5.3.5 may not be unique.

As an example consider the hyperbolic system

$$\frac{\partial^2\phi(x,t)}{\partial t^2} = \frac{\partial^2\phi(x,t)}{\partial x^2} - \alpha\,.\frac{\partial\phi(x,t)}{\partial t} + \int_0^1 k(x,y)\phi(y,t)\,\mathrm{d}y + u\phi(x,t)$$
$$\phi(0) = \phi(1) = 0. \tag{5.3.24}$$

Then if $\Phi = (\phi, \partial\phi/\partial t)$, we can write the equation in the form

$$\frac{\mathrm{d}\Phi}{\mathrm{d}t} = \begin{bmatrix} 0 & I \\ A+K & -\alpha \end{bmatrix}\Phi + u\begin{bmatrix} 0 & 0 \\ I & 0 \end{bmatrix}\Phi, \tag{5.3.25}$$

where

$$(K\phi)(x) = \int_0^1 k(x,y)\phi(y,t)\,\mathrm{d}y, \qquad A\phi = \frac{\partial^2\phi}{\partial x^2}.$$

Equation (5.3.25) is defined on the Hilbert space $H = H_0^1\,(0,1) \oplus L^2(0,1)$ which is given the inner product

$$\langle(\phi_1,\psi_1),(\phi_2,\psi_2)\rangle_H = \langle(-A)^{1/2}\phi_1,(-A)^{1/2}\phi_2\rangle_{L^2} + \langle y_1,\psi_2\rangle_{L^2}.$$

If

$$\mathscr{A} = \begin{pmatrix} 0 & I \\ A & -\alpha \end{pmatrix},$$

then $D(\mathscr{A}) = (H_0^1(0,1) \cap H^2(0,1)) \oplus H_0^1(0,1)$. Also, we have

$$\langle\Phi,\mathscr{A}\Phi\rangle_H = -\alpha\,\|\psi\|^2$$
$$\langle\Phi,\mathscr{A}^*\Phi\rangle_H = -\alpha\,\|\psi\|^2, \tag{5.3.26}$$

where $\Phi = (\phi,\psi) \in D(\mathscr{A})$. Then (Banks (1983)) $\mathscr{A}$ generates a stable semigroup. Moreover, the operator

$$\mathscr{A} + \begin{bmatrix} 0 & 0 \\ K & 0 \end{bmatrix}$$

is a bounded perturbation of $\mathscr{A}$ and so it also generates a (not necessarily stable) semigroup. Since the dual of

$$\begin{bmatrix} 0 & 0 \\ K & 0 \end{bmatrix}$$

is

$$\begin{bmatrix} 0 & 0 \\ K^* & 0 \end{bmatrix},$$

condition (5.3.15) holds trivially, because

$$P_\alpha^B = \begin{bmatrix} 0 & 0 \\ I & 0 \end{bmatrix}$$

with $\alpha = 1$. Hence Theorem 5.3.5 applies and the system is stabilizable.

5.4 DISTRIBUTED BILINEAR SYSTEMS AND CONTROLLABILITY

5.4.1 Approximate controllability

We shall consider, in this final section (following Ball *et al.* (1982)), the bilinear system

$$\dot{x}(t) = Ax(t) + u(t)Bx(t), \qquad x(0) = x_0 \tag{5.4.1}$$

where A generates a C^0 semigroup $T(t)$ on a Hilbert space H and $B \in \mathscr{L}(H)$. Recall from (5.1.3) that the Fréchet derivative of the solution $x(t; u, x_0)$ of (5.4.1) at $u = 0$ is given by

$$(D_u x(t; u, x_0)|_{u=0}).\, u = \int_0^t T(t-s)u(s)BT(s)x_0 \, ds. \tag{5.4.2}$$

Introduce the operator $L_\tau : Z(\tau) \to H$, where

$$L_\tau u = \int_0^\tau T(\tau - s)u(s)BT(s)x_0 \, ds. \tag{5.4.3}$$

where we recall that $Z(\tau)$ is any Banach space which is dense and continuously included in $L^1([0, \tau]; \mathbb{R})$.

If L_τ is surjective then, by the inverse function theorem (and (5.4.2)), it follows that (5.4.1) is locally exactly controllable in the sense that if $h \in H$ is given then there exists $\varepsilon > 0$ such that $x(\tau; u, x_0) = h$ for some $u \in Z(\tau)$ provided $\| h - T(\tau) \| < \varepsilon$. However, L_τ is not surjective in general as we shall see. The best we can hope for is approximate controllability (see Banks (1983)) in the sense that L_τ may have dense range. In fact we have

Lemma 5.4.1
If

$$\langle h, T(\tau - s)BT(s)x_0 \rangle = 0$$

for all $s \in [0, \tau]$ implies $h = 0$, then the range of L_τ is dense in H.

Proof
We have

$$\langle h, L_\tau u\rangle = \int_0^\tau \langle h, T(\tau - s)BT(s)x_0\rangle u(s)\, ds. \tag{5.4.4}$$

The range of L_τ is dense in H if and only if $\langle h, L_\tau u\rangle = 0$ for all $u \in Z(\tau)$ implies that $h = 0$. However, if $\langle h, L_\tau u\rangle = 0$ for all $u \in Z(\tau)$, then by (5.4.4) we must have

$$\langle h, T(\tau - s)BT(s)x_0\rangle = 0, \qquad s \in [0, \tau]$$

since $Z(\tau)$ is dense in $L^1([0, \tau]; \mathbb{R})$. Then $h = 0$ by assumption. □

Of course, if A is bounded, then

$$T(-s)BT(s)x_0 = Bx_0 + s[A, B]x_0 + \frac{s^2}{2}[A, [A, B]]x_0 + \cdots$$

from the Campbell–Hausdorff formula, and we then obtain the familiar accessibility condition

$$\dim \operatorname{span}\{Bx_0, [A, B]x_0, [A, [A, B]]x_0, \ldots\} = n$$

if $H = \mathbb{R}^n$ (see Chapter 2).

In order to prove that L_τ is not generally surjective (if $\dim H = \infty$) we shall need the following technical result:

Lemma 5.4.2
Suppose that $u_n \rightharpoonup u$ as $n \to \infty$ in $L^1([0, \tau]; \mathbb{R})$. Then

$$\zeta_n \triangleq \sup_{t \in [0, \tau]} \left\| \int_0^t (u_n(s) - u(s))T(t - s)Bx(s)\, ds \right\| \to 0$$

as $n \to \infty$, where $x(t) = x(t; u, x_0)$ is the unique solution of (5.4.1).

Proof
See Ball *et al.* (1982). □

If, again, $x(t; u, x_0)$ denotes the unique solution of (5.4.1) with the control u, then we define the *reachable set from* x_0 by

$$R(x_0) = \{x(t; u, x_0) : t \geqslant 0, u \in L^r_{\text{loc}}([0, \infty); \mathbb{R}), r > 1\}.$$

Theorem 5.4.3
Let $x_0 \in H$. Then $R(x_0)$ is contained in a countable union of compact subsets of H and so has dense complement.

Proof
First let $u_n \rightharpoonup u$ in $L^1([0, \tau]; \mathbb{R})$ and put $x_n(t) = x(t; u_r, x_0)$,

$x(t) = x(t; u, x_0)$. Then

$$z_n(t) = \int_0^t (u_n(s) - u(s))T(t-s)Bx(s)\,\mathrm{d}s + \int_0^t u_n(s)T(t-s)Bz_n(s)\,\mathrm{d}s,$$

where $z_n(t) = x_n(t) - x(t)$. Hence,

$$\| z_n(t)\| \leqslant \zeta_n + \int_0^t |u_n(s)|\ \|T(t-s)\|\ \|B\|\ \|z_n(s)\|\,\mathrm{d}s$$

$$\leqslant \zeta_n + C\int_0^t |u_n(s)|\ \|z_n(s)\|\,\mathrm{d}s,$$

for some constant C. Hence, Gronwall's inequality implies that

$$\| z_n(t)\| \leqslant \zeta_n \exp\left(C\int_0^t |u_n(s)|\,\mathrm{d}s\right)$$

$$\to 0$$

uniformly in $[0, \tau]$ as $n \to \infty$, by Lemma 5.4.2. Hence,

$$x(.; u_n, x_0) \to x(.; u, x_0) \qquad \text{in } C([0,\tau]; H). \tag{5.4.5}$$

Now define the subset $R_{mnr}(x_0)$ of $R(x_0)$, for any positive integers m, n, r by

$$R_{mnr}(x_0) = \{ x(t; u, x_0) : t \in [0, m], \| u\|_{L^{1+1/r}([0,m];\mathbb{R})} \leqslant n\}.$$

To show that each set $R_{mnr}(x_0)$ is precompact we choose a sequence $\{x(t_j; u_j, x_0)\}$ in $R_{mnr}(x_0)$ and show that

$$x(t_\mu; u_\mu, x_0) \to x(t; u, x_0) \qquad \text{in } H \tag{5.4.6}$$

for some subsequence $\{x(t_\mu; u_\mu, x_0)\}$. However, we can find subsequences $\{t_\mu\} \subseteq [0, m]$ and $\{u_\mu\} \subseteq L^{1+1/r}([0,m];\mathbb{R})$ such that $t \to t$ and $u_\mu \rightharpoonup u$ in $L^{1+1/r}([0,m];\mathbb{R})$, by the reflexivity of the latter space (since in such a space, any bounded sequence has a weakly convergent subsequence). By (5.4.5) it follows that (5.4.6) is true.

Finally note that

$$R(x_0) \subseteq \bigcup_{m,n,r=1}^{\infty} R_{mnr}(x_0)$$

and so $R(x_0)$ is contained in a countable union of compact sets. Since $\dim H = \infty$, $R_{mnr}(x_0)$ is nowhere dense and by the category theorem, $R(x_0)$ has dense complement. □

5.4.2 Controlling a finite number of modes

Although it is not possible to control exactly to every element of H, we may guess that it is possible to control to a given finite-dimensional subspace. This is in fact the case and we have

Theorem 5.4.4
If A and B satisfy the same assumptions as before and $G:H\to\mathbb{R}^n$ is a bounded linear map, then if

$$\langle h, GT(\tau-s)BT(s)x_0\rangle = 0 \tag{5.4.7}$$

for all $s\in[0,\tau]$, $\tau>0$, implies that $h=0$, we can find an $\varepsilon\,(=\varepsilon(\tau))>0$ such that, given $y\in\mathbb{R}^n$ with $\| y-GT(\tau)x_0\|_{\mathbb{R}^n}<\varepsilon$, we have $Gx(\tau;u,x_0)=y$, for some $u\in Z(\tau)$.

Proof
This is proved in the same way as Lemma 5.4.1, using the remarks preceding the lemma. □

Note that if G is surjective, then the condition that (5.4.7) implies $h=0$ follows from the condition

$$\langle h, T(\tau-s)BT(s)x_0\rangle = 0, \qquad s\in[0,\tau] \Rightarrow h=0 \tag{5.4.8}$$

of Lemma 5.4.1, since if (5.4.8) holds, then

$$\langle h, GT(\tau-s)BT(s)x_0\rangle = 0 \Rightarrow \langle G^*h, T(\tau-s)BT(s)x_0\rangle = 0$$

and so by (5.4.8), $G^*h=0$. If G is surjective, then G^* is injective, so $h=0$.

The next result shows that if one tries to control an increasing number of modes, then the controls must increase in size indefinitely.

Theorem 5.4.5
Let H_n be a subspace of H of dimension n with

$$\overline{\left(\bigcup_{n=1}^{\infty} H_n\right)} = H,$$

and let $G_n:H\to H_n$ be the corresponding projection. Let $\mathscr{U}_r$, $r>1$, be the set of bounded subsequences of $L^r([0,\tau];\mathbb{R})$, i.e.

$$\mathscr{U}_r = \{\{u_n\}\subseteq L^r([0,\tau];\mathbb{R}) : \|u_n\|_{L^r([0,\tau];\mathbb{R})} \leqslant \text{const. depending on } u\}$$

and let H' be the set

$$H' = \{h\in H : \exists\tau>0, r>1,\ \{u_n\}\in\mathscr{U}_r \text{ such that } G_n x(\tau;u_n,x_0) = G_n h, n\geqslant 1\},$$

i.e. H' is the set of points whose projections on H_n can be reached by using a bounded sequence of controls. Then

$$\overline{H\backslash H'} = H.$$

Proof
We may suppose that the projections G_n are uniformly bounded. If $h \in H'$, then we can find a bounded sequence $\{u_n\} \subseteq L^r([0,\tau];\mathbb{R})$ for some $r > 1$ as in the definition of H'. Then (possibly for some subsequence), we can find u such that $u_n \rightharpoonup u$ in $L^1([0,\tau];\mathbb{R})$. Then,

$$\begin{aligned}\| x(\tau;u,x_0) - h \| \leqslant & \| x(\tau;u,x_0) - G_n x(\tau;u,x_0)\| \\ & + \| G_n x(\tau;u,x_0) - G_n x(\tau;u_n,x_0)\| \\ & + \| G_n x(\tau;u_n,x_0) - G_n h \| + \| G_n h - h \| \\ \to & \, 0\end{aligned}$$

as $n \to \infty$, by (5.4.5). Hence $H' \subseteq R(x_0)$ and the latter has dense complement by Theorem 5.4.3. □

5.4.3 Riesz bases

In order to apply the above methods to hyperbolic systems it is convenient to consider a generalization of the notion of orthonormal basis of a Hilbert space, called a Riesz basis. If H is a real or complex Hilbert space, then a sequence $\{\omega_i\}_{i=1}^{\infty} \subseteq H$ is called a *Riesz basis* of H if every $h \in H$ has a unique representation

$$h = \sum_{i=1}^{\infty} a_i \omega_i$$

where the series is convergent in H and, moreover,

$$C_1 \sum_{i=1}^{\infty} | a_i |^2 \leqslant \| h \|^2 \leqslant C_2 \sum_{i=1}^{\infty} | a_i |^2$$

for some constants C_1, C_2 (independent of h).
The following facts are elementary:

Proposition 5.4.6
If $\{\omega_i\}$ is a Riesz basis of H, and $\{e_i\}$ is a complete orthonormal basis of H, then

(a) there is an isomorphism $L: H \to H$ given by

$$L(\Sigma_{i=1}^{\infty} a_i e_i) = \Sigma_{i=1}^{\infty} a_i \omega_i.$$

(b) $\displaystyle\sum_{i=1}^{\infty} |\langle h, \omega_i\rangle|^2 \leqslant \| L^* \|^2 \| h \|^2$ for all $h \in H$.

(c) given any sequence $\{a_i\} \in l^2$, there exists a unique solution $h \in H$ of the equations

$$\langle h, \omega_i \rangle = a_i, \qquad i = 1, 2, \ldots. \tag{5.4.9}$$

□

The only other result we need concerning Riesz bases is the following, which (partially) specifies a basis for $L^2([0, \tau]; \mathbb{C})$.

Theorem 5.4.7
If $\{\mu_k\}_{k=-\infty}^{\infty}$ is a sequence for which

$$0 = \mu_0 < \mu_1 < \mu_2 < \cdots, \qquad \mu_{-k} = -\mu_k, \quad k > 0$$

and

$$\lim_{k \to \infty} (\mu_{k+1} - \mu_k) \geqslant \gamma > 0,$$

then the functions $\{e^{i\mu_k t}\}_{k=-\infty}^{\infty}$ may be extended to a Riesz basis of $L^2([0, \tau]; \mathbb{C})$, provided $\tau > 2\pi/\gamma$.

Proof
See Ball *et al.* (1982). □

If $S \subseteq L^2([0, \tau]; \mathbb{C})$ is the closed linear span of the set $\{e^{i\mu_k t}\}$ then we can extend this to a Riesz basis of $L^2([0, \tau]; \mathbb{C})$ by choosing any orthonormal basis of $S^{\perp}$.

5.4.4 Hyperbolic systems

Consider now the abstract hyperbolic system

$$\begin{aligned} \ddot{x} + Ax + u(t)Bx &= 0 \\ x(0) = x_0 \in D(A^{1/2}), \qquad \dot{x}(0) &= x_1 \in H \end{aligned} \tag{5.4.10}$$

for some real Hilbert space H (where, of course, A is assumed to be positive and self-adjoint with $\overline{D(A)} = H$). We shall assume also that $B: D(A^{1/2}) \to H$ is a bounded linear operator and that A^{-1} is compact and A has simple eigenvalues λ_n^2, $n = 1, 2, \ldots$ with $0 < \lambda_1 < \lambda_2 < \cdots$. Let $\{\phi_n\}$ be the corresponding complete orthonormal basis of eigenfunctions.

We have seen before how to write (5.4.10) in the 'phase space' $D(A^{1/2}) \times H$. However, we shall see below that it is more usefully expressed in the Hamiltonian formulation which we now discuss. We first introduce the Hilbert space $\mathscr{H} = H \oplus iH$ which is the complexification of H and define the inner product on $\mathscr{H}$ by

$$\langle x_1 + iy_1, x_2 + iy_2 \rangle_{\mathscr{H}} = \langle x_1, x_2 \rangle + \langle y_1, y_2 \rangle + i[\langle y_1, x_2 \rangle - \langle x_1, y_2 \rangle],$$

for all $x_1, x_2, y_1, y_2 \in H$. Note that $D(A^{1/2}) \times H$ and $\mathscr{H}$ are isometric under the map

$$(x_1, x_2) \to A^{1/2}x_1 + ix_2,$$

and if we define $z = A^{1/2}x + i\dot{x}$, then (5.4.10) becomes

$$\begin{aligned} i\dot{z} &= iA^{1/2}\dot{x} - (-Ax - u(t)Bx) \\ &= A^{1/2}z + u(t)BA^{-1/2}\text{Re}\,(z) \qquad (5.4.11) \\ z(0) &= z_0 \end{aligned}$$

where

$$z_0 = A^{1/2}x_0 + ix_1 \in \mathscr{H}.$$

Since $\{\phi_n\}$ is a basis of H we may also regard $\{\phi_n\}$ as a basis of $\mathscr{H}$, where the expansion coefficients of any element are complex, i.e.

$$z = \sum_{n=1}^{\infty} z_n\phi_n, \qquad z_n \in \mathbb{C}, \quad \{z_n\} \in l^2, \quad z \in \mathscr{H}.$$

If we write $\mathscr{A} = -iA^{1/2}$, then, if $T(t)$ denotes the group generated by $\mathscr{A}$, we have

$$T(t)z = \sum_{n=1}^{\infty} z_n\, e^{-i\lambda_n t}\phi_n.$$

Hence, if $\mathscr{B} = -iBA^{-1/2}\,\text{Re}$, then

$$\mathscr{B}\,T(t)z = -i \sum_{n,m=1}^{\infty} \frac{B_{mn}}{\lambda_m}\,\text{Re}(e^{-i\lambda_m t}z_m)\phi_n,$$

where

$$B\phi_m = \sum_{n=1}^{\infty} B_{mn}\phi_n.$$

Thus we have

$$T(-t)\mathscr{B}\,T(t)z = -\frac{i}{2} \sum_{m,n=1}^{\infty} \frac{B_{mn}}{\lambda_m}\,(e^{i(\lambda_n-\lambda_m)t}z_m + e^{i(\lambda_n+\lambda_m)t}\bar{z}_m)\phi_n. \qquad (5.4.12)$$

If $h = \sum_{n=1}^{\infty} h_n\phi_n$, $h_n \in \mathbb{C}$ is an element of $\mathscr{H}$, then by (5.4.12) we have

$$\begin{aligned} 2i\langle T(-t)\mathscr{B}\,T(t)z_0, h\rangle &= \sum_{n=1}^{\infty} \bar{h}_n(z_{0n} + \bar{z}_{0n}\,e^{2i\lambda_n t})\frac{B_{nn}}{\lambda_n} \\ &+ \sum_{\substack{m \neq n \\ m,n=1}}^{\infty} \bar{h}_n(z_{0m}\,e^{i(\lambda_n-\lambda_m)t} + \bar{z}_{0m}\,e^{i(\lambda_n+\lambda_m)t})\frac{B_{mn}}{\lambda_m} \qquad (5.4.13) \end{aligned}$$

where

$$z_0 = \sum_{n=1}^{\infty} z_{0n}\phi_n = A^{1/2}x_0 + ix_1,$$

and

$$z_{0n} = \lambda_n\langle x_0, \phi_n\rangle + i\langle x_1, \phi_n\rangle.$$

Let β denote the set of pairs (p,q) such that $B_{pq} \neq 0$ and $p \neq q$ and assume that the set

$$\Gamma = \{e^{2i\lambda_n t}\}_{n=1}^{\infty} \cup \{e^{i(\lambda_p - \lambda_q)t}, e^{i(\lambda_p+\lambda_q)t}, (p,q) \in \beta\}$$

can be extended to a Riesz basis of $L^2([0,\tau];\mathbb{C})$. Then the coefficients of these functions on the right-hand side of (5.4.13) must be zero if the right-hand side is zero. Hence

$$\frac{\bar{h}_n \bar{z}_{0n} B_{nn}}{\lambda_n} = 0, \qquad n \geqslant 1. \tag{5.4.14}$$

Theorem 5.4.8

Suppose that the initial values x_0, x_1 satisfy

$$B_{nn}[\langle x_0, \phi_n\rangle^2 + \langle x_1, \phi_n\rangle^2] \neq 0, \qquad n \geqslant 1 \tag{5.4.15}$$

and that $\tau > 0$ is chosen so that the set Γ can be extended to a Riesz basis of $L^2([0,\tau];\mathbb{C})$. Then for any $\tau_1 \geqslant \tau$ and any bounded surjective maps

$$G_1 : D(A^{1/2}) \to \mathbb{R}^m, G_2 : H \to \mathbb{R}^n,$$

there exists ε_{τ_1} such that if

$$\| y_1 - G_1 x(\tau_1; 0, x_0, x_1)\|_{\mathbb{R}^m} < e_{\tau_1}, \| y_2 - G_2 x(\tau_1; 0, x_0, x_1)\|_{\mathbb{R}^n} < \varepsilon_{\tau_1},$$

then there exists $u \in Z(\tau_1)$ with

$$G_1 x(\tau_1; u, x_0, x_1) = y_1, G_2 \dot{x}(\tau_1; u, x_0, x_1) = y_2.$$

Proof

Consider the expression (5.4.13) and suppose that

$$\langle T(-t)\mathscr{B}T(t)z_0, h\rangle_{\mathscr{H}} = 0 \text{ for } t \in [0, \tau_1].$$

By (5.4.14) and (5.4.15) we have $h_n = 0$ for $n \geqslant 1$ and so $h = 0$. Then the result follows from Theorem 5.4.4 and the following remark. □

The problem of showing that the set Γ can be extended to a Riesz basis is difficult in general and so we assume now that B is diagonal, i.e. $B_{mn} = 0$ if $m \neq n$. In this case the equation (5.4.11) may be reduced to a set of ordinary differential equations

$$\dot{z}_n = -i\lambda_n z_n - iu(t)\frac{b_n}{\lambda_n}\operatorname{Re} z_n, \qquad n \geqslant 1, \quad z_n(0) = z_{0n}. \tag{5.4.16}$$

where $z = \sum_{n=1}^{\infty} z_n \phi_n$. It is not difficult to show that, when B is diagonal, Theorem 5.4.3 is true with L^1 controls, i.e. $R(x_0)$ can be replaced by the set

$$R(z_0) = \{z_n(t; u, z_0) : t \geqslant 0, u \in L^1_{\text{loc}}([0,\infty); \mathbb{R})\}.$$

To proceed any further and prove that the reachable set is dense in $\mathscr{H}$ we change coordinates to remove the operator 'Re' in (5.4.16). In fact, if we define

$$\zeta_n = \frac{\lambda_n}{b_n}\left[\frac{z_n}{z_{0n}} \exp i\left(\lambda_n t + \frac{b_n}{2\lambda_n} u(t)\right) - 1\right]$$

it follows from (5.4.16) that ζ satisfies the equation

$$\dot{\zeta}_n(t) = \frac{-iu(t)}{2}\frac{\bar{z}_{0n}}{z_{0n}}\left(\frac{b_n}{\lambda_n}\bar{\zeta}_n(t)+1\right)\exp\left[2i\left(\lambda_n t + \frac{b_n}{2\lambda_n} v(t)\right)\right] \quad (5.4.17)$$

$$\zeta_n(0) = 0.$$

where $v(t) = \int_0^t u(s)\, ds$. Using the methods of Section 5.1 we can then prove

Theorem 5.4.9
Let $\{z_{0n}\} \in l^2$, with $z_{0n} \neq 0, n \geqslant 1$, and assume that $\{e^{2i\lambda_n t}\}$ can be extended to a Riesz basis of $L^2([0, \tau_1]; \mathbb{R})$. Suppose that $b_n/\lambda_n = c + \gamma_n$ for some $c \in \mathbb{R}$ and $\{\gamma_n\} \in l^2$. Then, if $u \in L^2_{loc}([0, \infty); \mathbb{R})$, equation (5.4.17) has a unique absolutely continuous solution $\zeta_n = \zeta_n(t; u)$ for all $t \geqslant 0$ and $\{\zeta_n(.,u)\} \in C([0, \tau]; l^2)$ where $\tau \in [0, \tau_1]$. Moreover, the map $u \to \{\zeta_n(\tau; u)\}$ is C^1 from $L^2([0, \tau]; \mathbb{R})$ to l^2 for $\tau \in [0, \tau_1]$ and the Fréchet derivative of the solution with respect to u is given by

$$D_u\{\zeta_n(\tau; 0)\} . u = -\frac{i}{2}\frac{\bar{z}_{0n}}{z_{0n}}\int_0^\tau u(t)\exp(2i\lambda_n t)\, dt. \quad (5.4.18)$$

Consider the map $Q: L^2([0, \tau]; \mathbb{R}) \to l^2 \times \mathbb{R}$ (for some $\tau > 0$) given by

$$Q(u) = \left(\{\zeta_n(\tau; u)\}; \int_0^\tau u(t)\, dt\right).$$

Then, by (5.4.18), we have

$$D_u Q(0) . u = \left(\left\{\frac{-i}{2}\frac{\bar{z}_{0n}}{z_{0n}}\int_0^\tau u(t)\exp(2i\lambda_n t)\, dt\right\}, \int_0^\tau u(t)\, dt\right).$$

By Theorem 5.4.9, Q is C^1; we show that $D_u Q(0)$ is surjective, provided $\{1, e^{\pm 2i\lambda_n t}\}$ can be extended to a Riesz basis of $L^2([0, \tau]; \mathbb{C})$. In fact, if $\{a_n\} \in l^2$ and $\alpha \in \mathbb{R}$, then, by Proposition 5.4.6 (c) we can solve the equations

$$\int_0^\tau v(t)\exp(2i\lambda_n t)\, dt = c_n, \quad \int_0^\tau v(t)\exp(-2i\lambda_n t)\, dt = \bar{c}_n, \quad \int_0^\tau v(t)\, dt = \alpha$$

for $v \in L^2([0, \tau]; \mathbb{C})$, where $c_n = 2i(z_{0n}/\bar{z}_{0n})a_n$. Clearly,

$$\int_0^\tau (\text{Im } v(t))\exp(2i\lambda_n t)\, dt = \int_0^\tau \text{Im } v(t)\, dt = 0.$$

Hence if we set $u = \text{Re}(v)$ it follows that $D_u Q(0)$ is surjective. Applying Theorem 5.4.7, we have therefore proved

Theorem 5.4.10
Suppose that $\{z_{0n}\} \in l^2$ with $b_n \neq 0, z_{0n} \neq 0$ for $n \geqslant 1$ and assume that

$$\lim_{n \to \infty} (\lambda_{n+1} - \lambda_n) \geqslant \nu > 0.$$

Then if $\tau > \pi/\nu$ there exists $\epsilon_\tau > 0$ such that, for any $h \in l^2$, $\theta \in \mathbb{R}$, with $\| h \|_{l^2} + | \theta | < \epsilon_\tau$, there exists $u \in L^2([0, \tau]; \mathbb{R})$ such that

$$\frac{\lambda_n}{b_n} \left(\frac{z_n(\tau)}{z_{0n}} \exp\left[\mathrm{i}\left(\lambda_n \tau + \frac{b_n}{2\lambda_n} \right) \theta \right] - 1 \right) = h_n, \qquad n = 1, 2, \ldots \quad (5.4.19)$$

and $\int_0^\tau u(t)\, \mathrm{d}t = \theta$. □

Note that, if λ_n/σ is an integer for some $\sigma > 0$ and all n, then (5.4.19) becomes

$$\frac{z_n(2\pi/\sigma)}{z_{0n}} = \exp\left(\frac{-\mathrm{i} b_n \theta}{2\lambda_n} \right) \left(1 + \frac{b_n h_n}{\lambda_n} \right), \qquad n \geqslant 1$$

if we take $\tau = 2\pi/\sigma \ (> \pi/\nu)$.

Theorem 5.4.10 is a local approximate controllability result. It can be extended to the following global result:

Theorem 5.4.11
Let $z_0 = \{z_{0n}\} \in l^2$ with $z_{0n} \neq 0$, $n \geqslant 1$ and let λ_n/σ be an integer for all n and some $\sigma > 0$. Then, given $h \in l^2$ such that $1 + (b_n/\lambda_n)h_n \neq 0$ for all $\theta \in \mathbb{R}$, there exist an integer $m > 0$ and a control $u \in L^2([0, 2m\pi/\sigma]; \mathbb{R})$ such that

$$\frac{z_n(2m\pi/\sigma)}{z_{0n}} = \exp\left(\frac{-\mathrm{i} b_n \theta}{2\lambda_n} \right) \left(1 + \frac{b_n}{\lambda_n} h_n \right), \qquad n \geqslant 1$$

where $\theta = \displaystyle\int_0^{2m\pi/\sigma} u(t)\, \mathrm{d}t$.

Proof
See Ball *et al.* (1982). □

Since the set $\{h \in l^2 : 1 + (b_n/\lambda_n)h_n \neq 0, n \geqslant 0\}$ is dense in l^2, we have

Corollary 5.4.12
If the hypotheses of Theorem 5.4.11 hold, then the reachable set

$$R(z_0) = \{z(t; u, z_0) : t \geqslant 0, u \in L^2_{\mathrm{loc}}([0, \infty); \mathbb{R})\}$$

is dense in $\mathscr{H}$. □

Example
Consider the wave equation

$$\psi_{tt} - \psi_{xx} + u(t)\psi = 0, \qquad 0 < x < 1.$$
$$\psi(0, t) = \psi(1, t) = 0$$
$$\psi(x, 0) = \psi_0(x), \quad \psi_t(x, 0) = \psi_1(x), \qquad 0 < x < 1.$$

Define

$$A = \frac{-\mathrm{d}^2}{\mathrm{d}x^2}, \qquad D(A) = H^2(0, 1) \cap H_0^1(0, 1)$$

$$B = I.$$

(Note that $D(A^{1/2}) = H_0^1(0, 1)$.)
Then

$$\lambda_n = n\pi, \quad \phi_n = \sqrt{2} \sin n\pi x, \qquad n = 1, 2, \ldots, b_n = 1.$$

Hence the operator B is diagonal and $\{b_n/\lambda_n\} = \{1/n\pi\} \in l^2$. The space $\mathscr{H}$ is given by

$$\mathscr{H} = L^2(0, 1) \oplus \mathrm{i}L^2(0, 1).$$

Moreover, note that $\{1, \mathrm{e}^{\pm 2\mathrm{i}\lambda_n t}\}$ is a Riesz basis of $L^2([0, 1]; \mathbb{C})$ and it can be extended to a Riesz basis of $L^2([0, \tau]; \mathbb{C})$ if $\tau \geqslant 1$. If we define the reachable set

$$R_r(\{\psi_0, \psi_1\}) = \{(\psi(t; u, \psi_0, \psi_1), \psi_t(t; u, \psi_0, \psi_1)) : t \geqslant 0, u \in L^r_{\mathrm{loc}}([0, \infty); \mathbb{R})\},$$

then the above results show that $R_r(\{\psi_0, \psi_1\})$ has dense complement in $H_0^1(0, 1) \times L^2(0, 1)$ and also $R_2(\{\psi_0, \psi_1\})$ is dense in this space if $\lambda_n(\psi_0, \phi_n) + \mathrm{i}(\psi_1, \phi_n) \neq 0$ for each n and the time interval has length at least 1.

Further examples are given in Ball *et al.* (1982).

REFERENCES

R. Abraham and J. E. Marsden (1978). *Foundations of Mechanics*, (2nd edn), Benjamin/Cummings.

R. A. Adams (1975). *Sobolev Spaces*, Academic Press.

V. M. Alekseev (1961). An estimate for the perturbations of the solutions of ordinary differential equations (Russian). *Vestnik Moskov. Univ. Ser. I Mat. Mek.* **2**, 28–36.

L. Auslander and R. MacKenzie (1963). *Introduction to Differentiable Manifolds*, McGraw-Hill.

J. M. Ball (1977). Strongly continuous semigroups, weak solutions, and the variation of constants formula. *Proc. Amer. Math. Soc.*, **63**, 370–373.

J. M. Ball, 1978. On the asymptotic behaviour of generalized processes, with applications to nonlinear evolution equations, *J. Diff. Eqns.*, **27**, 224–265.

J. M. Ball, J. E. Marsden and M. Slemrod (1982). Controllability for distributed bilinear systems, *SIAM J. Contr.*, **20**, 575–597.

J. M. Ball and M. Slemrod (1979a). Feedback stabilization of distributed semilinear control systems, *Appl. Math. Optim.*, **5**, 169–179.

J. M. Ball and M. Slemrod (1979b). Nonharmonic Fourier series and the stabilization of distributed semi-linear control systems, *Comm. Pure Appl. Maths.*, **32**, 555–587.

S. P. Banks (1983). *State-space and Frequency Domain Methods in the Control of Distributed Parameter Systems*. Peter Peregrinus.

S. P. Banks (1984a). Controllability and optimal control of partial differential equations on compact manifolds, *Int. J. Sys. Sci.*, **15**, 543–562.

S. P. Banks (1984b). On nonlinear perturbations of nonlinear dynamical systems and applications to control, *IMA J. Math. Cont. Inf.*, **1**, 67–81.

S. P. Banks (1985a). On nonlinear systems and algebraic geometry, *Int. J. Contr.*, **42**, 333–352.

S. P. Banks (1985b). On nonlinear perturbations of dynamical systems, *IMA J. Math. Cont. Inf.*, **2**, 61–70.

S. P. Banks (1986a). *Control Systems Engineering*, Prentice-Hall.

S. P. Banks (1986b). Tensor operators and limit cycles in nonlinear systems, *Int. J. Control*, **43**, 883–890.

S. P. Banks (1986c). On the control of partial differential equations and algebraic geometry. In P. A. Cook (ed.), *Proc. 4th IMA conference on control theory*, Cambridge.

S. P. Banks and A. Ashtiani (1985). Linear and bilinear infinite-dimensional representations of finite-dimensional nonlinear systems, *Int. J. Sys. Sci.*, **16**, 841–853.

S. P. Banks and A. S. Morris (1986). Global stabilization of nonlinear systems and its application to robotics, *Proc. of IMACS-IFAC symp, on modelling and simulation for control of lumped and distributed parameter systems*, Lille, France.

S. P. Banks and M. K. Yew (1985). On a class of suboptimal controls for infinite-dimensional bilinear systems, *Sys. Contr. Letts.*, **5**, 327–333.

S. P. Banks and M. K. Yew (1986). On the optimal control of bilinear systems and its relation to Lie algebras, *Int. J. Control*, **43**, 891–900.

V. Barbu (1976). *Nonlinear Semigroups and Differential Equations in Banach Spaces*, Noordhoff Int. Publ.

B. Bonnard (1984). Controllabilité de systèmes méchaniques sur les groupes de Lie, *SIAM J. Contr.*, **33**, 711–722.

W. M. Boothby (1975). A transitivity problem from control theory. *J. Diff. Eqns.*, **17**, 296–307.

W. M. Boothby (1982). Some comments on positive orthant controllability of bilinear systems, *SIAM J. Contr.*, **20**, 634–644.

W. M. Boothby and E. N. Wilson (1979). Determination of the transitivity of bilinear systems, *SIAM J. Contr.*, **17**, 212–221.

F. Brauer (1966). Perturbations of nonlinear systems of differential equations, *J. Math. Anal. Appl.*, **14**, 198–206.

R. W. Brockett (1972). System theory on a group manifold and coset spaces, *SIAM J. Contr.*, **10**, 265–284.

R. W. Brockett (1973). Control theory on Lie groups, *Proc. NATO Advanced Study Institute*, Imperial College, London.

R. W. Brockett (1976a). Volterra series and geometric control theory, *Automatica*, **12**, 167–176.

R. W. Brockett (1976b). Nonlinear systems and differential geometry, *Proc. IEEE*, **64**, 61–72.

R. W. Brockett and A. Rahimi (1972). Lie algebras and linear differential equations, in L. Weiss (ed.), *Ordinary Differential Equations*, Academic Press.

T. Carleman (1932). Application de la théorie des équations integrales singulière aux équations differentiales de la dynamique, *T. Ark. Mat. Astron. Fys.*, **22B** (1).

C. Chevalley (1946). *Theory of Lie Groups*, Vol. I, Princeton Univ. Press.

W. L. Chow (1939). Uber systeme von linearen partiellen differential-gleichungen erster ordnung. *Math. Ann.*, **117**, 98–105.

E. A. Coddington and N. Levinson (1955). *Theory of Ordinary Differential Equations*, McGraw-Hill.

P. C. Collingwood (1985). *Algebraic Estimation and Graded Polynomial Systems*, Ph.D. thesis, Warwick Univ.

P. E. Crouch (1981). Dynamical realizations of finite Volterra series, *SIAM J. Contr.*, **19**, 177–202.

P. E. Crouch (1984). Solvable approximations to control systems, *SIAM J. Contr.*, **22**, 40–54.

L. A. Crum and J. A. Heinen (1974). Simultaneous reduction and expansion of multi-dimensional Laplace transform kernels, *SIAM J. Appl. Math.*, **26**, 753–771.

R. F. Curtain and A. J. Pritchard (1978). *Infinite-dimensional Linear Systems Theory*, Springer-Verlag.

C. M. Dafermos (1972). Uniform processes and semicontinuous Liapunov functionals, *J. Diff. Eqns.*, **11**, 401–415.

P. D'Alessandro, A. Isidori and A. Ruberti (1973). A new approach to the theory of canonical decomposition of linear dynamical systems, *SIAM J. Contr.*, **11**, 148–158.

P. D'Alessandro, A. Isidori and A. Ruberti (1974). Realization and structure theory of bilinear dynamical systems, *SIAM J. Contr.*, **12**, 517–535.

H. D'Angelo (1970). *Linear Time-varying Systems: Analysis and Synthesis*. Allyn and Bacon, Boston.

E. J. Davison and E. G. Kunze (1970). Some sufficient conditions for the global and local controllability of nonlinear time-varying systems, *SIAM J. Contr.*, **8**, 489–497.

M. D. Di Benedetto and A. Isidori (1978). Triangular canonical forms for bilinear systems, *IEEE Trans. Aut. Contr.*, AC-**23**, 877–880.

G. Di Pillo, C. Bruni and G. Koch (1974). Bilinear systems: an appealing class of 'nearly linear' systems in theory and applications, *IEEE Trans. Aut. Contr.*, **19**, 334–343.

S. K. Donaldson (1983). An application of gauge theory to the topology of 4-manifolds, *J. Diff. Geom.*, **18**, 269–316.

N. Dunford and J. T. Schwartz (1958). *Linear Operators*, Part I, Wiley-Interscience.

N. Dunford and J. T. Schwartz (1963). *Linear Operators*, Part II, Wiley-Interscience.

M. E. Evans (1983). Bilinear systems with homogeneous input–output maps, *IEEE Trans. Aut. Contr.*, AC-**28**, 113–115.

P. L. Falb and W. A. Wolovich (1967). Decoupling in the design and synthesis of multivariable control systems, *IEEE Trans. Aut. Contr.*, AC-**12**, 651–659.

M. Fliess (1980). A Note on Volterra series expansions for nonlinear differential systems, *IEEE Trans. Aut. Contr.*, AC-**25**, 116–117.

O. Forster (1981). *Reimann Surfaces*, GTM 81, Springer-Verlag.

A. E. Frazho (1980). A shift operator approach to bilinear system theory, *SIAM J. Contr.*, **18**, 640–658.

D. Freed and K. Uhlenbeck (1984). *Topology and 4-manifolds*, Springer-Verlag.

A. Friedman (1969). *Partial Differential Equations*, Holt, Rinehart and Winston.

J. P. Gauthier and G. Bornard (1981). Observability for any $u(t)$ of a class of non-linear systems, *IEEE Trans. Aut. Contr.*, AC-**26**, 922–924.

K. Grasse (1984). On accessibility and normal accessibility: The openness of controllability in the fine C^0 topology, *J. Diff. Eqns.*, **53**, 387–414.

H. Grauert and K. Fritzsche (1976). *Several Complex Variables*, GTM 38, Springer-Verlag.

W. Greub (1978). *Multilinear Algebra* (2nd edn), Springer-Verlag.

J. W. Grizzle and S. I. Marcus (1985). The structure of nonlinear control systems possessing symmetries, *IEEE Trans. Aut. Contr.*, AC-**30**, 248–258.

P. O. Gutman (1981). Stabilizing controllers for bilinear systems, *IEEE Trans. Aut. Contr.*, AC-**26**, 917–922.

S. Helgason (1962). *Differential Geometry and Symmetric Spaces*, Academic Press.

G. Helmberg (1969). Introduction to Spectral Theory in Hilbert Spaces, North Holland.

D. Henry (1981). *Geometric Theory of Semilinear Parabolic Equations*, Lecture notes in mathematics no. 840, Springer-Verlag.

R. Hermann (1968). *Differential Geometry and the Calculus of Variations*, Academic Press.

R. Hermann and A. Krener (1977). Nonlinear controllability and observability, *IEEE Trans. Aut. Contr.*, AC-**22**, 728–740.

H. Hermes (1979). Controllability of nonlinear delay differential equations, *Nonlin. Anal, Theory App. Meth.*, **3**, 483–493.

R. M. Hirschorn (1973). Topological semigroups, sets of generators and controllability, *Duke Math. J.*, **40**, 937–947.

R. M. Hirschorn (1975). Controllability in nonlinear systems, *J. Diff. Eqns.*, **19**, 46–61.

R. M. Hirschorn (1979a). Invertibility of nonlinear control systems, *SIAM J. Contr.*, **17**, 289–297.

R. M. Hirschorn (1979b). Invertibility of multivariable nonlinear control systems, *IEEE Trans. Aut. Contr.*, AC-**24**, 855–865.

R. M. Hirschorn (1981). (A, B)-Invariant distributions and disturbance decoupling of nonlinear systems, *SIAM J. Contr.*, **19**, 1–19.

J. M. Holtzman (1970). *Nonlinear System Theory*, Prentice-Hall.

L. R. Hunt, M. Luksic and R. Su (1986). Exact linearizations of input-output systems, *Int. J. Contr.*, **43**, 247–255.

L. R. Hunt, R. Su and G. Meyer (1983). Global transformations of nonlinear systems, *IEEE Trans. Aut. Contr.*, AC-**28**, 24–31.

A. Isidori (1985). The matching of a prescribed linear input-output behaviour in a nonlinear sytstem, *IEEE Trans. Aut. Contr.*, AC-**30**, 258–265.

A. Isidori, A. J. Krener, C. Gori-Giorgi and S. Monaco (1981). Nonlinear decoupling via feedback: a differential-geometric approach, *IEEE Trans. Aut. Contr.*, AC-**26**, 331–345.

A. Isidori and A. Ruberti (1984). On the synthesis of linear input-output responses for nonlinear systems, *Sys. & Contr. Letts.*, **4**, 17–22.

B. Jakubczyk (1980). Existence and uniqueness of realizations of nonlinear systems, *SIAM J. Contr.*, **18**, 455–471.

V. Jurdjevic and J. P. Quinn (1978). Controllability and stability, *J. Diff. Eqns.*, **28**, 381–389.

V. Jurdjevic and H. Sussman (1972). Control systems on Lie groups, *J. Diff. Eqns.*, **12**, 313–329.

R. E. Kalman (1965). Irreducible realizations and the degree of a rational matrix, *SIAM J. Appl Math.*, **13**, 520–543.

R. E. Kalman, P. L. Falb and M. A. Arbib (1969). *Topics in Mathematical System Theory*, McGraw-Hill.

K. Kendig (1977). *Elementary Algebraic Geometry*, GTM 44, Springer-Verlag.

A. A. Kirillov (1976). *Elements of the Theory of Representations*, Springer-Verlag.

J. Klamka (1976). Relative controllability of nonlinear systems with delays in the control, *Automatica*, **12**, 633–634.

S. Kobayashi and K. Nomizu (1969). *Foundations of Differential Geometry*, Wiley-Interscience.

S. R. Kou, D. L. Elliot and T. J. Tarn (1973). Observability of nonlinear systems, *Inform. Contr.*, **22**, 89–99.

E. Kreindler and P. E. Sarachik (1964). On the concepts of controllability and observability of linear systems, *IEEE Trans. Aut. Contr.*, AC-**9**, 129–136.

A. J. Krener (1973). On the equivalence of control systems and the linearization of nonlinear systems, *SIAM J. Contr.*, **11**, 670–676.

A. J. Krener (1977). A decomposition theory for differential systems, *SIAM J. Contr.*, **15**, 813–829.

K. B. Krohn and J. L. Rhodes (1965). Algebraic theory of machines 1. The main decomposition theorem, *Trans. Amer. Math. Soc.*, **116**, 450–464.

M. Kuranishi (1951). On everywhere dense imbeddings of free groups in a Lie group, *Nagoya Math J.*, **2**, 63–71.

S. Lang (1972). *Differentiable Manifolds*, Addison–Wesley.

E. B. Lee and L. Marcus (1961). Optimal control for nonlinear processes, *Arch. Rat. Mech Anal.*, **8**, 36–58.

S. Lefschetz (1977). *Differential Equations: Geometric Theory*, Dover.

C. Lesiak and A. J. Krener (1978). The existence and uniqueness of Volterra series for nonlinear systems, *IEEE Trans. Aut. Contr.*, AC-**23**, 1090–1095.

C. Lobry (1970). Controllabilité des systèmes nonlinéaires, *SIAM J. Contr.*, **8**, 573–605.

C. Lobry (1973). Dynamical polysystems and control theory, in D. Q. Mayne and R. W. Brockett (eds.), *Geometric methods in systems theory*, D. Riedel, Dordrecht, Holland.

R. Longchamp (1980). Stable feedback control of bilinear systems, *IEEE Trans. Aut. Contr.*, AC-**25**, 302–306.

J. K. Lubbock (1969). Multidimensional Laplace transforms for solution of nonlinear equations, *Proc. IEE*, **116**, 2075–2082.

D. L. Lukes (1974). Global controllability of distributed nonlinear equations, *SIAM J. Contr.*, **12**, 695–704.

S. I. Marcus (1984). Algebraic and geometric methods in nonlinear filtering, *SIAM J. Contr.*, **22**, 817–844.

A. A. Markov (1958). The problem of homeomorphy. *Proc. of Internat. Congress of Mathematicians* (Russian) English Translation, Cambridge University Press (1960).

Y. Matsushima (1951). On the discrete subgroups and homogeneous spaces of nilpotent Lie groups, *Nagoya Math. J.*, **12**, 95–100.

J. Milnor (1956). On manifolds homeomorphic to the 7-sphere, *Ann. Math.*, **64**, 399–405.

G. E. Mitzel, S. J. Clancy and W. J. Rugh (1979). On transfer function representations for homogeneous nonlinear systems, *IEEE Trans. Aut. Contr.*, AC-**24**, 242–249.

R. R. Mohler (1973). *Bilinear Control Preocesses*, Academic Press.

D. Montgomery and L. Zippin (1955). *Topological Transformation Groups*, Interscience.

T. Nagano (1966). Linear differential systems with singularities and an application to transitive Lie algebras, *J. Math. Soc. Japan*, **18**, 398–404.

H. Nijmeijer (1981). Controlled invariance for affine control systems, *Int. J. Contr.*, **34**, 825–833.

H. Nijmeijer (1983). Feedback decomposition of nonlinear control systems, *IEEE Trans. Aut. Contr.*, AC-**28**, 861–862.

H. Nijmeijer and J. M. Schumacher (1985). Zeros at infinity for affine nonlinear control systems, *IEEE Trans. Aut. Contr.*, AC-**30**, 566–573.

H. Nijmeijer and A. Van der Schaft (1982). Controlled invariance for nonlinear systems, *IEEE Trans. Aut. Contr.*, AC-**27**, 904–914.

H. Nijmeijer and A. Van der Schaft (1983). The disturbance decoupling problem for nonlinear control systems, *IEEE Trans. Aut. Contr.*, AC-**28**, 621–623.

R. S. Palais (1957). A global formulation of the Lie theory of transformation groups *Amer. Math. Soc.*, no. 22.

R. S. Palais (1968). *Foundations of Global Nonlinear Analysis*, Benjamin.

J. G. Pearlman (1978). Canonical forms for bilinear input/output maps. *IEEE Trans. Aut. Contr.*, AC-**23**, 595–602.

J. G. Pearlman and M. Denham (1979). Canonical realization of bilinear input/output maps, *SIAM J. Contr.*, **17**, 451–468.

I. R. Porteous (1969). *Topological Geometry*, Van Nostrand Reinhold.

D. Rebhuhn (1977). On the set of attainability of nonlinear nonautonomous control systems, *SIAM J. Contr.*, **15**, 803–812.

D. Rebhuhn (1980). Invertibility of C^∞ multivariable input–output systems, *IEEE Trans. Aut. Contr.*, AC-**25**, 207–212.

F. Riesz and B. Sz. Nagy (1955). *Functional Analysis*, F. Ungar.

R. E. Rink and R. R. Mohler (1968). Completely controllable bilinear systems, *SIAM J. Contr.*, **6**, 477–486.

W. J. Rugh (1981). *Nonlinear System Theory: The Volterra–Wiener Approach*, John Hopkins Univ. Press.

E. P. Ryan and N. J. Buckingham (1983). On asymptotically stabilizing feedback control of bilinear systems, *IEEE Trans. Aut. Contr.*, AC-**28**, 863–864.

A. A. Sagle and R. E. Walde (1973). *Introduction to Lie Groups and Lie Algebras*, Academic Press.

J. T. Schwartz (1969). *Nonlinear Functional Analysis*, Gordon and Breach.

I. E. Segal (1963). Nonlinear semigroups, *Ann. Math.*, **78**, 339–364.

L. M. Silverman (1969). Inversion of multivariable linear systems, *IEEE Trans. Aut. Contr.*, AC-**14**, 270–276.

S. N. Singh (1982a). Functional reproducibility of multivariable nonlinear systems, *IEEE Trans. Aut. Contr.*, AC-**27**, 270–272.

S. N. Singh (1982b). Invertibility of observable multivariable nonlinear systems, *IEEE Trans. Aut. Contr.*, AC-**27**, 487–489.

M. Slemrod (1978). Stabilization of bilinear control systems with applications to nonconservative problems in elasticity, *SIAM J. Contr.*, **16**, 131–141.

E. H. Spanier (1966). *Algebraic Topology*. McGraw-Hill.

R. Su (1982). On the linear equivalents of nonlinear systems, *Sys. & Contr. Letts.*, **2**, 48–52.

H. J. Sussmann (1973). Orbits of families of vector fields and integrability of distributions, *Trans. Amer. Math. Soc.*, **180**, 171–188.

H. J. Sussmann (1976). Some properties of vector field systems that are not altered by small perturbations, *J. Diff. Eqns.*, **20**, 292–315.

H. J. Sussmann (1977). Existence and uniqueness of minimal realizations of nonlinear systems, *Math. Systems Theory*, **10**, 263–284.

H. J. Sussmann (1983). Lie brackets and local controllability: a sufficient condition for scalar-input systems, *SIAM J. Contr.*, **21**, 686–713.

H. J. Sussmann and V. Jurdjevic (1972). Controllability of nonlinear systems. *J. Diff. Eqns.*, **12**, 95–116.

H. Takata (1979). Transformation of a nonlinear system into an augmented linear system, *IEEE Trans. Aut. Contr.*, AC-**24**, 736–741.
F. Takens (1976). *Variational and Conservative Systems*, Rapport ZW-7603, Maths. Institute, Groningen, the Netherlands.
M. Takesaki (1979). *Theory of Operator Algebras I*, Springer-Verlag.
H. Tanabe (1979). *Equations of Evolution*, Pitman.
Y-S. Tang, A. I. Mees and L. O. Chua (1983). Hopf bifurcation via Volterra series, *IEEE Trans. Aut. Contr.*, AC-**28**, 42–53.
A. E. Taylor (1958). *Introduction to Functional Analysis*, Wiley.
S. G. Tzafestas, K. E. Anagnostou and T. G. Pimenides (1984). Stabilizing optimal control of bilinear systems with a generalized cost, *Optimal Contr. Apps. & Methods*, **5**, 111–117.

A. J. Van der Schaft (1982a). Observability and controllability for smooth nonlinear systems, *SIAM J. Contr.*, **20**, 338–354.
A. J. Van der Schaft (1982b). Controllability and observability for affine nonlinear Hamiltonian systems, *IEEE Trans. Aut. Contr.*, AC-**27**, 490–492.
P. M. Van Dooren, P. Dewilde and J. Wandewalle (1979). On the determination of the Smith–MacMillan form of a rational matrix from its Laurent expansion, *IEEE Trans. Aut. Contr.*, CAS-**26**, 180–189.
V. S. Varadarajan (1974). *Lie groups, Lie Algebras and their Representations*, Prentice-Hall.

Z-X. Wan (1975). *Lie Algebras*, Pergamon Press.
R. O. Wells (1979). *Differential Analysis on Complex Manifolds*, GTM 65, Springer-Verlag.
E. H. Wichmann (1961). Note on the algebraic aspect of the integration of a system of ordinary linear differential equations, *J. Math. Phys.*, **2**, 876–880.
J. C. Willems (1979). Systems theoretic models for the analysis of physical systems, *Richerche di Automatica*, **10**, 71–106.
D. Williamson (1977). Observation of bilinear systems with application to biological control, *Automatica*, **13**, 243–254.
W. M. Wonham (1979). *Linear Multivariable Control: A Geometric Approach (2nd edn)*, Vol. 10, applications of maths. Series, Springer-Verlag.

K. Yosida (1974). *Functional Analysis*, Springer-Verlag.
D. C. Youla (1966). The synthesis of linear dynamical systems from prescribed weighting patterns, *SIAM J. Appl. Math.*, **14**, 527–549.

INDEX